Fault-Tolerant Process Control

Prashant Mhaskar • Jinfeng Liu •
Panagiotis D. Christofides

Fault-Tolerant Process Control

Methods and Applications

Springer

Prashant Mhaskar
Department of Chemical Engineering
McMaster University
Hamilton, Ontario, Canada

Jinfeng Liu
Dept. of Chemical & Mat. Engineering
University of Alberta
Edmonton, Alberta, Canada

Panagiotis D. Christofides
Dept. of Chemical & Biomolecular Engin.
University of California
Los Angeles, CA, USA

ISBN 978-1-4471-4807-4 ISBN 978-1-4471-4808-1 (eBook)
DOI 10.1007/978-1-4471-4808-1
Springer London Heidelberg New York Dordrecht

Library of Congress Control Number: 2012953998

Printed on acid-free paper

Springer is part of Springer Science+Business Media (www.springer.com)

Preface

The chemical industry is a vital sector of the global economy. Increasingly faced with the requirements of safety, environmental sustainability, energy efficiency, and profitability, chemical process operation is relying extensively on automated process control systems involving a large number of control actuators and measurement sensors. While process automation is critical in achieving the above requirements, the increasing reliance on actuators and sensors tends to increase the vulnerability of the process to faults (for example, defects/malfunctions in process equipment, sensors and actuators, failures in the controllers or in the control loops), leading to the failure of the control system and potentially causing a host of economic, environmental, and safety problems that can seriously degrade the operating efficiency of the process. Problems due to faults may include physical damage to the process equipment, raw material and energy waste, increase in process downtime, resulting in significant production losses, and jeopardizing personnel and environmental safety. Management of abnormal situations resulting from actuator and sensor malfunctions is a challenge in the chemical industry since abnormal situations account for tens of billions of dollars in annual lost revenue in the US alone.

The above considerations provide a strong motivation for the development of methods and strategies for the design of advanced fault-tolerant control systems that ensure an efficient and timely response to enhance fault recovery, prevent faults from propagating or developing into total failures, and reduce the risk of safety hazards. To this end, this book presents methods for the design of advanced fault-tolerant control systems for chemical processes which explicitly deal with actuator/controller failures and sensor data losses. Specifically, the book proposes: (i) a fault-detection, isolation, and diagnosis framework for handling actuator and sensor faults for nonlinear systems; (ii) reconfiguration and safe-parking based fault-handling methodologies; (iii) integrated data and model based fault-detection and isolation and fault-tolerant control methods; (iv) methods for handling sensor malfunctions; and (v) methods for monitoring the performance of low-level proportional-integral-derivative (PID) control loops. The proposed methods employ tools ranging from nonlinear systems analysis, Lyapunov techniques, optimization, statistical methods, and hybrid systems theory and are predicated upon the idea of integrating fault-detection, local feedback control, and supervisory control. The applicability and performance of the proposed methods are demonstrated through a number of chemical process examples.

Application of the proposed fault-tolerant control methods to processes subject to actuator and sensor malfunctions is expected to significantly improve their operation and performance, increase process safety and reliability, and minimize the negative economic impact of failures on overall process operation.

The book requires basic knowledge of differential equations, linear and nonlinear control theory, and optimization methods, and is intended for researchers, graduate students, and process control engineers. Throughout the book, practical implementation issues are discussed to help engineers and researchers understand the application of the methods in greater depth.

Finally, we would like to thank all the people who contributed in some way to this project. In particular, we would like to thank our colleagues at McMaster University, the University of Alberta, and UCLA for creating a pleasant working environment. Last, but not least, we would like to express our deepest gratitude to our families for their dedication, encouragement and support over the course of this project. We dedicate this book to them.

Hamilton, Ontario, Canada — Prashant Mhaskar
Edmonton, Alberta, Canada — Jinfeng Liu
Los Angeles, CA, USA — Panagiotis D. Christofides

Contents

List of Figures

List of Tables

Chapter 1
Introduction

1.1 Motivation

The operation of chemical processes is characterized both by the complexity of the individual units and the intricate interconnection of these geographically distributed units via a network of material and energy streams, and control loops. The nonlinear behavior exhibited by most chemical processes, together with the presence of constraints on the operating conditions, modeling uncertainty and disturbances, and the lack of availability of state measurements has motivated several research results in the area of nonlinear process control focusing on these issues. The development of the advanced control algorithms (alongside developments in sensing, communication, and computing technologies) has led to extensive automation of plant operation. Increased automation, however, also makes the plant susceptible to faults (e.g., defects/malfunctions in process equipment, sensors and actuators, failures in the controllers or in the control loops), which, if not appropriately handled in the control system design, can potentially cause a host of undesired economic, environmental, and safety problems. These considerations provide a strong motivation for the development of advanced fault-tolerant control methods that ensure an efficient and timely response to enhance fault recovery, prevent faults from propagating or developing into total failures, and reduce the risk of safety hazards.

The area of fault-tolerant control stands on three key pegs: (i) fault detection and isolation methods, (ii) robust and nonlinear control designs, and (iii) fault-handling mechanisms. While there have been significant contributions in these three individual areas, the key to a successful fault-tolerant control method lies on a seamless integration of the above three in a way that accounts for system complexities such as nonlinearity, uncertainty, and constraints and provides a mechanism for an efficient and timely response to enhance fault recovery. Motivated by the above, this book presents methods for integrated fault-detection and isolation and fault-tolerant control, accompanied by their application to nonlinear process systems.

P. Mhaskar et al., *Fault-Tolerant Process Control*, DOI 10.1007/978-1-4471-4808-1_1,

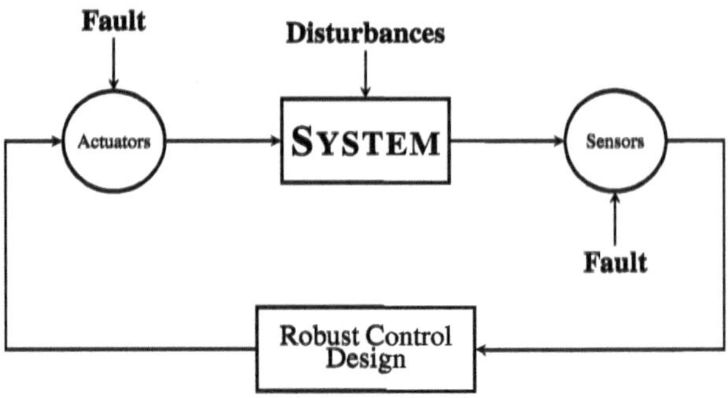

Fig. 1.1 A traditional fault-tolerant control structure

1.2 Background

Over the past 15 years, fault-tolerant control has become an active area of research within control engineering as a means for avoiding disaster in the case of a fault; see, for example, [44, 48, 111, 113, 114]. Many research studies can be found in the field of aerospace control engineering [17, 131, 182] as well as within chemical process control [13, 111, 113]. Fault-tolerant control works on the basic premise that there still exist some degrees of freedom/partial controllability in the presence of a fault, which inherently stems from some form of actuator/sensor redundancy. Fault-tolerant control solutions typically exist as an integrated fault-detection/isolation and diagnosis and fault-tolerant control framework, with process control algorithms being integral to the success of the framework. In the remainder of this section, we will briefly review the state-of-the-art in the key components of a fault-tolerant control framework.

The highly nonlinear behavior of many chemical processes has motivated extensive research on nonlinear process control. Chemical process nonlinearities can arise from the first principles process model, bounds on manipulated inputs, controller elements, or complex process interactions. Excellent reviews of results in the area of nonlinear process control can be found, for example, in [15, 67]; for a more recent review, see [28]. The problems caused by input constraints have motivated numerous studies on the dynamics and control of systems subject to input constraints. Important contributions in this area include results on optimization-based control methods such as model predictive control (e.g., [60, 100]), Lyapunov-based control (e.g., [45, 46, 78, 85, 156]), and hybrid predictive control (e.g., [50, 109]).

The traditional approach to handling faults has been to design robust control structures. The control designs essentially rely on availability of sufficient control effort in the presence of faults that allows the controller to implement control action to counter the effect of faults (that are treated essentially as disturbances). Such designs can be categorized as "passive" designs in that no explicit action is taken based on the occurrence of faults. The benefit of this approach is that it does not require an explicit fault-detection and isolation mechanism, with the obvious limitation being the conservativeness of the control design. Figure 1.1 shows a schematic of the traditional fault-tolerant control structure.

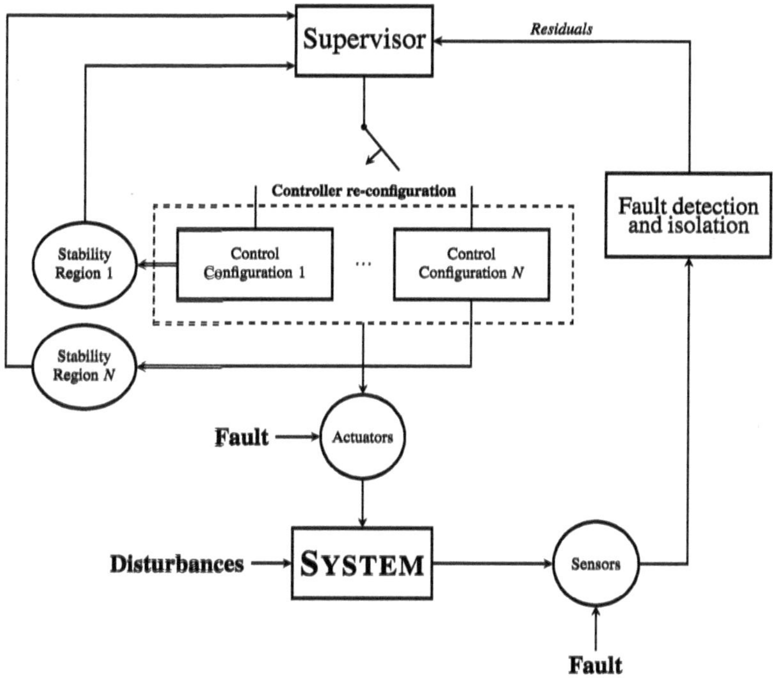

Fig. 1.2 An active fault tolerant control structure

More recent results have focussed on designing active fault tolerant control structures where the specific action taken is triggered by the detection (and isolation and, where applicable, diagnosis) of the fault. Figure 1.2 shows a schematic of the traditional fault-tolerant control structure. Methods for fault detection and isolation fall into two broad categories: model-based methods [54, 164] and data-based methods [163]. Model-based methods utilize a mathematical model of the process to build, under appropriate assumptions, dynamic filters that use process measurements to compute residuals that relate directly to specific faults; in this way, fault detection and isolation can be accomplished for specific model and fault structures (see, for example, [54, 164]). The problem of using fundamental process models for the purpose of detecting faults has been studied extensively in the context of linear systems [21, 22, 39, 54, 55, 63, 98, 106, 132, 164]; and more recently, some results in the context of nonlinear and distributed parameter systems have been derived [8, 18, 38, 44, 48, 115, 134, 146, 153, 177, 179, 180]. In this approach, fault detection and isolation (FDI) is often achieved by generating residuals through the system model and input/output data. Under fault-free conditions, the magnitudes of residuals are small. A fault is reported when a residual breaches the user-specified threshold. Due to the presence of plant–model mismatch, residuals that are sensitive to faults but insensitive to uncertainty and disturbances are desired. On the other hand, data-based methods are primarily based on past measured plant-data to construct indicators that identify deviations from normal operation to detect faults (e.g., [7, 36, 42, 43, 82, 123, 128, 142, 155, 178]). Analyzing measured data gives a picture of the location and direction of the system in the state-space. It is then possible to extract information about the fault by comparing the location and/or direction of the system in the state-space with past behavior under faulty operation (e.g., [144, 174]) or with expected behavior as predicted by the structure or model of the system. Several methods have been developed that process the measured data to reduce their dimension and extract information from the data with respect to actuator/sensor faults using principle component analysis (PCA) or partial least squares (PLS) techniques (e.g., [90, 123, 139, 171]). These methods reduce the dimensionality of the data by eliminating directions in the state-space with low common-cause variance. Many methods use this reduced space and consequent null space to gain further information about the process behavior as well as about actuator/sensor faults, including techniques such as contribution plots (e.g., [80]) or multi-scale statistical process control using wavelets (e.g., [6, 7, 12]). One of the main drawbacks of these data-based methods is that in order to accomplish fault isolation, they commonly require fault-specific historical data that may be costly to obtain. Furthermore, due to the nature of the chemical process, its structure and/or how it is instrumented, in practice, it is often hard to distinguish between regions/directions corresponding to operation in the presence of different faults due to overlap, making fault isolation difficult. For a comprehensive review of model-based and data-based fault detection and isolation methods, the reader may refer to [163, 164]. In general, most of the FDI methods mentioned thus far rely on measurements that are continuously or synchronously sampled, and they do not account for measurements that arrive asynchronously. Recently, research has been done on the topic of feedback control with asynchronous measurements [113, 120]. These efforts provide a starting framework for control subject to asynchronous measurements, but they do not include FDI.

Unknown input observers are developed in [22] to decouple the effect of unknown inputs, such as disturbances, from that of the faults for linear systems. A fault detection filter can then be developed to make the residuals directional for the purpose of fault isolation by using the remaining design freedom. For nonlinear systems, the problem has been studied by using uniform thresholds in [115] (and adaptive thresholds in [177, 179, 180]), where the isolation of faults relies on the existence of a state variable such that its evolution is directly and uniquely affected by the potential fault. For systems modeled by polynomial differential algebraic equations, analytical redundancy relations, which are constructed through a successive derivation of the system inputs and outputs to eliminate the unknown state variables, are used to generate structured residuals for FDI (e.g., [153]). Furthermore, a geometric approach is explored in [38], where a nonlinear FDI filter is designed to solve the fundamental problem of residual generation. Recently, a feedback control law was designed to decouple the dependency between certain system state variables to allow fault isolation using the structure of the closed-loop system (e.g., [129, 130]).

Compared to the problem of actuator faults, relatively fewer results on the handling of sensor faults for nonlinear systems are available. In one line of work, the problem of sensor FDI has been studied for Lipschitz nonlinear systems (see, e.g.,

[140, 162, 176, 179]). In [140], a nonlinear state observer is designed to generate state estimates by using a single sensor, and residuals are defined as the differences between the measurements and the corresponding state estimates. The fault isolation logic, however, is only limited to systems with three or more outputs. The method developed in [176, 179] utilizes adaptive estimation techniques to account for unstructured but bounded uncertainty, which requires knowledge of Lipschitz constants in the generation of the adaptive thresholds. A bank of fault isolation estimators are activated after the detection of a fault, and fault mismatch functions are used to describe the faults that are isolable. Linear matrix inequality techniques are used to design observers [133], which can be used to identify the fault vector, thereby achieving detection and estimation at the same time. A sliding mode observer is designed [172] to reconstruct or estimate faults by transforming sensor faults into pseudo-actuator faults. However, this approach requires a special structure of the system, and there is a limitation on the nonlinearities that can be handled. A bank of nonlinear observers are used to generate residuals that are sensitive to faults in all the sensors except for the one under consideration [99]. However, the design of the observer gain is based on a linearized model. In addition to sensor bias faults, the effect of intermittent unavailability of measurements has also been studied (see, e.g., [113, 120]). Despite the above methods, there exist limited results that consider the problem of detecting and isolating sensor faults for nonlinear systems.

The occurrence of faults in chemical processes and subsequent switching to fallback control configurations naturally leads to the superposition of discrete events on the underlying continuous process dynamics thereby making a hybrid systems framework a natural setting for the analysis and design of fault-tolerant control structures. Proper coordination of the switching between multiple (or redundant) actuator/sensor configurations provides a means for fault-tolerant control. However, at this stage, despite the large and growing body of research work on a diverse array of hybrid system problems (e.g., [37, 47, 51, 61, 62, 68]), the use of a hybrid system framework for the study of fault-tolerant control problems for nonlinear systems subject to constraints has received limited attention.

In summary, a close examination of the existing literature indicates a lack of general and practical methods for the design of integrated fault-detection and fault-tolerant control structures for chemical plants accounting explicitly for actuator/controller and sensor failures, process nonlinearities exhibited by most chemical processes, input constraints, modeling uncertainty and disturbances, and the lack of availability of state measurements.

1.3 Objectives and Organization of the Book

Motivated by the lack of a comprehensive integrated fault-detection and isolation and fault-tolerant control approach for nonlinear process systems, the broad objectives of this book are as follows:

1. To develop integrated fault-detection, isolation and diagnosis frameworks for handling actuator and sensor faults for nonlinear systems

2. To develop reconfiguration and safe-parking based fault-handling methodologies
3. To develop integrated data and model based fault-detection and isolation and fault-tolerant control methods
4. To develop methods for handling sensor malfunctions and methods for monitoring the performance of low-level proportional-integral-derivative (PID) control loops
5. To illustrate the applications of the developed fault-detection and isolation and fault-tolerant control methods to nonlinear process systems

The book is organized as follows. In Chap. 2, we first review some basic results on nonlinear systems and control, including definitions of stability of nonlinear systems, Lyapunov-based control, feedback linearization, model predictive control, Lyapunov-based model predictive control and hybrid systems.

In Chap. 3, we focus on actuator faults for single-input nonlinear systems and present a methodology to detect and handle the actuator fault through controller reconfiguration. First, the problem is considered under the assumption that state feedback is available; and then the approach is extended to the case where only certain outputs are available for measurement. Simulations of a chemical reactor example are carried out to illustrate the effectiveness of the presented approaches.

In Chap. 4, we generalize the results of Chap. 3 to include multi-input multi-output nonlinear systems subject to multiple faults in the control actuators and constraints on the manipulated inputs. We present a framework for integrated fault detection and isolation and fault-tolerant control. Similar to Chap. 3, we consider the case that state feedback is available first and then consider the case of output feedback. Applications of the methods to a chemical reactor and a reverse osmosis water desalination process are presented to demonstrate the applicability and effectiveness of the methods.

In Chap. 5, we move away from the assumption of availability of a redundant control configuration and present a "safe-parking" approach to handle faults. The safe-parking approach dictates driving the system to a (appropriately chosen) temporary operating point (the so-called safe-park point) until the fault is rectified. The choice of the safe-park point is based on stability and performance considerations, and also necessitates fault diagnosis (estimating the magnitude of the fault), going beyond FDI. A comprehensive mechanism for Fault-detection and Diagnosis (FDD) and safe-parking is presented and illustrated through chemical process examples.

In Chap. 6, we relax the assumption on the knowledge of the location and magnitude of the fault made in Chap. 5 and consider the problem of designing an integrated fault diagnosis and fault-handling framework to deal with actuator faults in nonlinear systems. A model-based fault diagnosis design is first proposed, which cannot only identify the failed actuator, but also estimate the fault magnitude. The efficacy of the integrated fault diagnosis and safe-parking framework is demonstrated through a chemical reactor example.

In Chap. 7, we demonstrate the use of FDI considerations in both control design and performance monitoring. We first develop a data-based method of fault detection and isolation that utilizes the design of the controller to enhance the isolability of the faults in the closed-loop system. It is demonstrated in this chapter that a data-based FDI scheme is able to isolate a given set of faults if the nonlinear closed-loop system satisfies certain isolability conditions in the presence of common-cause process variation. This is achieved through the use of appropriate nonlinear control laws that effectively decouple the dependency between certain process state variables. The theoretical results are applied to a continuous stirred tank reactor example and to a polyethylene reactor example. Next, we focus on the problem of monitoring and retuning of low-level PID control loops used to regulate control actuators to the values computed by a model-based controller. Under the assumption that the real-time measurement of the actuation level is unavailable, we use process state measurements and process models to carry out PID controller monitoring and compute appropriate residuals. Once a poorly-tuned PID controller is detected and isolated, a PID tuning method based on the estimated transfer function of the control actuator is applied to retune the PID controller. The presented method is applied to a nonlinear reactor–separator process operating under model predictive control with low-level PID controllers regulating the control actuators.

In Chap. 8, we consider the problem of sensor FDI and FTC for nonlinear systems subject to input constraints. The key idea of the presented method is to exploit model-based sensor redundancy through state observer design. An output feedback control design using high-gain observers is first presented; and then an FDI scheme is presented, which comprises of a bank of high-gain observers. Residuals are defined as the discrepancies between these state estimates and their predicted values based on previous estimates. A fault is identified when all the residuals breach their thresholds except for the one generated without using the measurements provided by the faulty sensor. Upon FDI, the state estimate generated using measurements from the remaining healthy sensors is used to preserve practical stability of the closed-loop system. The implementation of the sensor FDI and fault-handling framework subject to uncertainty and measurement noise is illustrated using a chemical reactor example.

Finally, in Chap. 9, we address the problem of control and fault-handling subject to asynchronous measurements and data losses. First, we develop an approach for handling sensor data losses via Lyapunov-based model predictive control. Specifically, in this control scheme, when feedback is lost due to sensor data losses, the actuators implement the last optimal input trajectory evaluated by the controller. This control scheme allows for an explicit characterization of the stability region and guarantees practical stability in the absence of sensor data losses. Application of the control scheme to a continuous crystallization process subject to sensor malfunctions is presented to illustrate the robustness of the control scheme when the process is subject to measurement unavailability, asynchronous sampling and parametric model uncertainties. Next, an integrated fault detection, isolation and fault-tolerant control framework is applied to a polyethylene reactor system where several process measurements are not available synchronously. First, an FDI scheme that employs model-based techniques is designed that allows for the isolation of the faults. This scheme employs model-based FDI filters in addition to observers that estimate the fault-free evolution of the asynchronously measured states during times when they are unmeasured. The FDI scheme provides detection and isolation for a fault where the fault entered into the differential equation of only synchronously measured states, and grouping of faults where the fault entered into the differential equation of any asynchronously measured state.

Chapter 2
Background on Nonlinear Systems and Control

In this chapter, we review some basic results on the analysis and control of nonlinear systems. This review is not intended to be exhaustive but to provide the reader with the necessary background for the results presented in the subsequent chapters. The results presented in this chapter are standard in the nonlinear systems and control literature. For detailed discussion and proofs of the results, the reader may refer to the classic books [72, 76].

2.1 Notation

Throughout this book, the operator $|\cdot|$ is used to denote the absolute value of a scalar and the operator $\|\cdot\|$ is used to denote Euclidean norm of a vector, while we use $\|\cdot\|_Q$ to denote the square of a weighted Euclidean norm, i.e., $\|x\|_Q = x^T Qx$ for all $x \in \mathbb{R}^n$. The symbol Ω_r is used to denote the set $\Omega_r := \{x \in \mathbb{R}^n : V(x) \leq r\}$ where V is a scalar positive definite, continuous differentiable function and $V(0) = 0$, and the operator '$/$' denotes set subtraction, that is, $A/B := \{x \in \mathbb{R}^n : x \in A, x \notin B\}$. The notation $R = [r_1\ r_2]$ is used to denote the augmented vector $R \in \mathbb{R}^{m+n}$ comprising the vectors $r_1 \in \mathbb{R}^m$ and $r_2 \in \mathbb{R}^n$. The notation $x(T^+)$ denotes the limit of the trajectory $x(t)$ as T is approached from the right, i.e., $x(T^+) = \lim_{t \to T^+} x(t)$. The notation $L_f h$ denotes the standard Lie derivative of a scalar function $h(\cdot)$ with respect to the vector function $f(\cdot)$, i.e., $L_f h(x) = \frac{\partial h}{\partial x} f(x)$.

2.2 Nonlinear Systems

In this book, we deal with a class of time invariant nonlinear systems that can be described by the following state-space model:

$$\dot{x} = f(x, u), \tag{2.1}$$

P. Mhaskar et al., *Fault-Tolerant Process Control*, DOI 10.1007/978-1-4471-4808-1_2,

where $x \in \mathbb{R}^n$ denotes the vector of state variables, $u \in \mathbb{R}^m$ denotes the vector of control (manipulated) input variables, and f is a locally Lipschitz vector function on $\mathbb{R}^n \times \mathbb{R}^m$ such that $f(0,0) = 0$. This implies that the origin is an equilibrium point for the unforced system. The input vector is restricted to be in a nonempty convex set $U \subseteq \mathbb{R}^m$ which is defined as follows:

$$U := \{u \in \mathbb{R}^m : \|u\| \leq u^{\max}\}, \tag{2.2}$$

where $u^{\max}$ is the magnitude of the input constraint. Another version of the set that we will use is

$$U_{con} := \{u \in \mathbb{R}^m : u_i^{\min} \leq u_i \leq u_i^{\max}, i = 1, \dots, m\}, \tag{2.3}$$

where $u_i^{\min}$ and $u_i^{\max}$ denote the constraints on the minimum and maximum value of the ith input.

In many chapters, we will restrict our analysis to a special case of the system of Eq. (2.1) where the input vector u enters the dynamics of the state x in an affine fashion as follows:

$$\dot{x} = f(x) + G(x)u, \tag{2.4}$$

where f is a locally Lipschitz vector function on $\mathbb{R}^n$ such that $f(0) = 0$ and G is an $n \times m$ matrix of locally Lipschitz vector functions on $\mathbb{R}^n$.

2.3 Stability of Nonlinear Systems

For all control systems, stability is the primary requirement. One of the most widely used stability concepts in control theory is that of *Lyapunov stability*, which we employ throughout the book. In this section, we briefly review basic facts from Lyapunov's stability theory. To begin with, we note that Lyapunov stability and asymptotic stability are properties not of a dynamical system as a whole, but rather of its individual solutions. We restrict our attention to the class of time-invariant nonlinear systems:

$$\dot{x} = f(x), \tag{2.5}$$

where the control input u does not appear explicitly. This does not necessarily mean that the input to the system is zero. It could be that the input u has been specified as a given function of the state x, $u = u(x)$, and could be considered as a special case of the system of Eq. (2.1).

The solution of Eq. (2.5), starting from x_0 at time $t_0 \in \mathbb{R}$, is denoted as $x(t; x_0, t_0)$, so that $x(t_0; x_0, t_0) = x_0$. Because the solutions of Eq. (2.5) are invariant under a translation of t_0, that is, $x(t+T; x_0, t_0+T) = x(t; x_0, t_0)$, the stability properties of $x(t; x_0, t_0)$ are *uniform*, i.e., they do not depend on t_0. Therefore, without loss of generality, we assume $t_0 = 0$ and write $x(t; x_0)$ instead of $x(t; x_0, 0)$.

Lyapunov stability concepts describe continuity properties of $x(t; x_0, t_0)$ with respect to x_0. If the initial state x_0 is perturbed to $\tilde{x}_0$, then, for stability, the perturbed solution $\tilde{x}(t; x_0)$ is required to stay close to $x(t; x_0)$ for all $t \geq 0$. In addition, for asymptotic stability, the error $\tilde{x}(t; x_0) - x(t; x_0)$ is required to vanish as $t \rightarrow \infty$. Some solutions of Eq. (2.5) may be stable and some unstable. We are particularly interested in studying and characterizing the stability properties of *equilibria*, that is, constant solutions $x(t; x_e) \equiv x_e$ satisfying $f(x_e) = 0$.

For convenience, we state all definitions and theorems for the case when the equilibrium point is at the origin of $\mathbb{R}^n$; that is, $x_e = 0$. There is no loss of generality in doing so since any equilibrium point under investigation can be translated to the origin via a change of variables. Suppose $x_e \neq 0$, and consider the change of variables, $z = x - x_e$. The derivative of z is given by:

$$\dot{z} = \dot{x} = f(x) = f(z + x_e) := g(z),$$

where $g(0) = 0$. In the new variable z, the system has an equilibrium point at the origin. Therefore, for simplicity and without loss of generality, we will always assume that $f(x)$ satisfies $f(0) = 0$ and confine our attention to the stability properties of the origin $x_e = 0$.

2.3.1 Stability Definitions

The origin is said to be a *stable* equilibrium point of the system of Eq. (2.5), in the sense of Lyapunov, if for every $\varepsilon > 0$ there exists a $\delta > 0$ such that we have:

$$\|x(0)\| \leq \delta \quad \Longrightarrow \quad \|x(t)\| \leq \varepsilon, \quad \forall t \geq 0. \tag{2.6}$$

In this case, we will also simply say that the system of Eq. (2.5) is stable. A similar convention will apply to other stability concepts introduced below. The origin is said to be *unstable* if it is not stable. The ε–δ requirement for stability takes a challenge–answer form. To demonstrate that the origin is stable, for every value of ε that a challenger may care to design, we must produce a value of δ, possibly dependent on ε, such that a trajectory starting in a δ neighborhood of the origin will never leave the ε neighborhood.

The origin of the system of Eq. (2.5) is said to be *asymptotically stable* if it is stable and δ in Eq. (2.6) can be chosen so that (attractivity property of the origin):

$$\|x(0)\| \leq \delta \quad \Longrightarrow \quad x(t) \rightarrow 0 \quad \text{as } t \rightarrow \infty. \tag{2.7}$$

When the origin is asymptotically stable, we are often interested in determining how far from the origin the trajectory can be and still converge to the origin as t approaches ∞. This gives rise to the definition of the *region of attraction* (also called *region of asymptotic stability*, *domain of attraction*, and *basin*). Let $\phi(t; x)$ be the solution of Eq. (2.5) that starts at initial state x at time $t = 0$. Then the region of attraction is defined as the set of all points x such that $\lim_{t\to\infty}\phi(t;x)=0$. If the origin is a stable equilibrium and its domain of attraction is the entire state-space, then the origin is called *globally asymptotically stable*.

If the system is not necessarily stable but has the property that all solutions with initial conditions in some neighborhood of the origin converge to the origin, then it is called (locally) attractive. We say that the system is *globally attractive* if its solutions converge to the origin from all initial conditions.

The system of Eq. (2.5) is called *exponentially stable* if there exist positive real constants δ, c, and λ such that all solutions of Eq. (2.5) with $\|x(0)\| \le \delta$ satisfy the inequality:

$$\|x(t)\| \le c\|x(0)\|e^{-\lambda t}, \quad \forall t \ge 0. \tag{2.8}$$

If this exponential decay estimate holds for any $x(0) \in \mathbb{R}^n$, the system is said to be *globally exponentially stable*.

2.3.2 Stability Characterizations Using Function Classes $\mathcal{K}$, $\mathcal{K}_\infty$, and $\mathcal{KL}$

Scalar comparison functions, known as class $\mathcal{K}$, $\mathcal{K}_\infty$, and $\mathcal{KL}$, are important stability analysis tools that are frequently used to characterize the stability properties of a nonlinear system.

Definition 2.1 A function $\alpha : [0,a) \to [0,\infty)$ is said to be of class $\mathcal{K}$ if it is continuous, strictly increasing, and $\alpha(0)=0$. It is said to belong to class $\mathcal{K}_\infty$ if $a=\infty$ and $\alpha(r)\to\infty$ as $r\to\infty$.

Definition 2.2 A function $\beta : [0,a)\times[0,\infty) \to [0,\infty)$ is said to be of class $\mathcal{KL}$ if, for each fixed $t \ge 0$, the mapping $\beta(r,t)$ is of class $\mathcal{K}$ with respect to r and, for each fixed r, the mapping $\beta(r,t)$ is decreasing with respect to t and $\beta(r,t)\to 0$ as $t\to\infty$.

We will write $\alpha \in \mathcal{K}$ and $\beta \in \mathcal{KL}$ to indicate that α is a class $\mathcal{K}$ function and β is a class $\mathcal{KL}$ function, respectively. As an immediate application of these function classes, we can rewrite the stability definitions of the previous section in a more compact way. For example, stability of the system of Eq. (2.5) is equivalent to the property that there exist a $\delta > 0$ and a class $\mathcal{K}$ function, α, such that all solutions with $\|x(0)\| \le \delta$ satisfy:

$$\|x(t)\| \le \alpha(\|x(0)\|), \quad \forall t \ge 0. \tag{2.9}$$

Asymptotic stability is equivalent to the existence of a $\delta > 0$ and a class $\mathcal{KL}$ function, β, such that all solutions with $\|x(0)\| \le \delta$ satisfy:

$$\|x(t)\| \le \beta(\|x(0)\|, t), \quad \forall t \ge 0. \tag{2.10}$$

Global asymptotic stability amounts to the existence of a class $\mathcal{KL}$ function, β, such that the inequality of Eq. (2.10) holds for all initial conditions. Exponential stability means that the function β takes the form $\beta(r,s) = cre^{-\lambda s}$ for some $c, \lambda > 0$.

2.3.3 Lyapunov's Direct (Second) Method

Having defined stability and asymptotic stability of equilibrium points, the next task is to find ways to determine stability. To be of practical interest, stability conditions must not require that we explicitly solve Eq. (2.5). The direct method of Lyapunov aims at determining the stability properties of an equilibrium point from the properties of $f(x)$ and its relationship with a positive-definite function $V(x)$.

Definition 2.3 Consider a C^1 (i.e., continuously differentiable) function $V : \mathbb{R}^n \to \mathbb{R}$. It is called *positive-definite* if $V(0) = 0$ and $V(x) > 0$ for all $x \neq 0$. If $V(x) \to \infty$ as $\|x\| \to \infty$, then V is said to *be radially unbounded.*

If V is both positive-definite and radially unbounded, then there exist two class $\mathcal{K}_\infty$ functions α_1, α_2 such that V satisfies:

$$\alpha_1(\|x\|) \leq V(x) \leq \alpha_2(\|x\|) \tag{2.11}$$

for all x. We write $\dot{V}$ for the derivative of V along the solutions of the system of Eq. (2.5), i.e.:

$$\dot{V}(x) = \frac{\partial V}{\partial x} f(x). \tag{2.12}$$

The main result of Lyapunov's stability theory is expressed by the following statement.

Theorem 2.1 (Lyapunov) *Let $x = 0$ be an equilibrium point for the system of Eq.* (2.5) *and $D \subset \mathbb{R}^n$ be a domain containing $x = 0$ in its interior. Suppose that there exists a positive-definite C^1 function $V : \mathbb{R}^n \to \mathbb{R}$ whose derivative along the solutions of the system of Eq.* (2.5) *satisfies*:

$$\dot{V}(x) \leq 0, \quad \forall x \in D \tag{2.13}$$

then $x = 0$ of the system of Eq. (2.5) *is stable. If the derivative of V satisfies*:

$$\dot{V}(x) < 0, \quad \forall x \in D/\{0\} \tag{2.14}$$

then $x = 0$ of the system of Eq. (2.5) *is asymptotically stable. If in the latter case, V is also radially unbounded, then $x = 0$ of the system of Eq.* (2.5) *is globally asymptotically stable.*

A continuously differentiable positive-definite function $V(x)$ satisfying Eq. (2.13) is called a *Lyapunov function*. The surface $V(x) = c$, for some $c > 0$, is called a *Lyapunov surface* or a level surface. The condition $\dot{V} \leq 0$ implies that when a trajectory crosses a Lyapunov surface $V(x) = c$, it moves inside the set $\Omega_c = \{x \in \mathbb{R}^n : V(x) \leq c\}$ and can never come out again. When $\dot{V} < 0$, the trajectory moves from one Lyapunov surface to an inner Lyapunov surface with smaller c. As c decreases, the Lyapunov surface $V(x) = c$ shrinks to the origin, showing that the trajectory approaches the origin as time progresses. If we only know that $\dot{V}(x) \leq 0$, we cannot be sure that the trajectory will approach the origin, but we can conclude that the origin is stable since the trajectory can be contained inside any ball, B_ε, by requiring that the initial state x_0 lie inside a Lyapunov surface contained in that ball.

The utility of a Lyapunov function arises from the need (or difficulty) of specifying a unique (necessary and sufficient) direction of movement of states for stability. To understand this, consider any scalar system (whether linear or nonlinear). The necessary and sufficient condition for stability is that, for any value of the state x, the value of $\dot{x}$ should be opposite in sign to x, and greater than zero in magnitude (unless $x = 0$). A Lyapunov function that allows readily capturing this requirement is $V(x) = \frac{x^2}{2}$, resulting in $\dot{V}(x) = x\dot{x}$. If and only if the origin of the systems is stable (i.e., x is opposite in sign to $\dot{x}$), it will result in $\dot{V}(x) < 0$.

For non-scalar systems, this 'unique' direction of movement of states, while possible for linear systems (see Remark 2.1), is in general difficult to identify for nonlinear systems. For instance, if one considers a simple two state system, and restricts the choice of the Lyapunov function to quadratic forms, it is clear that the square of the distance to the origin (resulting in 'circles' as level curves) is not necessarily the only choice of the Lyapunov-function, and there is no unique way to find a necessary and sufficient direction of the movement of states to achieve stability. This is the problem that lies at the core of the Lyapunov-stability theory—the inability to define (and/or construct) a unique Lyapunov function for a given system that is necessary and sufficient to establish stability. Having recognized this limitation, it is important to note that the Lyapunov-based analysis at least provides sufficient conditions to ascertain stability.

In this direction, various converse Lyapunov theorems show that the conditions of Theorem 2.1 are also necessary. For example, if the system is asymptotically stable, then there exists a positive-definite C^1 function V that satisfies the inequality of Eq. (2.14). The theorems, however, do not provide a way of constructing this Lyapunov function.

Remark 2.1 It is well-known that for the linear time-invariant system

$$\dot{x} = Ax \tag{2.15}$$

asymptotic stability, exponential stability, and their global versions are all equivalent and amount to the property that A is a Hurwitz matrix, i.e., all eigenvalues of A have negative real parts. Fixing an arbitrary positive-definite symmetric matrix Q and finding the unique positive-definite symmetric matrix P that satisfies the Lyapunov equation

$$A^T P + PA = -Q,$$

one obtains a quadratic Lyapunov function $V(x) = x^T Px$ whose time derivative along the solutions of the system of Eq. (2.15) is $\dot{V} = -x^T Qx$. The explicit formula for P is

$$P = \int_0^\infty e^{A^T t} Q e^{At}\, dt.$$

Indeed, we have

$$A^T P + PA = \int_0^\infty \frac{d}{dt}\left(e^{A^T t} Q e^{At}\right) dt = -Q,$$

because A is Hurwitz.

2.3.4 *LaSalle's Invariance Principle*

With some additional knowledge about the behavior of solutions, it is possible to prove asymptotic stability using a Lyapunov function which satisfies the nonstrict inequality of Eq. (2.13). This is facilitated by *LaSalle's invariance principle*. To state this principle, we first recall the definition of an invariant set.

Definition 2.4 A set M is called (positively) invariant with respect to the given system if all solutions starting in M remain in M for all future times.

We now state a version of LaSalle's theorem.

Theorem 2.2 (LaSalle) *Suppose that there exists a positive-definite $\mathcal{C}^1$ function $V : \mathbb{R}^n \to \mathbb{R}$ whose derivative along the solutions of the system of Eq.* (2.5) *satisfies the inequality of Eq.* (2.13). *Let M be the largest invariant set contained in the set $\{x : \dot{V}(x) = 0\}$. Then the system of Eq.* (2.5) *is stable and every solution that remains bounded for $t \geq 0$ approaches M as $t \to \infty$. In particular, if all solutions remain bounded and $M = \{0\}$, then the system of Eq.* (2.5) *is globally asymptotically stable.*

To deduce global asymptotic stability with the help of this result, one needs to check two conditions. First, all solutions of the system must be bounded. This property follows automatically from the inequality of Eq. (2.13) if V is chosen to be radially unbounded; however, radial boundedness of V is not necessary when boundedness of solutions can be established by other means. The second condition is that V be not identically zero along any nonzero solution. We also remark that if one only wants to prove asymptotic convergence of bounded solutions to zero and is not concerned with Lyapunov stability of the origin, then positive-definiteness of V is not needed (this is in contrast to Theorem 2.1).

While Lyapunov's stability theorem readily generalizes to time-varying systems, for LaSalle's invariance principle this is not the case. Instead, one usually works with the weaker property that all solutions approach the set $\{x : \dot{V}(x) = 0\}$.

2.3.5 Lyapunov's Indirect (First) Method

Lyapunov's indirect method allows one to deduce stability properties of the nonlinear system of Eq. (2.5), where f is $\mathcal{C}^1$, from stability properties of its *linearization*, which is the linear system of Eq. (2.15) with

$$A := \frac{\partial f}{\partial x}(0). \tag{2.16}$$

By the mean value theorem, we can write

$$f(x) = Ax + g(x)x,$$

where g is given componentwise by $g_i(x) := \frac{\partial f_i}{\partial x}(z_i) - \frac{\partial f_i}{\partial x}(0)$ for some point, z_i, on the line segment connecting x to the origin, $i = 1, \dots, n$. Since $\frac{\partial f}{\partial x}$ is continuous, we have $\|g(x)\| \to 0$ as $x \to 0$. From this it follows that if the matrix A is Hurwitz (i.e., all its eigenvalues lie in the open left half of the complex plane), then a quadratic Lyapunov function for the linearization serves—locally—as a Lyapunov function for the original nonlinear system. Moreover, its rate of decay in a neighborhood of the origin can be bounded below by a quadratic function, which implies that stability is, in fact, exponential. This is summarized by the following result.

Theorem 2.3 *If f is $\mathcal{C}^1$ and the matrix of Eq.* (2.16) *is Hurwitz, then the system of Eq.* (2.5) *is locally exponentially stable.*

It is also known that if the matrix A has at least one eigenvalue with a positive real part, the origin of the nonlinear system of Eq. (2.5) is not stable. If A has eigenvalues on the imaginary axis but no eigenvalues in the open right half-plane, the linearization test is inconclusive. However, in this critical case, the system of Eq. (2.5) cannot be exponentially stable since exponential stability of the linearization is not only a sufficient but also a necessary condition for (local) exponential stability of the nonlinear system.

2.3.6 Input-to-State Stability

It is of interest to extend stability concepts to systems with disturbance inputs. In the linear case represented by the system

$$\dot{x} = Ax + B\theta,$$

it is well known that if the matrix A is Hurwitz, i.e., if the unforced system, $\dot{x} = Ax$, is asymptotically stable, then bounded inputs θ lead to bounded states while inputs converging to zero produce states converging to zero. Now, consider a nonlinear system of the form

$$\dot{x} = f(x, \theta), \tag{2.17}$$

where θ is a measurable bounded disturbance input. In general, global asymptotic stability of the unforced system $\dot{x} = f(x, 0)$ does not guarantee input-to-state stability with respect to θ of the kind mentioned above. For example, the scalar system

$$\dot{x} = -x + x\theta \tag{2.18}$$

has unbounded trajectories under the bounded input $\theta \equiv 2$. This motivates the following important concept, introduced by Sontag [151].

Definition 2.5 The system of Eq. (2.17) is called *input-to-state stable* (ISS) with respect to θ if for some functions $\gamma \in \mathcal{K}_\infty$ and $\beta \in \mathcal{KL}$, for every initial state $x(0)$, and every input θ, the corresponding solution of the system of Eq. (2.17) satisfies the inequality

$$\|x(t)\| \le \beta(\|x(0)\|, t) + \gamma(\|\theta\|^s_{[0,t]}), \tag{2.19}$$

where $\|\theta\|^s_{[0,t]} := \text{ess.sup}\{\|\theta(s)\| : s \in [0,t]\}$ (supremum norm on $[0,t]$ except for a set of measure zero).

Since the system of Eq. (2.17) is time-invariant, the same property results if we write

$$\|x(t)\| \le \beta(\|x(t_0)\|, t - t_0) + \gamma(\|\theta\|^s_{[t_0,t]}), \quad \forall t \ge t_0 \ge 0. \tag{2.20}$$

The ISS property admits the following Lyapunov-like equivalent characterization: The system of Eq. (2.17) is ISS if and only if there exists a positive-definite radially unbounded C^1 function $V : \mathbb{R}^n \to \mathbb{R}$ such that for some class $\mathcal{K}_\infty$ functions α and χ we have

$$\frac{\partial V}{\partial x} f(x, \theta) \le -\alpha(\|x\|) + \chi(\|\theta\|), \quad \forall x, \theta. \tag{2.21}$$

This is, in turn, equivalent to the following "gain margin" condition:

$$\|x\| \ge \rho(\|\theta\|) \quad \Longrightarrow \quad \frac{\partial V}{\partial x} f(x, \theta) \le -\alpha(\|x\|), \tag{2.22}$$

where $\alpha, \rho \in \mathcal{K}_\infty$. Such functions V are called *ISS-Lyapunov functions*. If the system of Eq. (2.17) is ISS, then $\theta(t) \to 0$ implies $x(t) \to 0$.

The system of Eq. (2.17) is said to be *locally input-to-state stable* (locally ISS) if the bound of Eq. (2.19) is valid for solutions with sufficiently small initial conditions and inputs, i.e., if there exists a $\delta > 0$ such that Eq. (2.19) is satisfied whenever $\|x(0)\| \le \delta$ and $\|\theta\|^s_{[0,t]} \le \delta$. It turns out that (local) asymptotic stability of the unforced system $\dot{x} = f(x, 0)$ implies local ISS.

2.4 Stabilization of Nonlinear Systems

This book is primarily about control *design*. Our objective is to create closed-loop systems with desirable stability and performance properties, rather than analyze the properties of a given system. For this reason, we are interested in an extension of the Lyapunov function concept, called a *control Lyapunov function* (CLF).

Suppose that our problem for the time-invariant system

$$\dot{x} = f(x, u), \tag{2.23}$$

where $x \in \mathbb{R}^n$, $u \in \mathbb{R}$ (i.e., we consider the unconstrained problem), $f(0, 0) = 0$, is to design a feedback control law $\alpha(x)$ for the control variable u such that the equilibrium $x = 0$ of the closed-loop system

$$\dot{x} = f\bigl(x, \alpha(x)\bigr) \tag{2.24}$$

is globally asymptotically stable. We can pick a function $V(x)$ as a Lyapunov function candidate, and require that its derivative along the solutions of the system of Eq. (2.24) satisfies $\dot{V} \le -W(x)$, where $W(x)$ is a positive-definite function. We therefore need to find $\alpha(x)$ to guarantee that for all $x \in \mathbb{R}^n$

$$\frac{\partial V}{\partial x}(x) f\bigl(x, \alpha(x)\bigr) \le -W(x). \tag{2.25}$$

This is a difficult task. A stabilizing control law for the system of Eq. (2.23) may exist, but it may fail to satisfy Eq. (2.25) because of a poor choice of $V(x)$ and $W(x)$. A system for which a good choice of $V(x)$ and $W(x)$ exists is said to possess a CLF. This notion is made more precise below.

Definition 2.6 A smooth positive-definite radially unbounded function $V : \mathbb{R}^n \to \mathbb{R}$ is called a control Lyapunov function (CLF) for the system of Eq. (2.23) if

$$\inf_{u \in \mathbb{R}} \left\{ \frac{\partial V}{\partial x}(x) f(x, u) \right\} < 0, \quad \forall x \neq 0. \tag{2.26}$$

The CLF concept of Artstein [9] is a generalization of Lyapunov design results by Jacobson and Judjevic and Quinn. Artstein showed that Eq. (2.26) is not only necessary, but also sufficient for the existence of a control law satisfying Eq. (2.25), that is, the existence of a CLF is equivalent to global asymptotic stabilizability.

For systems affine in the control, namely,

$$\dot{x} = f(x) + g(x)u, \quad f(0) = 0, \tag{2.27}$$

the CLF inequality of Eq. (2.25) becomes

$$L_f V(x) + L_g V(x)u \le -W(x). \tag{2.28}$$

If V is a CLF for the system of Eq. (2.27), then a particular stabilizing control law $\alpha(x)$, smooth for all $x \neq 0$, is given by Sontag's formula [150]:

$$u = \alpha_s(x) = \begin{cases} -\frac{L_f V(x) + \sqrt{(L_f V)^2(x) + (L_g V)^4(x)}}{(L_g V)^2(x)} L_g V(x), & L_g V(x) \neq 0, \\ 0, & L_g V(x) = 0. \end{cases} \tag{2.29}$$

It should be noted that Eq. (2.28) can be satisfied only if

$$L_g V(x) = 0 \quad \Longrightarrow \quad L_f V(x) < 0, \quad \forall x \neq 0. \tag{2.30}$$

The intuitive interpretation of the existence of a CLF is as follows: For any x such that $L_g V(x) \neq 0$, since there are no constraints on the input, $\dot{V}$ can be made negative by picking a 'large enough' control action, with an appropriate sign, to counter the effect of possibly positive $L_f V(x)$ term. For all x such that $L_g V(x) = 0$, the control action has no effect on the Lyapunov-function derivative. For it to be possible to show stability using the CLF V, it should therefore be true that whenever $L_g V(x) = 0$, we also have that $L_f V(x) < 0$. This is the requirement that is formalized in Eq. (2.30). With such a CLF, Eq. (2.29) results in

$$W(x) = \sqrt{(L_f V)^2(x) + (L_g V)^4(x)} > 0, \quad \forall x \neq 0. \tag{2.31}$$

A further characterization of a stabilizing control law $\alpha(x)$ for the system of Eq. (2.27) with a given CLF V is that $\alpha(x)$ is continuous at $x = 0$ if and only if the CLF satisfies the *small control property*: For each $\varepsilon > 0$ there is a $\delta(\varepsilon) > 0$ such that, if $x \neq 0$ satisfies $|x| \le \delta$, then there is some u with $|u| < \varepsilon$ such that

$$L_f V(x) + L_g V(x)u < 0. \tag{2.32}$$

The main deficiency of the CLF concept as a design tool is that for most nonlinear systems a CLF is not known. The task of finding an appropriate CLF maybe as complex as that of designing a stabilizing feedback law. In the next section, we review one commonly used tool for designing a Lyapunov-based control law that utilizes coordinate transformations. We also note that in the presence of input constraints, the concept of a CLF needs to be revisited, and this issue is discussed in Sect. 2.6.

2.5 Feedback Linearization and Zero Dynamics

One of the popular methods for nonlinear control design (or alternatively, one way to construct a Lyapunov-function for the purpose of control design) is *feedback linearization*, which employs a change of coordinates and feedback control to transform a nonlinear system into a system whose dynamics are linear (at least partially). This transformation allows the construction and use of a Lyapunov function for the control design utilizing results from linear systems analysis. A great deal of research has been devoted to this subject over the last four decades, as evidenced by the comprehensive books [72, 126] and the references therein. In this section, we briefly review some of the basic geometric concepts that will be used in subsequent chapters. While this book does not require the formalism of differential geometry, we will employ Lie derivatives only for notational convenience. If $f : \mathbb{R}^n \to \mathbb{R}^n$ is a vector field and $h : \mathbb{R}^n \to \mathbb{R}$ is a scalar function, the notation $L_f h$ is used for $\frac{\partial h}{\partial x} f(x)$. It is recursively extended to

$$L_f^k h(x) = L_f\big(L_f^{k-1} h(x)\big) = \frac{\partial}{\partial x}\big(L_f^{k-1} h(x)\big) f(x).$$

Let us consider the following nonlinear system:

$$\begin{aligned} \dot{x} &= f(x) + g(x)u, \\ y &= h(x), \end{aligned} \tag{2.33}$$

where $x \in \mathbb{R}^n$, $u \in \mathbb{R}$, $y \in \mathbb{R}$, f, g, h are analytic (i.e., infinitely differentiable) vector functions. The derivative of the output $y = h(x)$ is given by

$$\begin{aligned} \dot{y} &= \frac{\partial h}{\partial x}(x) f(x) + \frac{\partial h}{\partial x}(x) g(x) u \\ &= L_f h(x) + L_g h(x) u. \end{aligned} \tag{2.34}$$

If $L_g h(x_0) \neq 0$, then the system of Eq. (2.33) is said to have *relative degree one at* x_0 (note that since the functions are smooth $L_g h(x_0) \neq 0$ implies that there exists a neighborhood of x_0 on which $L_g h(x) \neq 0$). In our terminology, this implies that the output y is separated form the input u by one integration only. If $L_g h(x_0) = 0$, there are two cases:

(i) If there exist points arbitrarily close to x_0 such that $L_g h(x) \neq 0$, then the system of Eq. (2.33) does not have a well-defined relative degree at x_0.
(ii) If there exists a neighborhood B_0 of x_0 such that $L_g h(x) = 0$ for all $x \in B_0$, then the relative degree of the system of Eq. (2.33) may be well-defined.

In case (ii), we define

$$\psi_1(x) = h(x), \qquad \psi_2(x) = L_f h(x) \tag{2.35}$$

and compute the second derivative of y

$$\begin{aligned}\ddot{y} &= \frac{\partial \psi_2}{\partial x}(x)f(x) + \frac{\partial \psi_2}{\partial x}(x)g(x)u \\ &= L_f^2 h(x) + L_g L_f h(x)u.\end{aligned} \tag{2.36}$$

If $L_g L_f h(x_0) \neq 0$, then the system of Eq. (2.33) is said to have *relative degree two at* x_0. If $L_g L_f h(x) = 0$ in a neighborhood of x_0, then we continue the differentiation procedure.

Definition 2.7 The system of Eq. (2.33) is said to have relative degree r at the point x_0 if there exists a neighborhood B_0 of x_0 on which

$$L_g h(x) = L_g L_f h(x) = \cdots = L_g L_f^{r-2} h(x) = 0, \tag{2.37}$$

$$L_g L_f^{r-1} h(x) \neq 0. \tag{2.38}$$

If Eq. (2.37)–(2.38) are valid for all $x \in \mathbb{R}^n$, then the relative degree of the system of Eq. (2.33) is said to be globally defined.

Suppose now that the system of Eq. (2.33) has relative degree r at x_0. Then we can use a change of coordinates and feedback control to locally transform this system into the *cascade interconnection* of an r-dimensional linear system and an $(n-r)$-dimensional nonlinear system. In particular, after differentiating r times the output $y = h(x)$, the control appears:

$$y^{(r)} = L_f^r h(x) + L_g L_f^{r-1} h(x)u. \tag{2.39}$$

Since $L_g L_f^{r-1} h(x) \neq 0$ in a neighborhood of x_0, we can linearize the input–output dynamics of the system of Eq. (2.33) using feedback to cancel the nonlinearities in Eq. (2.39):

$$u = \frac{1}{L_g L_f^{r-1} h(x)} \left[-L_f^r h(x) + v \right]. \tag{2.40}$$

Then the dynamics of y and its derivatives are governed by a chain of r integrators: $y^{(r)} = v$. Since our original system of Eq. (2.33) has dimension n, we need to account for the remaining $n - r$ states. Using differential geometry tools, it can be shown that it is always possible to find $n - r$ functions $\psi_{r+1}, \ldots, \psi_n(x)$ with $\frac{\partial \psi_i}{\partial x}(x)g(x) = 0$, for $i = r+1, \ldots, n$ such that the change of coordinates

$$\begin{aligned}&\zeta_1 = y = h(x), \qquad \zeta_2 = \dot{y} = L_f h(x), \ldots, \zeta_r = y^{(r-1)} = L_f^{r-1} h(x), \\ &\eta_1 = \psi_{r+1}, \ldots, \eta_{n-r} = \psi_n(x)\end{aligned} \tag{2.41}$$

is locally invertible and transforms, along with the feedback law of Eq. (2.40), the system of Eq. (2.33) into

$$\begin{aligned}
\dot{\zeta}_1 &= \zeta_2, \\
&\vdots \\
\dot{\zeta}_r &= v, \\
\dot{\eta}_1 &= \Psi_1(\zeta, \eta), \\
&\vdots \\
\dot{\eta}_{n-r} &= \Psi_{n-r}(\zeta, \eta), \\
y &= \zeta_1,
\end{aligned} \tag{2.42}$$

where $\Psi_1(\zeta, \eta) = L_f^{r+1}h(x)$, $\Psi_{n-r}(\zeta, \eta) = L_f^n h(x)$.

The states $\eta_1, \ldots, \eta_{n-r}$ have been rendered *unobservable* from the output y by the control of Eq. (2.40). Hence, feedback linearization in this case is the nonlinear equivalent of placing $n - r$ poles of a linear system at the origin and canceling the r zeros with the remaining poles. Of course, to guarantee stability, the canceled zeros must be stable. In the nonlinear case, using the new control input v to stabilize the linear subsystem of Eq. (2.42) does not guarantee stability of the whole system, unless the stability of the nonlinear part of the system of Eq. (2.42) has been established separately.

When v is used to keep the output y equal to zero for all $t > 0$, that is, when $\zeta_1 \equiv \cdots \equiv \zeta_r \equiv 0$, the dynamics of $\eta_1, \ldots, \eta_{n-r}$ are described by

$$\begin{aligned}
\dot{\eta}_1 &= \Psi_1(0, \eta), \\
&\vdots \\
\dot{\eta}_{n-r} &= \Psi_{n-r}(0, \eta).
\end{aligned} \tag{2.43}$$

They are called the zero dynamics of the system of Eq. (2.33) because they evolve on the subset of the state-space on which the output of the system is identically zero. If the equilibrium at $\eta_1 = \cdots = \eta_{n-r} = 0$ of the zero dynamics of Eq. (2.43) is asymptotically stable, the system of Eq. (2.33) is said to be *minimum phase*.

Remark 2.2 Most nonlinear analytical controllers emanating from the area of geometric control are input–output linearizing and induce a linear input–output response in the absence of constraints [72, 81]. For the class of processes modeled by equations of the form of Eq. (2.33) with relative order r and under the minimum phase assumption, the appropriate linearizing state feedback controller is given by

$$u = \frac{1}{L_g L_f^{r-1} h(x)} \big(v - L_f^r h(x) - \beta_1 L_f^{r-1} h(x) - \cdots - \beta_{r-1} L_f h(x) - \beta_r h(x)\big) \tag{2.44}$$

and induces the linear rth order response

$$\frac{d^r y}{dt^r} + \beta_1 \frac{d^{r-1} y}{dt^{r-1}} + \cdots + \beta_{r-1} \frac{dy}{dt} + \beta_r y = v, \tag{2.45}$$

where the tunable parameters, $\beta_1, \ldots, \beta_r$, are essentially closed-loop time constants that influence and shape the output response. The nominal stability of the process is guaranteed by placing the roots of the polynomial $s^r + \beta_1 s^{r-1} + \cdots + \beta_{r-1} s + \beta_r$ in the open left-half of the complex plane.

2.6 Input Constraints

The presence of input constraints requires revisiting the concept of the CLF for both linear and nonlinear systems. To understand this, consider a scalar linear system of the form $\dot{x} = \alpha x + \beta u$, with $u^{\min} \le u \le u^{\max}$. For the sake of simplicity and without loss of generality, let us assume $u^{\min} < 0 < u^{\max}$ and $\beta > 0$. For the case of scalar systems, it is possible to determine the entire set of initial conditions from where the system can be driven to the origin subject to input constraints (regardless of the choice of the control law). This set is generally referred to as the null controllable region (NCR). An explicit computation of the NCR is possible in this case because for scalar systems (as discussed earlier) there exists a unique direction in which the system states needs to move to achieve stability.

To determine this set, one can simply analyze the system trajectory to the left and right of zero. Consider first $x > 0$, and the requirement that for $x > 0$, $\dot{x} < 0$. If $\alpha < 0$, $\dot{x} < 0$ $\forall x > 0$ (and also $\dot{x} > 0$ $\forall x < 0$). On the other hand, if $\alpha > 0$, $\dot{x} < 0$ can only be achieved for $x < \frac{-u^{\min}\beta}{\alpha}$. Similarly, $\dot{x} > 0$ can only be achieved for $x > \frac{-u^{\max}\beta}{\alpha}$. The analysis reveals what was perhaps intuitive to begin with: For linear systems, if the steady state is open-loop stable, the NCR is the entire state space, while if the steady state is open-loop unstable, it has a finite NCR, which in this case is $\{x : \frac{-u^{\max}\beta}{\alpha} < x < \frac{-u^{\min}\beta}{\alpha}\}$. The same result for the NCR can also be obtained using a CLF $V(x) = \frac{x^2}{2}$ and determining the states for which $\dot{V} < 0$ is achievable using the available control action. Furthermore, it points to the requirement of additional considerations when defining CLFs for systems with constrained inputs. In particular, requiring that $\dot{V}(x) < 0$ $\forall x$ is simply not achievable for certain cases, at best what is achievable is that $\dot{V}(x) < 0$ $\forall x \in \mathrm{NCR} - \{0\}$. The definition of a CLF (or more appropriately, a constrained CLF) then becomes intricately linked with the characterization of the NCR. The characterization of the NCR, however, is an increasingly difficult (although possible, see [71]) problem when considering non-scalar linear systems, and currently an open problem for nonlinear systems.

To understand the impact of the lack of availability of constrained CLFs (CCLFs), let us first consider again the linear scalar system under a feedback law of the form $u_c(x) = -kx$, with $k > 0$ such that $(\alpha - k\beta) < 0$ under two possible scenarios: (i) $\alpha < 0$ (i.e., for the unforced system, there is an isolated equilibrium point at the origin and the system is stable at that operating point) and (ii) $\alpha > 0$ (i.e., for the unforced system, there is an isolated equilibrium point at the origin and the system is unstable at that operating point). Due to the presence of input constraints, the closed-loop system is no longer a linear system, but operates in three 'modes', depending on the state, described by the following set of equations:

$$\begin{aligned}
\frac{dx}{dt} &= \alpha x + \beta u_c, \quad u^{\min} \le u_c \le u^{\max}, \\
\frac{dx}{dt} &= \alpha x + \beta u^{\max}, \quad u_c > u^{\max}, \\
\frac{dx}{dt} &= \alpha x + \beta u^{\min}, \quad u^{\min} > u_c.
\end{aligned} \tag{2.46}$$

Let us analyze the three possible modes of operation of the closed-loop system for scenario (i). For $-\frac{|u^{\max}|}{k} \le x \le \frac{|u^{\min}|}{k}$, we have that $\frac{dx}{dt} = \alpha x + \beta u_c = (\alpha - k\beta)x$, which establishes that for all initial conditions x_0 such that $-\frac{|u^{\max}|}{k} \le x_0 \le \frac{|u^{\min}|}{k}$, the prescribed control action u_c is within the constraints and the system state will be driven to the origin. For $\frac{|u^{\min}|}{k} < x \le \frac{-u^{\min}\beta}{\alpha}$, $u_c > u^{\max}$ resulting in $u = u^{\max}$, in turn resulting in $\dot{x} < 0$. A similar result is obtained for $\frac{-u^{\max}\beta}{\alpha} < x < -\frac{|u^{\max}|}{k}$. The analysis shows that for scalar systems, while the region of unconstrained operation for a particular control law might depend on the specific control law chosen, the stability region under the control law might still possibly be the entire NCR.

The issue of directionality again crops up when considering non-scalar systems. While it is relatively easy to determine the region of unconstrained operation for a particular control law, and, in certain cases, the region of attraction for the closed-loop system, it is not necessary that the region of attraction for the closed-loop system match the NCR. This happens due to the fact that it is in general difficult to determine, for a particular value of the state, the unique direction in which the inputs should saturate to achieve closed-loop stability. To achieve this objective, recent control designs have utilized the explicit characterization of the NCR [71] in designing CCLF based control laws that ensure stabilization from all initial conditions in the NCR [93, 94]. For nonlinear systems, where the characterization of the NCR is still an open problem, a meaningful control objective is to be able to explicitly account for the constraints in the control design and provide an explicit characterization of the closed-loop stability region.

2.7 Model Predictive Control

One of the control methods useful for accounting for constraints and optimality simultaneously is that of model predictive control (MPC). MPC is an approach which accounts for optimality considerations explicitly and is widely adopted in industry as an effective approach to deal with large multivariable constrained optimal control problems. The main idea of MPC is to choose control actions by repeatedly solving an online a constrained optimization problem, which aims at minimizing a performance index over a finite prediction horizon based on predictions obtained by a system model. In general, an MPC design is composed of three components:

1. A model of the system. This model is used to predict the future evolution of the system in open-loop and the efficiency of the calculated control actions of an MPC depends highly on the accuracy of the model.
2. A performance index over a finite horizon. This index is minimized subject to constraints imposed by the system model, restrictions on control inputs and system state, and other considerations at each sampling time to obtain a trajectory of future control inputs.
3. A receding horizon scheme. This scheme introduces the notion of feedback into the control law to compensate for disturbances and modeling errors, whereby only the first piece of the future input trajectory is implemented and the constrained optimization problem is resolved at the next sampling instance.

Consider the control of the system of Eq. (2.1) and assume that the state measurements of the system of Eq. (2.1) are available at synchronous sampling time instants $\{t_{k\geq 0}\}$, a standard MPC is formulated as follows [60]:

$$\min_{u\in S(\Delta)} \int_{t_k}^{t_{k+N}} \left[\|\tilde{x}(\tau)\|_{Q_c} + \|u(\tau)\|_{R_c}\right] d\tau + F\big(x(t_{k+N})\big) \tag{2.47}$$

$$\text{s.t.}\quad \dot{\tilde{x}}(t) = f\big(\tilde{x}(t), u(t)\big), \tag{2.48}$$

$$u(t) \in U, \tag{2.49}$$

$$\tilde{x}(t_k) = x(t_k), \tag{2.50}$$

where $S(\Delta)$ is the family of piece-wise constant functions with sampling period Δ, N is the prediction horizon, Q_c and R_c are strictly positive definite symmetric weighting matrices, $\tilde{x}$ is the predicted trajectory of the system due to control input u with initial state $x(t_k)$ at time t_k, and $F(\cdot)$ denotes the terminal penalty.

The optimal solution to the MPC optimization problem defined by Eq. (2.47)–(2.50) is denoted as $u^*(t|t_k)$ which is defined for $t \in [t_k, t_{k+N})$. The first step value of $u^*(t|t_k)$ is applied to the closed-loop system for $t \in [t_k, t_{k+1})$. At the next sampling time t_{k+1}, when a new measurement of the system state $x(t_{k+1})$ is available, and the control evaluation and implementation procedure is repeated. The manipulated input of the system of Eq. (2.1) under the control of the MPC of Eq. (2.47)–(2.50) is defined as follows:

$$u(t) = u^*(t|t_k), \quad \forall t \in [t_k, t_{k+1}), \tag{2.51}$$

which is the standard receding horizon scheme.

In the MPC formulation of Eq. (2.47)–(2.50), Eq. (2.47) defines a performance index or cost index that should be minimized. In addition to penalties on the state and control actions, the index may also include penalties on other considerations; for example, the rate of change of the inputs. Equation (2.48) is the model of the system of Eq. (2.1) which is used in the MPC to predict the future evolution of the system. Equation (2.49) takes into account the constraint on the control input, and Eq. (2.50) provides the initial state for the MPC which is a measurement of the actual system state. Note that in the above MPC formulation, state constraints are not considered but can be readily taken into account.

It is well known that the MPC of Eq. (2.47)–(2.50) is not necessarily stabilizing. To understand this, let us consider a discrete time version of the MPC implementation, for a scalar system described by $x(k+1) = \alpha x(k) + u(k)$, in the absence of input constraints. Also, let $N = 1$, q and r denote the horizon, penalty on the state deviation and input deviation, respectively. The objective function then simplifies to $q(\alpha^2 x(k)^2 + u(k)^2 + 2\alpha x(k)u(k)) + ru(k)^2$, and the minimizing control action is $u(k) = \frac{-q\alpha x(k)}{q+r}$, resulting in the closed-loop system $x(k+1) = \frac{r\alpha x(k)}{q+r}$. The minimizing solution will result in stabilizing control action only if $q > r(\alpha - 1)$. Note that for $\alpha < 1$, this trivially holds (i.e., the result trivially holds for stabilization around an open-loop stable steady state). For $\alpha > 1$, the result establishes how large the penalty on the set point deviation should be compared to the penalty on the control action for the controller to be stabilizing. The analysis is meant to bring out the fact that generally speaking, the stability of the closed-loop system in the MPC depends on the MPC parameters (penalties and the control horizon) as well as the system dynamics. Note also that even though we have analyzed an unconstrained system, the prediction horizon we used was finite (in comparison to linear quadratic regulator designs, where the infinite horizon cost is essentially captured in computing the control action, and therefore results in stabilizing controller in the absence of constraints). Finally, also note that for the case of infinite horizon, the optimum solution is also the stabilizing one, and it can be shown that such an MPC will stabilize the system with the NCR as the stability region (albeit at an impractical computational burden).

To achieve closed-loop stability without relying on the objective function parameters, different approaches have been proposed in the literature. One class of approaches is to use well-designed terminal penalty terms that capture infinite horizon costs; please, see [16, 100] for surveys of these approaches. Another class of approaches is to impose stability constraints in the MPC optimization problem [3, 14, 100]. There are also efforts focusing on getting explicit stabilizing MPC laws using offline computations [92]. However, the implicit nature of MPC control law makes it very difficult to explicitly characterize, a priori, the admissible initial conditions starting from where the MPC is guaranteed to be feasible and stabilizing. In practice, the initial conditions are usually chosen in an ad hoc fashion and tested through extensive closed-loop simulations.

2.8 Lyapunov-Based MPC

In this section, we introduce Lyapunov-based MPC (LMPC) designs proposed in [93, 108, 110] which allow for an explicit characterization of the stability region and guarantee controller feasibility and closed-loop stability.

For the predictive control of the system of Eq. (2.1), the key idea in LMPC-based designs is to utilize a Lyapunov-function based constraint and achieve immediate decay of the Lyapunov function. The set of initial conditions for which it is possible to achieve an instantaneous decay in the Lyapunov function value can be computed explicitly, and picking the (preferably largest) level curve contained in this set can provide the explicitly characterized feasibility and stability region for the LMPC.

The following example of the LMPC design is based on an existing explicit control law $h(x)$ which is able to stabilize the closed-loop system [108, 110]. The formulation of the LMPC is as follows:

$$\min_{u \in S(\Delta)} \int_{t_k}^{t_{k+N}} \left[\left\| \tilde{x}(\tau) \right\|_{Q_c} + \left\| u(\tau) \right\|_{R_c} \right] d\tau \tag{2.52}$$

$$\text{s.t.} \quad \dot{\tilde{x}}(t) = f\big(\tilde{x}(t), u(t)\big), \tag{2.53}$$

$$u(t) \in U, \tag{2.54}$$

$$\tilde{x}(t_k) = x(t_k), \tag{2.55}$$

$$\frac{\partial V(x(t_k))}{\partial x} f\big(x(t_k), u(t_k)\big) \le \frac{\partial V(x(t_k))}{\partial x} f\big(x(t_k), h\big(x(t_k)\big)\big), \tag{2.56}$$

where $V(x)$ is a Lyapunov function associated with the nonlinear control law $h(x)$. The optimal solution to this LMPC optimization problem is denoted as $u_l^*(t|t_k)$ which is defined for $t \in [t_k, t_{k+N})$. The manipulated input of the system of Eq. (2.1) under the control of the LMPC of Eq. (2.52)–(2.56) is defined as follows:

$$u(t) = u_l^*(t|t_k), \quad \forall t \in [t_k, t_{k+1}), \tag{2.57}$$

which implies that this LMPC also adopts a standard receding horizon strategy.

In the LMPC defined by Eq. (2.52)–(2.56), the constraint of Eq. (2.56) guarantees that the value of the time derivative of the Lyapunov function, $V(x)$, at time t_k is smaller than or equal to the value obtained if the nonlinear control law $u = h(x)$ is implemented in the closed-loop system in a sample-and-hold fashion. This is a constraint that allows one to prove (when state measurements are available every synchronous sampling time) that the LMPC inherits the stability and robustness properties of the nonlinear control law $h(x)$ when it is applied in a sample-and-hold fashion; please, see [30, 125] for results on sampled-data systems.

Let us denote the stability region of $h(x)$ as Ω_ρ. The stability properties of the LMPC implies that the origin of the closed-loop system is guaranteed to be stable and the LMPC is guaranteed to be feasible for any initial state inside Ω_ρ when the sampling time Δ is sufficiently small. Note that the region Ω_ρ can be explicitly characterized; please, refer to [110] for more discussion on this issue. The main advantage of the LMPC approach with respect to the nonlinear control law $h(x)$ is that optimality considerations can be taken explicitly into account (as well as constraints on the inputs and the states [110]) in the computation of the control actions within an online optimization framework while improving the closed-loop performance of the system. Since the closed-loop stability and feasibility of the

LMPC of Eq. (2.52)–(2.56) are guaranteed by the nonlinear control law $h(x)$, it is unnecessary to use a terminal penalty term in the cost index (see Eq. (2.52) and compare it with Eq. (2.47)) and the length of the horizon N does not affect the stability of the closed-loop system but it affects the closed-loop performance.

2.9 Hybrid Systems

Hybrid systems are characterized by the co-existence of continuous modes of operation along with discrete switches between the distinct modes of operation and arise frequently in the design and analysis of fault-tolerant control systems. The class of hybrid systems of interest to the focus of this book–switched systems–can be described by

$$\dot{x} = f_{i(x,t)}(x) + g_{i(x,t)}(x)u_{i(x,t)}, \tag{2.58}$$

where $x \in \mathbb{R}^n$, $u \in \mathbb{R}^n$ are the continuous variables and $i \in N$ are the discrete variables indexing the mode of operation. The nature of the function $i(x,t)$ and, in particular, its two specific forms $i(x)$ and $i(t)$ result in the so-called state-dependent and time-dependent switching. What is of more interest from a stability analysis and design point of view (both when considering the design of control laws and, in the case of time-dependent switching, the switching signal) is the possibility of infinitely many switches where it becomes crucial to explicitly consider the switched nature of the system in the stability analysis. In particular, when the possibility of infinitely many switches exists, establishing stability in the individual modes of operation is not sufficient [19], and additional conditions on the behavior of the Lyapunov-functions (used to establish stability in the individual modes of operation) during the switching (as well as of sufficient dwell-time [68]) need to be satisfied for the stability of the switched system. For the case of finite switches, the considerations include ensuring stability requirements at the onset of a particular mode are satisfied and, in particular, satisfied for the terminal (last) mode of operation.

2.10 Conclusions

In this chapter, some fundamental results on nonlinear systems analysis and control were briefly reviewed. First, the class of nonlinear systems that will be considered in this book was presented; then the definitions of stability of nonlinear systems were introduced; and following that, techniques for stabilizing nonlinear systems, for example, Lyapunov-based control, feedback linearization, handling constraints, model predictive control and Lyapunov-based model predictive control and stability of hybrid (switched) systems were discussed.

Chapter 3
Integrated Fault-Detection and Fault-Tolerant Control

3.1 Introduction

In this chapter, we consider the problem of implementing fault-detection and fault-tolerant control to single-input nonlinear processes with input constraints subject to control actuator failures. An approach predicated upon the idea of integrating fault-detection, feedback and supervisory control is presented and demonstrated. To illustrate the main idea behind the approach, we first assume availability of measurements of all the process state variables. For the processes under consideration, a family of candidate control configurations, characterized by different manipulated inputs, is first identified. For each control configuration, a Lyapunov-based controller that enforces asymptotic closed-loop stability in the presence of constraints is designed, and the constrained stability region associated with it is explicitly characterized. A fault-detection filter is used to compute the expected closed-loop behavior in the absence of faults. Deviations of the process states from the expected closed-loop behavior are used to detect faults. A switching policy is then derived on the basis of the stability regions to orchestrate the activation/deactivation of the constituent control configurations in a way that guarantees closed-loop stability in the event that a failure is detected. Often, in chemical process applications, all state variables are not available for measurement. To deal with the problem of lacking process state measurements, a nonlinear observer is designed to generate estimates of the states, which are then used to implement the state feedback controller and the fault-detection filter. A switching policy is then derived to orchestrate the activation/deactivation of the constituent control configurations in a way that accounts for the estimation error. Finally, simulation studies are presented to demonstrate the implementation and evaluate the effectiveness of the proposed fault-tolerant control scheme.

3.2 Process Description

We consider a class of continuous-time, single-input nonlinear processes with constraints on the manipulated input, represented by the following state-space descrip-

P. Mhaskar et al., *Fault-Tolerant Process Control*, DOI 10.1007/978-1-4471-4808-1_3,

tion:

$$\begin{aligned} &\dot{x}(t) = f\big(x(t)\big) + g_{k(t)}\big(x(t)\big)\big(u_{k(t)} + m_{k(t)}\big), && y_m = h_m(x), \\ &k(t) \in \mathcal{K} = \{1, \dots, N\}, N < \infty, && |u_{k(t)}| \le u^k_{\max}, \end{aligned} \tag{3.1}$$

where $x(t) \in \mathbb{R}^n$ denotes the vector of process state variables, $y_m \in \mathbb{R}$ denotes the measured variable, $u_k(t) \in [-u^k_{\max}, u^k_{\max}] \subset \mathbb{R}$ denotes the constrained manipulated input associated with the kth control configuration and $m_{k(t)} \in \mathbb{R}$ denotes the fault in the kth control configuration. For each value that k assumes in $\mathcal{K}$, the process is controlled via a different manipulated input which defines a given control configuration.

It is assumed that the origin is the equilibrium point of the nominal process (i.e., $f(0) = 0$), $g_k(x) \neq 0\ \forall x \in \mathbb{R}^n$ and that the vector functions $f(\cdot)$ and $g_k(\cdot)$ are sufficiently smooth, for all k, on $\mathbb{R}^n$. It is also assumed that for any $|u_k| \le u^k_{\max}$ the solution of the system of Eq. (3.1) exists and is continuous for all t.

3.3 Motivating Example

To motivate our fault-tolerant control design methodology, we introduce in this section a benchmark chemical reactor example that will be used to illustrate the design and implementation of the fault-tolerant control structure. To this end, consider a well-mixed, non-isothermal continuous stirred tank reactor (see Fig. 3.1) where three parallel irreversible elementary exothermic reactions of the form $\mathrm{A} \xrightarrow{k_1} \mathrm{B}$, $\mathrm{A} \xrightarrow{k_2} \mathrm{U}$, and $\mathrm{A} \xrightarrow{k_3} \mathrm{R}$ take place, where A is the reactant species, B is the desired product and U, R are undesired byproducts. Under standard modeling assumptions, a mathematical model of the process can be derived from material and energy balances and takes the following form:

$$\begin{aligned} \frac{dT}{dt} &= \frac{F}{V}(T_{\mathrm{A}0} - T) + \sum_{i=1}^{3} \frac{(-\Delta H_i)}{\rho c_p} k_{i0} \exp\left(\frac{-E_i}{RT}\right) C_{\mathrm{A}} + \frac{Q}{\rho c_p V}, \\ \frac{dC_{\mathrm{A}}}{dt} &= \frac{F}{V}(C_{\mathrm{A}0} - C_{\mathrm{A}}) - \sum_{i=1}^{3} k_{i0} \exp\left(\frac{-E_i}{RT}\right) C_{\mathrm{A}}, \\ \frac{dC_{\mathrm{B}}}{dt} &= -\frac{F}{V} C_{\mathrm{B}} + k_{10} \exp\left(\frac{-E_1}{RT}\right) C_{\mathrm{A}}, \end{aligned} \tag{3.2}$$

where C_{A} and C_{B} denote the concentrations of the species A and B, T denotes the temperature of the reactor, Q denotes the rate of heat input/removal from the reactor, V denotes the volume of the reactor, ΔH_i, k_i, E_i, $i = 1, 2, 3$, denote the enthalpies, pre-exponential constants and activation energies of the three reactions, respectively, and c_p and ρ denote the heat capacity and density of the reactor, respectively. The values of the process parameters and the corresponding steady-state

Table 3.1 Process parameters and steady-state values for the reactor of Eq. (3.2)

$V = 1000.0$ L	$R = 8.314$ J/mol K
$C_{A0s} = 4.0$ mol/L	$T_{A0s} = 300.0$ K
$\Delta H_1 = -5.0 \times 10^4$ J/mol	$\Delta H_2 = -5.2 \times 10^4$ J/mol
$\Delta H_3 = -5.4 \times 10^4$ J/mol	$k_{10} = 5.0 \times 10^4$ min^{-1}
$k_{20} = 5.0 \times 10^3$ min^{-1}	$k_{30} = 5.0 \times 10^3$ min^{-1}
$E_1 = 5.0 \times 10^4$ J/mol	$E_2 = 7.53 \times 10^4$ J/mol
$E_3 = 7.53 \times 10^4$ J/mol	$c_p = 0.231$ J/g K
$\rho = 1000.0$ g/L	$F = 83.3$ L/min
$T_s = 390.97$ K	$C_{A_s} = 3.58$ mol/L
$C_{D_s} = 0.42$ mol/L	

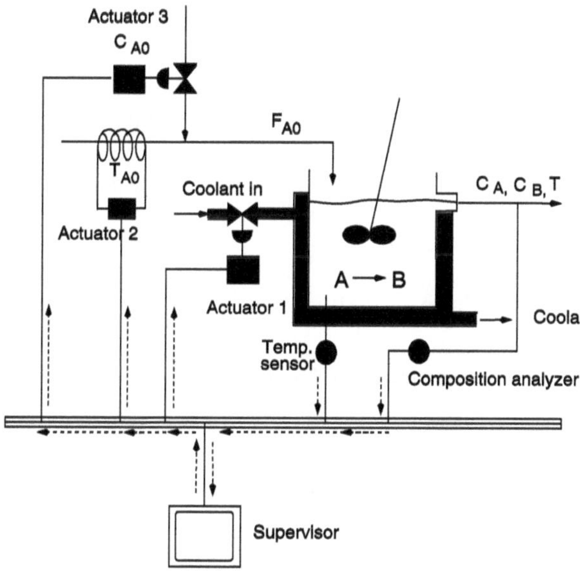

Fig. 3.1 A schematic of the CSTR showing the three candidate control configurations

values are given in Table 3.1. It was verified that these conditions correspond to an unstable equilibrium point of the process of Eq. (3.2).

The control objective considered here is the one of stabilizing the reactor at the (open-loop) unstable steady-state. Operation at this point is typically sought to avoid high temperature, while simultaneously achieving reasonable conversion. To accomplish this objective in the presence of control system failures, we consider as manipulated inputs the rate of heat input, $u_1 = Q$, subject to the constraints $|Q| \le u^1_{max} = 748$ kJ/s, the inlet stream temperature, $u_2 = T_{A0} - T_{A0s}$, subject to the constraints $|u_2| \le u^2_{max} = 100$ K, with $T_{A0s} = 300$ K and the inlet reactant concentration, $u_3 = C_{A0} - C_{A0s}$, subject to the constraints $|u_3| \le u^3_{max} = 4$ kmol/m^3, with $C_{A0s} = 4$ kmol/m^3.

Each of the above manipulated inputs, together with measurements of reactor temperature and/or concentration, represents a unique control configuration (or control-loop) that, by itself, can stabilize the reactor. In the event of some failure in the primary configuration (involving the heat input, Q), the important questions that arise include how can the supervisor detect this fault (note that measurements of the manipulated input variable are not available), and which control loop to activate once failure is detected in the active configuration. The answer to the first question involves the design of an appropriate fault-detection filter. The approach that we will utilize to answer the second question, i.e., that of deciding which backup controller should be activated in the event of a fault, will be based on the stability regions under the individual control configuration. To this end, we next review a state feedback control design that allows for characterizing the constrained stability region under each control configuration. Note that this particular choice of the controller is presented only as an example to illustrate our results, and that any other controller design that allows for an explicit characterization of the constrained stability region can be used instead. Note also that while the above example will be used to illustrate the main ideas behind the proposed fault-detection and fault-tolerant control method, we also investigate in the simulation studies below an application to a network of chemical reactors in the presence of uncertainty and measurement noise.

3.4 State Feedback Case

3.4.1 Bounded Lyapunov-Based Control

Consider the system of Eq. (3.1) for which a family of control Lyapunov functions (CLFs), $V_k(x)$, $k \in \mathcal{K} \equiv \{1, \dots, N\}$ has been found (see below for a discussion on the construction of CLFs). Using each control Lyapunov function, we construct, using the results in [85] (see also [45]), the following continuous bounded control law:

$$u_k(x) = -\frac{L_f^* V_k(x) + \sqrt{(L_f^* V_k(x))^2 + (u_{\max}^k \|(L_{g_k} V_k)(x)\|)^4}}{\|(L_{g_k} V_k)(x)\|^2 [1 + \sqrt{1 + (u_{\max}^k \|(L_{g_k} V_k)(x)\|)^2}]} (L_{g_k} V_k)(x) \quad (3.3)$$

when $(L_{g_k} V_k)(x) \neq 0$ and $u_k(x) = 0$ when $(L_{g_k} V_k)(x) = 0$, $L_f^* V_k(x) = \frac{\partial V_k(x)}{\partial x} f(x) + \rho_k V_k(x)$, $\rho_k > 0$ and $L_{g_k} V_k(x) = \frac{\partial V_k(x)}{\partial x} g_k(x)$. Let Π_k be the set defined by

$$\Pi_k(u_{\max}^k) = \{x \in \mathbb{R}^n : L_f^* V_k(x) \le u_{\max}^k \|(L_{g_k} V_k)(x)\|\} \quad (3.4)$$

and assume that

$$\Omega_k := \{x \in \mathbb{R}^n : V_k(x) \le c_k^{\max}\} \subseteq \Pi_k(u_{\max}^k) \quad (3.5)$$

for some $c_k^{\max} > 0$. It can be shown, using standard Lyapunov arguments, that in the absence of faults ($m_{k(t)} = 0$), Ω_k provides an estimate of the stability region, starting from where the control law of Eq. (3.3) guarantees asymptotic (and local exponential) stability of the origin of the closed-loop system under each control configuration. This implies that there exist class $\mathcal{KL}$ functions β_i, $i = 1, \dots, N$, such that $\|x(t)\| \leq \beta_i(\|x(0)\|, t)$. We will use this property later in the design of the output feedback controllers.

Referring to the above controller design, it is important to make the following remarks. First, a general procedure for the construction of CLFs for nonlinear systems of the form of Eq. (3.1) is currently not available. Yet, for several classes of nonlinear systems that arise commonly in the modeling of engineering applications, it is possible to exploit system structure to construct CLFs (see, for example, [56, 83]). Second, given that a CLF, V_k, has been obtained for the system of Eq. (3.1), it is important to clarify the essence and scope of the additional assumption that there exists a level set, Ω_k, of V_k that is contained in Π_k. Specifically, the assumption that the set, Π_k, contains an invariant subset around the origin, is necessary to guarantee the existence of a set of initial conditions for which closed-loop stability is guaranteed (note that even though $\dot{V}_k < 0\ \forall x \in \Pi_k \backslash \{0\}$, there is no guarantee that trajectories starting within Π_k remain within Π_k for all times). Moreover, the assumption that Ω_k is a level set of V_k is made only to simplify the construction of Ω_k. This assumption restricts the applicability of the proposed control method because a direct method for the construction of a CLF with level sets contained in Π_k is not available; see also Chap. 2. However, the proposed control method remains applicable if the invariant set Ω_k is not a level set of V_k but can be constructed in some other way (which, in general, is a difficult task). Note also that possibly larger estimates of the stability region can be computed using constructive procedures such as Zubov's method [41] or by using a combination of several Lyapunov functions.

3.4.2 State Feedback Fault-Tolerant Control

Consider the system of Eq. (3.1) where all process states are available as measurements, i.e., $h_m(x) = x$, and, without loss of generality, assume that it starts operating using control configuration i, under the controller of Eq. (3.3). At some unknown time, T_i^f, a fault occurs in the first control configuration such that for all $t \geq T_i^f$, $m_i = -u_i$, i.e., control configuration i fails. The problems at hand are those of detecting that a fault has occurred and, upon detection, to decide which of the available backup configurations should be implemented in the closed-loop to achieve fault-tolerant control. To this end, we consider a fault-detection filter and a switching logic of the form:

$$\dot{w}(t) = f_f(w, x), \qquad r(t) = h_f(w, x), \qquad k(t) = \varphi(r, w, x), \tag{3.6}$$

where $w \in \mathbb{R}^n$ is the state of the filter, $r(t) \in \mathbb{R}$ is a residual that indicates the occurrence of a fault and is the output of the filter, $f_f \in \mathbb{R}^n$ is the vector field describing the evolution of the filter state w, and $\varphi(r, w, x)$ is the switching logic that dictates which of the available control configurations should be activated.

The main idea behind the fault-tolerant control design is as follows: (i) use the available state measurements, the process model, and the computed control action to simulate the evolution of the closed-loop process in the absence of actuator faults, compare it with the actual evolution of the states, and use the difference between the two behaviors, if any, to detect faults, and (ii) having detected the fault, activate a backup control configuration for which the closed-loop state is within its stability region estimate. To formalize this idea, consider the constrained system of Eq. (3.1) for which a bounded controller of the form of Eq. (3.3) has been designed for each control configuration, and the stability region, Ω_j, $j = 1, \dots, N$ has been explicitly characterized. The fault-detection filter and the fault-tolerant control design are described in Theorem 3.1 below.

Theorem 3.1 *Let $k(0) = i$ for some $i \in \mathcal{K}$ and $x(0) := x_0 \in \Omega_i$. Set $w(0) = x(0)$ and consider the system*

$$\dot{w} = f(w) + g_i(w)u_i(w); \qquad r = \|w - x\|, \tag{3.7}$$

where $w \in \mathbb{R}^n$ is the filter state and $u_i(\cdot)$ is the feedback control law defined in Eq. (3.3). Let T_i^f be such that $m_i(t) = 0\ \forall 0 \le t \le T_i^f$, then $r(T_i^{f+}) > 0$ if and only if $m_i(T_i^f) \neq 0$. Furthermore, let T_i^s be the earliest time such that $r(t) > 0$, then the following switching rule:

$$k(t) = \begin{cases} i, & 0 \le t < T_i^s, \\ j \neq i, & t \ge T_i^s,\ x(T_i^s) \in \Omega_j \end{cases} \tag{3.8}$$

guarantees asymptotic stability of the origin of the closed-loop system.

Proof We split the proof of the theorem in two parts. In the first part, we show that the filter detects a fault if and only if one occurs, and in the second part we establish closed-loop stability under the switching rule of Eq. (3.8).
Part 1: Let $x(T_i^f) := x_{T_i^f}$ and $w(T_i^f) := w_{T_i^f}$, and consider

$$\begin{aligned} \dot{w}(T_i^f) - \dot{x}(T_i^f) &= f(x_{T_i^f}) + g(x_{T_i^f})\big(u_i(x_{T_i^f}) + m_i(T_i^f)\big) \\ &\quad - \big(f(w_{T_i^f}) + g(w_{T_i^f})u_i(w_{T_i^f})\big) \end{aligned} \tag{3.9}$$

with $m_i(T_i^f) \neq 0$. Since $w_{T_i^f} = x_{T_i^f}$, we have that

$$\begin{aligned} &f(x_{T_i^f}) + g(x_{T_i^f})\big(u_i(x_{T_i^f}) + m_i(T_i^f)\big) - \big(f(w_{T_i^f}) + g(w_{T_i^f})u_i(w_{T_i^f})\big) \\ &\quad = g(x_{T_i^f})m_i(T_i^f). \end{aligned} \tag{3.10}$$

Furthermore, since $g(x_{T_i^f}) \neq 0$, we have that

$$\dot{w}(T_i^f) - \dot{x}(T_i^f) = g(x_{T_i^f})m_i(T_i^f) \neq 0 \tag{3.11}$$

if and only if $m_i(T_i^f) \neq 0$. Since $w_{T_i^f} - x_{T_i^f} = 0$ and $\dot{w}(T_i^f) - \dot{x}(T_i^f) \neq 0$ if and only if $m_i(T_i^f) \neq 0$, we have that

$$w(T_i^{f+}) - x(T_i^{f+}) \neq 0 \tag{3.12}$$

or

$$r(T_i^{f+}) = \|w(T_i^{f+}) - x(T_i^{f+})\| > 0 \tag{3.13}$$

if and only if $m_i(T_i^f) \neq 0$.

Part 2: We prove closed-loop stability for the two possible cases; first, if no switching occurs, and second, if a switch occurs at a time T_i^s.

Case 1: The absence of a switch implies $r_i(t) = 0$. Furthermore, $r_i(t) = 0 \Rightarrow x(t) = w(t)$. Since $x(0) = w(0) \in \Omega_i$, and control configuration i is implemented for all times in this case, we have that asymptotic closed-loop stability is achieved.

Case 2: At time T_i^s, the supervisor switches to a control configuration j for which $x(T_i^s) \in \Omega_j$. From this time onwards, since configuration j is implemented in the closed-loop system for all times, and since $x(T_i^s) \in \Omega_j$, closed-loop asymptotic stability follows. This completes the proof of Theorem 3.1. □

The fault-detection filter and fault-tolerant controller are designed and implemented as follows (see also Fig. 3.2):

- Given any $x_0 \in \Omega_i$, initialize the filter states as $w(0) = x_0$ and integrate the filter dynamics using Eq. (3.7).
- Compute the norm of the difference between the filter states and the process states, $r(t) = \|w(t) - x(t)\|$, and if $r(t) = 0$, continue to implement control configuration i.
- At any time T_i^s that $r(T_i^s) > 0$, switch to a control configuration $j \neq i$, for which $x(T_i^s) \in \Omega_j$ to achieve asymptotic stability of the origin of the closed-loop system.

Note that the fault-detection filter uses a replica of the process dynamics, and that the state of the filter w is initialized at the same value as the process states $x(0)$. In the absence of faults, the evolution of $w(t)$ is identical to $x(t)$, and hence $r(t) = 0$. In the presence of faults, however, the effect of the fault is registered by a change in the evolution of the process, but not in that of the filter state (since the filter state dynamics include the computed control action, $u_i(w)$, and not the implemented control action, $u_i(w) + m_i$). This change is detected by a change in the value of $r(t)$ and declared as a fault. Note also that the fact that the faults m_i appear as additive terms to the manipulated input variable is a natural consequence

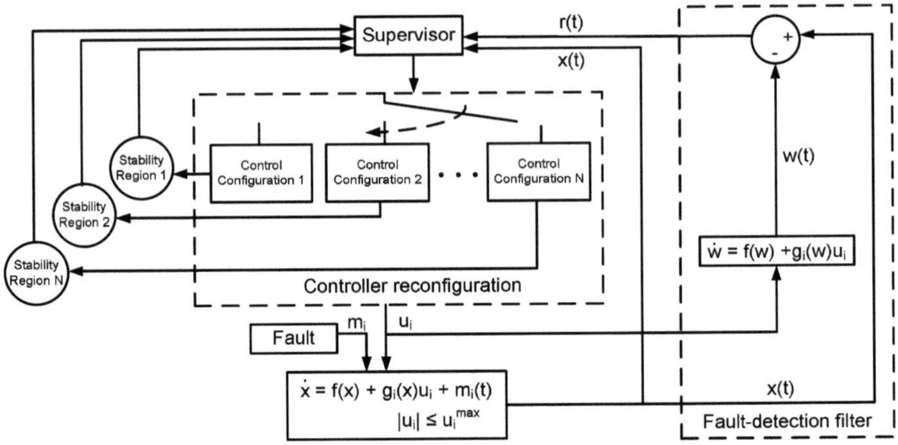

Fig. 3.2 Integrated fault-detection and fault-tolerant control design: state feedback case

of focusing on the problem of detecting (through the design of appropriate fault-detection filters) and dealing (via reconfiguration) with faults in control actuators. The approach employed in the design of the fault-detection filter can also be used to detect faults that do not necessarily appear in the control actuators, as long as they influence the evolution of the state variables.

Remark 3.1 Once a fault is detected, the switching logic ensures that the backup control configuration that is implemented in the closed-loop is one that can guarantee closed-loop stability in the presence of constraints, and this is achieved by verifying that the state of the process, at the time that a fault is detected, is present in the constrained stability region of the candidate control configuration. Note that while the bounded controller is used for a demonstration of the main ideas, other control approaches that provide an explicit characterization of the set of initial conditions for which closed-loop stability is guaranteed (achieved, for example, via the use of the hybrid predictive control approach [50] or via a Lyapunov-based model predictive control design [108]) can be used within the proposed framework. Note also that early detection of a fault enhances the chances that corrective action can be taken in time to achieve fault-tolerant control (Theorem 3.1 guarantees that a fault is detected as soon as it occurs). Specifically, it may happen that a fault occurs when the closed-loop state resides in the stability region of one of the backup configurations, but if the fault is not immediately detected, the destabilizing effect of the fault may drive the state outside the stability region of the backup configuration by the time a fault is detected (for a demonstration, see the simulation example in Sect. 3.4.3).

In the event that the process state, at the time of the failure of the primary control configuration, lies in the stability region of more than one backup control configuration, additional performance considerations such as ease and/or cost of implementing one control configuration over another can be used in choosing which control configuration should be implemented in the closed-loop system [111]. If the state at the time of a failure lies outside the stability region of all the backup controllers, then this indicates that the back up controllers do not have enough control action available and calls for increasing the allowable control action in the fall-back configurations. Recall that the set of initial conditions starting from where a given control configuration can stabilize a steady state—the so-called null-controllable region—is fundamentally limited by the constraints on the available control action, and that different control laws typically provide estimates of the stability region which are subsets of the null-controllable region; see also Chap. 2.

Remark 3.2 In the presence of plant-model mismatch or unknown disturbances, the value of $r(t)$ will be nonzero even in the absence of faults. The FDFTC problem in the presence of time-varying disturbances with known bounds on the disturbances can be handled by (i) redesigning the filter to account for the disturbances; specifically, requiring that a fault be declared only if the value of $r(t)$ increases beyond some threshold, δ, where δ accounts for the deviation of the plant dynamics from the nominal dynamics in the absence of faults (please, see the simulation example for a demonstration of this idea in an application to a network of chemical reactors in the presence of uncertainty and measurement noise) and (ii) by redesigning the controllers for the individual control configurations to mitigate the effect of disturbances on the process, and characterizing the robust stability regions and using them as criteria for deciding which backup controller should be implemented in the closed-loop system. Note that while Theorem 3.1 presents the fault-detection filter and fault-tolerant control (FDFTC) design for a fault in the primary control configuration, extensions to faults in successive backup configurations are straightforward and involve similar filter designs for the active control configuration and a switching logic that orchestrates switching to the remaining control configurations.

Remark 3.3 While we illustrate our approach using a single input, extensions to multi-input systems are possible, and fault-detection filters can be designed in the same way, using a replica of the process dynamics. The case of multi-input systems, however, introduces an additional layer of complexity due to the need of identifying which particular manipulated input has failed, i.e., the additional problem of fault-isolation. For the purpose of presenting the integrated fault-detection and fault-tolerant control structure, we focus here on multiple control configurations, where each control configuration comprises of a single input that does not require the filter to perform the additional task of fault-isolation. For a simple illustration of a fault-detection and isolation filter design, please, see the simulation example in Sect. 3.4.3. Please, also see Chap. 4 for a complete fault detection and isolation approach.

Remark 3.4 Note that the fault-detection filter presented in Theorem 3.1 detects the presence of both complete and partial failures. Once a fault is detected, the control reconfiguration strategy is the same for both cases, and that is to shut down the faulty configuration and switch to some well-functioning fall-back configuration. Note that in the case of a partial failure, unless the faulty configuration is shut down, the backup control configurations will have to be redesigned to be robust with respect to the bounded disturbance generated by the faulty configuration (for the backup control configuration, the unmeasured actuator action of the faulty control configuration will act as a disturbance and will be bounded because of the fact that the actuator itself has a limited capacity and, therefore, even if the implemented control action is not the same as that prescribed by the controller, it cannot exceed the physical limitations and will remain bounded). By shutting down the faulty configuration, however, the source of the disturbance is eliminated and no controller redesign is needed for the backup control configurations.

3.4.3 Simulation Results

In this subsection, we illustrate the implementation of the proposed fault-detection and fault-tolerant control methodology to the chemical reactor introduced as a motivating example in Sect. 3.3. We first describe the controller design for the individual control configurations. Note that our objective is full state stabilization; however, to facilitate the controller design and subsequent stability analysis, we use a state transformation to transform the system of Eq. (3.3) into the following one describing the input/output dynamics:

$$\dot{e} = Ae + l_k(e) + b\alpha_k u_k := \bar{f}_k(e) + \bar{g}_k(e)u_k \tag{3.14}$$

where $e \in \mathbb{R}^n$ is the variable in transformed coordinate (for the specific transformations used for each control configuration, please see below), $A = \left[\begin{smallmatrix} 0 & 1 \\ 0 & 0 \end{smallmatrix}\right]$, $b = \left[\begin{smallmatrix} 0 \\ 1 \end{smallmatrix}\right]$, $l_k(\cdot) = L_{f_k}^2 h_k(x)$, $\alpha_k(\cdot) = L_{g_k} L_{f_k} h_k(x)$, $h_k(x) = y_k$ is the output associated with the kth configuration, $x = [x_1\ x_2]^T$ with $x_1 = T - T_s$, $x_2 = C_A - C_{As}$, and the functions $f_k(\cdot)$ and $g_k(\cdot)$ can be obtained by rewriting the (T, C_A) model equations in Eq. (3.2) in the form of Eq. (3.1). The explicit forms of these functions are omitted for brevity. A quadratic Lyapunov function of the form $V_k = e^T P_k e$, where P_k is a positive-definite symmetric matrix that satisfies the inequality $A^T P_k + P_k A - P_k b b^T P_k < 0$, is used for controller design. In particular,

1. For the first configuration with $u_1 = Q$, we consider the controlled output $y_1 = C_A - C_{As}$. The coordinate transformation (in error variables form) takes the form $e_1 = C_A - C_{As}$, $e_2 = \frac{F}{V}(C_{A0} - C_A) - \sum_{i=1}^{3} k_{i0} e^{\frac{-E_i}{RT}} C_A$ and yields a relative degree of two with respect to the manipulated input.
2. For the second configuration with $u_2 = T_{A0} - T_{A0s}$, we choose the output $y_2 = C_A - C_{As}$ which yields the same relative degree as in the first configuration, $r_2 = 2$, and the same coordinate transformation.
3. For the third configuration with $u_3 = C_{A0} - C_{A0s}$, a coordinate transformation of the form used for configurations 1 and 2 above does not yield a sufficiently

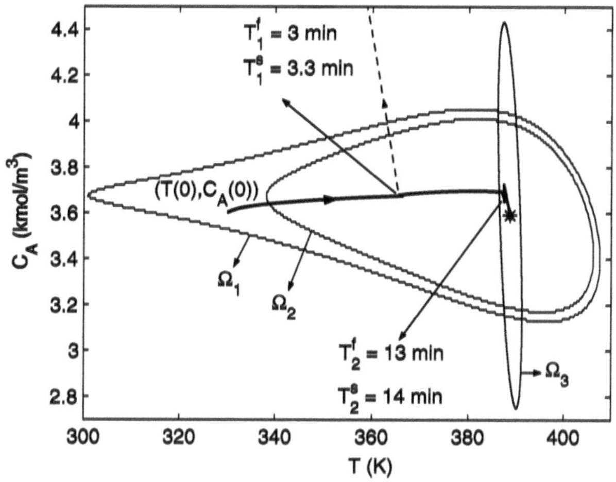

Fig. 3.3 Evolution of the closed-loop state profiles under the switching rule of Eq. (3.8) subject to failures in control systems 1 and 2 (*solid line*) and under arbitrary switching (*dashed line*)

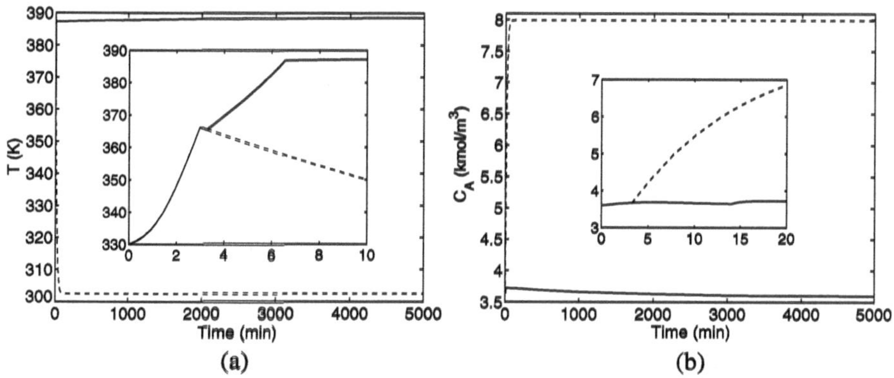

Fig. 3.4 Evolution of the closed-loop (**a**) temperature and (**b**) reactant concentration under the switching rule of Eq. (3.8) subject to failures in control systems 1 and 2 (*solid lines*) and under arbitrary switching (*dashed lines*)

large estimate of the stability region, we therefore choose a candidate Lyapunov function of the form $V_3(x) = x'Px$, where $P > 0$ and $x = [T - T_s \; C_A - C_{As}]'$ with $P = \left[\begin{smallmatrix} 0.011 & 0.019 \\ 0.019 & 0.101 \end{smallmatrix}\right]$.

Figure 3.3 depicts the stability region in the (T, C_A) space for each configuration. The desired steady-state is depicted with an asterisk that lies in the intersection of the three stability regions. The reactor as well as the fault-detection filter for the first control configuration is initialized at $T(0) = 330$ K, $C_A(0) = 3.6$ kmol/m^3, $C_B(0) = 0.0$ kmol/m^3, using the Q-control configuration, and the supervisor proceeds to monitor the evolution of the closed-loop trajectory.

As shown by the solid lines in Figs. 3.3–3.4, the controller proceeds to drive the closed-loop trajectory towards the desired steady-state, up until the Q-configuration fails after 3 minutes of reactor startup (see Fig. 3.6(a)). As can be seen in Fig. 3.5(a),

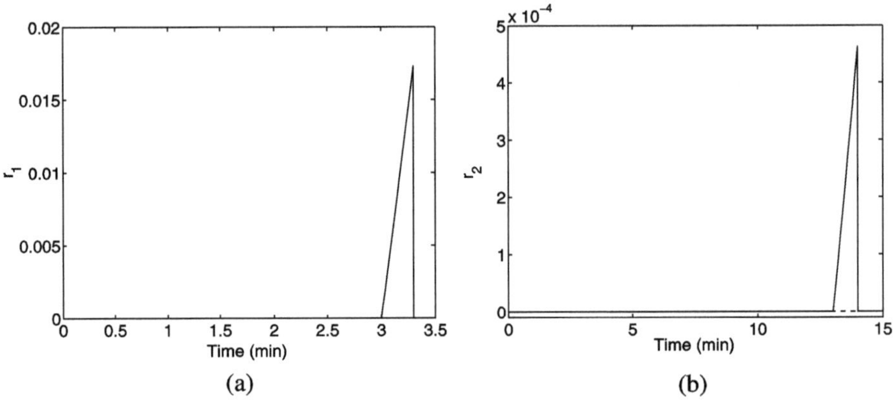

Fig. 3.5 Evolution of the closed-loop residual under the fault-detection filter for (**a**) control configuration 1 and (**b**) control configurations 2 and 3 under the switching rule of Eq. (3.8) subject to failures in control systems 1 and 2 (*solid lines*) and under arbitrary switching (*dashed lines*)

at this time the value of $r_1(t)$ becomes nonzero and the fault-detection filter detects this fault. If the supervisor switches arbitrarily and, in particular, switches to backup configuration 3, closed-loop stability is not achieved (dashed lines in Figs. 3.3–3.4). Note that this happens because the closed-loop state is outside the stability region of the third control configuration, and even though the third control configuration does not encounter a fault ($r_3(t) = 0$; see dashed line in Fig. 3.5(b)), the limited control action available in this configuration is unable to achieve closed-loop stability. On the basis of the switching logic of Eq. (3.8), the supervisor activates the second configuration (with T_{A0} as the manipulated input, see Fig. 3.6b), which continues to drive the state trajectory closer to the desired steady-state.

To demonstrate the implementation of the proposed FDFTC strategy when faults occur in successive control configurations, we consider the case when a second failure occurs (this time in the T_{A0}-configuration) at $t = 13$ minutes. Once again, the filter detects this failure via an increase in the value of $r_2(t)$ (solid line in Fig. 3.5(b)) using the fault-detection filter for control configuration 2. From Fig. 3.3, it is clear that the failure of the second control configuration occurs when the closed-loop trajectory is within the stability region of the third configuration. Therefore, the supervisor immediately activates the third control configuration (with C_{A0} as the manipulated input, see Fig. 3.6(c)) which finally stabilizes the reactor at the desired steady-state.

3.5 Handling Availability of Limited Measurements: The Output Feedback Case

The feedback controllers, the fault-detection filters, and the switching rules in the previous section were designed under the assumption of availability of measurements of all the process states. The unavailability of full state measurements has

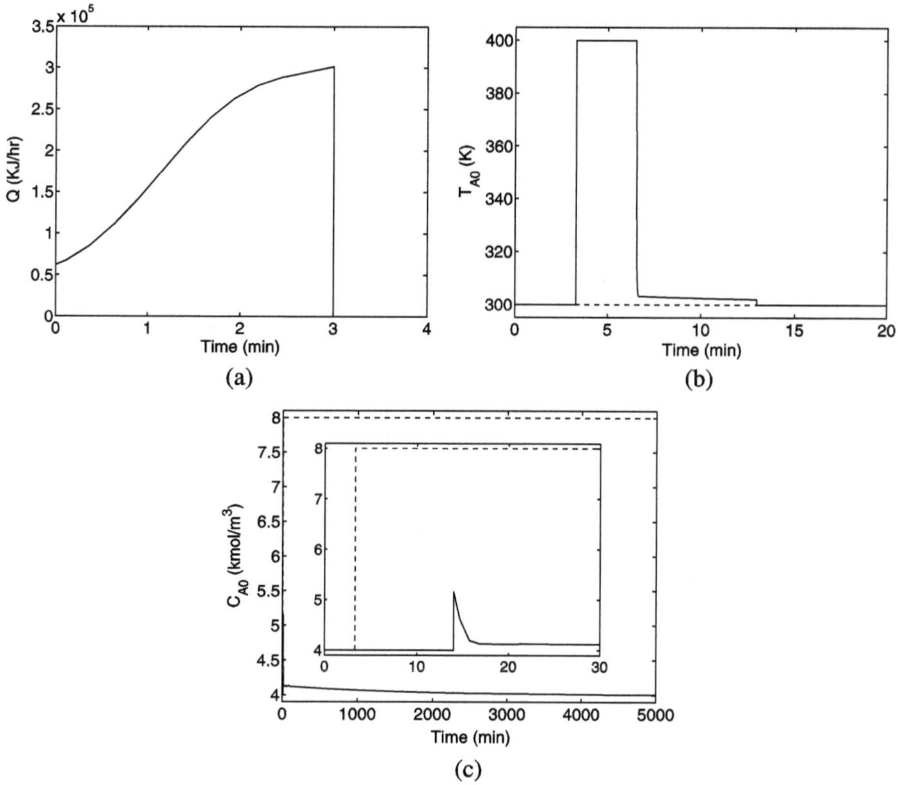

Fig. 3.6 Manipulated input profiles under (**a**) control configuration 1, (**b**) control configuration 2, and (**c**) control configuration 3 under the switching rule of Eq. (3.8) subject to failures in control systems 1 and 2 (*solid lines*) and under arbitrary switching (*dashed lines*)

several implications. First, it necessitates generating estimates of the states to be used in conjunction with both the state feedback controller and the fault-detection filter. The state estimates, however, contain errors, and this results in a difference between the expected closed-loop behavior of the measured variables (computed using the state estimates) and the evolution of the measured variables, even in the absence of actuator faults. The fault-detection filter has to be redesigned to account for this fact so that it does not treat this difference to be an indicator of an actuator fault (i.e., to prevent a false alarm). Also, the switching logic has to account for the fact that the supervisor can monitor only the state estimates and needs to make inferences about the true values of the states using the state estimates.

In the remainder of this section, we first review an output feedback controller design, proposed in [46], based on a combination of a high-gain observer and a state feedback controller (see also [26, 74, 75, 96, 152] for results on observer designs and output feedback control for unconstrained nonlinear systems) and characterize the stability properties of the closed-loop system under output feedback control. Then, we present the fault-detection filter and fault-tolerant controller and demonstrate its application via a simulation example.

3.5.1 Output Feedback Control

To facilitate the design of a state estimator with the required convergence properties, we make the following assumption:

Assumption 3.1 For each $i \in \mathcal{K}$, there exists a set of coordinates

$$[\xi_i] = \begin{bmatrix} \xi_i^1 \\ \xi_i^2 \\ \vdots \\ \xi_i^n \end{bmatrix} = \chi_i(x) = \begin{bmatrix} h_m(x) \\ L_f h_m(x) \\ \vdots \\ L_f^{n-1} h_m(x) \end{bmatrix} \tag{3.15}$$

such that the system of Eq. (3.1) takes the form

$$\begin{aligned} \dot{\xi}_i^1 &= \xi_i^2, \\ &\vdots \\ \dot{\xi}_i^{n-1} &= \xi_i^n, \\ \dot{\xi}_i^n &= L_f^n h_m\big(\chi_i^{-1}(\xi)\big) + L_{g_i} L_f^{n-1} h_m\big(\chi_i^{-1}(\xi)\big)(u_{i(t)} + m_{i(t)}), \end{aligned} \tag{3.16}$$

where $L_{g_i} L_f^{n-1} h_m(x) \neq 0$ for all $x \in \mathbb{R}^n$. Also, $\xi_i \to 0$ if and only if $x \to 0$.

We note that the change of variables is invertible since, for every x, the variable ξ_i is uniquely determined by the transformation $\xi_i = \chi_i(x)$. This implies that if one can estimate the values of ξ_i for all times, using an appropriate state observer, then we automatically obtain estimates of x for all times, which can be used to implement the state feedback controller. The existence of such a transformation will facilitate the design of high-gain observers which will be instrumental in preserving the same closed-loop stability properties achieved under full state feedback.

Proposition 3.1 below presents the output feedback controller used for each mode and characterizes its stability properties. The proof of the proposition, which invokes singular perturbation arguments (for a result on input-to-state stability with respect to singular perturbations, and further references, see [29]), is a special case of the proof of Theorem 3.2 in [46], and is omitted for brevity. To simplify the statement of the proposition, we first introduce the following notation. We define $\alpha_i(\cdot)$ as a class $\mathcal{K}$ function that satisfies $\alpha_i(\|x\|) \leq V_i(x)$. We also define the set $\Omega_{b,i} := \{x \in \mathbb{R}^n : V_i(x) \leq \delta_{b,i}\}$, where $\delta_{b,i}$ is chosen such that $\beta_i(\alpha_i^{-1}(\delta_{b,i}), 0) < \alpha_i^{-1}(c_i^{\max})$, where $\beta_i(\cdot,\cdot)$ is a class $\mathcal{KL}$ function and $c_i^{\max}$ is a positive real number defined in Eq. (3.5).

Proposition 3.1 *Consider the nonlinear system of Eq.* (3.1), *for a fixed mode*, $k(t) = i$, *and with* $m_i(t) \equiv 0$, *under the output feedback controller:*

$$\dot{\tilde{y}} = \begin{bmatrix} -L_i a_1^{(i)} & 1 & 0 & \dots & 0 \\ -L_i^2 a_2^{(i)} & 0 & 1 & \dots & 0 \\ \vdots & \vdots & \vdots & \ddots & \vdots \\ -L_i^n a_n^{(i)} & 0 & 0 & \dots & 0 \end{bmatrix} \tilde{y} + \begin{bmatrix} L_i a_1^{(i)} \\ L_i^2 a_2^{(i)} \\ \vdots \\ L_i^n a_n^{(i)} \end{bmatrix} y_m, \tag{3.17}$$

$$u_i = u_i^c(\hat{x}, u_i^{\max}),$$

where u_i^c *is defined in Eq.* (3.3), *the parameters,* $a_1^{(i)}, \dots, a_n^{(i)}$ *are chosen such that the polynomial* $s^n + a_1^{(i)} s^{n-1} + a_2^{(i)} s^{n-2} + \dots + a_n^{(i)} = 0$ *is Hurwitz,* $\hat{x} = \chi_i^{-1}(sat(\tilde{y}))$, $sat(\cdot) = \min\{1, \zeta_{\max,i}/|\cdot|\}(\cdot)$, *with* $\zeta_{\max,i} = \beta_\zeta(\delta_{\zeta,i}, 0)$ *where* β_ζ *is a class* $\mathcal{KL}$ *function and* $\delta_{\zeta,i}$ *is the maximum value of the norm of the vector* $[h_m(x), \dots, L_{f_i}^{n-1} h_m(x)]$ *for* $V_i(x) \le c_i^{\max}$ *and let* $\varepsilon_i = 1/L_i$. *Then, given* $\Omega_{b,i}$, *there exists* $\varepsilon_i^* > 0$ *such that if* $\varepsilon_i \in (0, \varepsilon_i^*]$, $x(0) \in \Omega_{b,i}$, *and* $\|\tilde{y}(0)\| \le \delta_{\zeta,i}$, *the origin of the closed-loop system is asymptotically (and locally exponentially) stable. Furthermore, given any positive real numbers,* $e_{m,i}$ *and* T_i^b, *there exists a positive real number* ε_i^{**} *such that if* $\varepsilon_i \in (0, \varepsilon_i^{**}]$ *then* $\|x(t) - \hat{x}(t)\| \le e_{m,i}$ *for all* $t \ge T_i^b$.

The state observer in Eq. (3.17) ensures sufficiently fast convergence that is necessary for the implementation of both the state feedback controller (and preserving its stability properties under output feedback control), and the fault-detection filter. The most important feature of this estimator (and one that will be used in the fault-detection filter design) is that the estimation error is guaranteed to fall below a certain value in a small period of time, T_i^b, which can be chosen arbitrarily small by sufficiently increasing the observer gain. This requirement or constraint on the error dynamics is needed even when other estimation schemes, such as moving horizon observers, are used (for example, see [116, 141]). For such observers, however, it is difficult in general to obtain a transparent relationship between the tunable observer parameters and the error decay rate.

Due to the lack of full state measurements, the supervisor can rely only on the available state estimates to decide whether switching at any given time is permissible, and, therefore, needs to make reliable inferences regarding the position of the states based upon the available state estimates. Proposition 3.2 below establishes the existence of a set, $\Omega_{s,l} := \{x \in \mathbb{R}^n : V_l(x) \le \delta_{s,l}\}$, such that once the state estimation error has fallen below a certain value (note that the decay rate can be controlled by adjusting L_l), the presence of the state within the output feedback stability region, $\Omega_{b,l}$, can be guaranteed by verifying the presence of the state estimates in the set $\Omega_{s,l}$. A similar approach was employed in the construction of the output feedback stability regions $\Omega_{b,l}$ and the regions for the state estimates $\Omega_{s,i}$ in the context of output feedback control of linear systems in [107].

Proposition 3.2 *Given any positive real number $\delta_{b,i}$, there exist positive real numbers $e^*_{m,i}$ and $\delta_{s,i}$ such that if $\|x - \hat{x}\| \leq e_{m,i}$, where $e_{m,i} \in (0, e^*_{m,i}]$, and $V_i(\hat{x}) \leq \delta_{s,i}$, then $V_i(x) \leq \delta_{b,i}$.*

Proof From the continuity of the function $V_i(\cdot)$, we have that for any positive real number $e_{m,i}$, there exists a positive real number γ_i such that $\|x - \hat{x}\| \leq e_{m,i} \Rightarrow |V_i(x) - V_i(\hat{x})| \leq \gamma_i \Rightarrow V_i(x) \leq V_i(\hat{x}) + \gamma_i$. Since γ_i can be made small by choosing $e_{m,i}$ small, it follows that given any positive real number $\delta_{b,i}$, there exists a positive real number, $e^*_{m,i}$, such that for all $e_{m,i} \in (0, e^*_{m,i}]$, $\gamma_i < \delta_{b,i}$. Now, let $\delta_{s,i}$ be any positive real number that satisfies $\delta_{s,i} + \gamma_i \leq \delta_{b,i}$. Then if $\|x - \hat{x}\| \leq e_{m,i} \leq e^*_{m,i}$ and $V_i(\hat{x}) \leq \delta_{s,i}$, we have $V_i(x) \leq V_i(\hat{x}) + \gamma_i \leq \delta_{s,i} + \gamma_i \leq \delta_{b,i}$. This completes the proof of the proposition. □

Note that for the inference that $\hat{x} \in \Omega_{s,i} \Rightarrow x \in \Omega_{b,i}$ to be useful in executing the switching, the set $\Omega_{s,i}$ needs to be contained within $\Omega_{b,i}$. From Proposition 3.2, this can be ensured if $e_{m,i}$ is sufficiently small, which in turn is ensured for all times greater than T_i^b provided that the observer gain is sufficiently large. In practice, the use of a sufficiently high observer gain leads to an $\Omega_{b,i}$ that is almost identical to Ω_i, and furthermore, once the error has sufficiently decreased, $\Omega_{s,i}$ can be taken to be almost equal to $\Omega_{b,i}$.

3.5.2 *Integrating Fault-Detection and Fault-Tolerant Output Feedback Control*

In this subsection, we will present a fault-tolerant controller that uses the estimates generated by the high-gain observer for the implementation of the fault-detection filter, the state feedback controllers and the switching logic (see Fig. 3.7). We proceed by first showing how the implementation of the design and implementation of the fault-detection filter should be modified to handle the absence of full state measurements. To this end, we consider the following system:

$$\begin{aligned} \dot{w}(t) &= f(w) + g_i(w)u_i(w), \\ r(t) &= \big\|\hat{x}(t) - w(t)\big\|. \end{aligned} \tag{3.18}$$

Note that, as in the full state feedback case, the state equation for the filter in Eq. (3.18) is a replica of the closed-loop state equation under full state feedback and in the absence of faults. However, because of the absence of full state measurements, the residual can only be defined in terms of the state estimates, not the actual states. The residual therefore provides a measure of the discrepancy between the evolution of the nominal closed-loop system (i.e., with no faults) under full state feedback and the evolution of the closed-loop state estimates under output feedback. Since the discrepancy can be solely due to estimation errors and not necessarily due

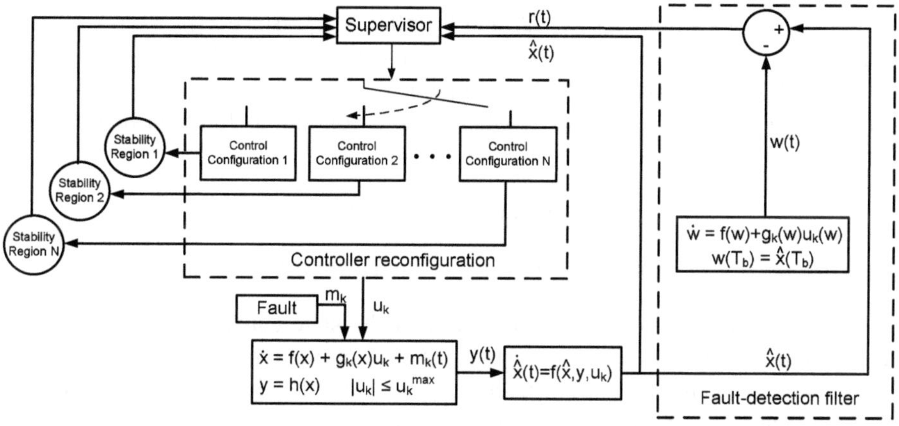

Fig. 3.7 Integrated fault-detection and fault-tolerant control design under output feedback

to faults, it is important to establish a bound on the residual which captures the expected difference in behavior in the absence of faults. This bound, which is given in Proposition 3.3 below, will be useful as a threshold to be used by the supervisor in declaring when a fault has occurred and consequently when switching becomes necessary.

Proposition 3.3 *Consider the nonlinear system of Eq.* (3.1), *for a fixed mode, $k(t) = i$, and with $m_i(t) \equiv 0$, under the output feedback controller of Eq.* (3.17). *Consider also the system of Eq.* (3.18). *Then, given the set of positive real numbers $\{\delta_{b,i}, \delta_{\zeta,i}, \delta_{m,i}, T_i^b\}$, there exists a positive real number, $\varepsilon_i' > 0$, such that if $\varepsilon_i \in (0, \varepsilon_i']$, $V_i(x(0)) \leq \delta_{b,i}$, $\|\tilde{y}(0)\| \leq \delta_{\zeta,i}$, $w(T_i^b) = \hat{x}(T_i^b)$, the residual satisfies a relation of the form $r(t) \leq \delta_{m,i}$ for all $t \geq T_i^b$.*

Proof Consider the system of Eq. (3.1) with $m_i(t) \equiv 0$ under the output feedback controller of Eq. (3.17). From the result of Proposition 3.1, we have that given $x(0) \in \Omega_{b,i}$ and any positive real number T_i^b, there exists a real positive number ε_i^{**} such that $\|x(t) - \hat{x}(t)\| \leq k_1 \varepsilon_i$, for all $t \geq T_i^b$, $\varepsilon_i \in (0, \varepsilon_i^{**}]$, for some $k_1 > 0$, i.e., $x(t) = \hat{x}(t) + O(\varepsilon_i)$, where $O(\varepsilon_i)$ is the standard order of magnitude notation. Now, consider the following two systems for $t \geq T_i^b$:

$$\dot{x}(t) = f\big(x(t)\big) + g_i\big(x(t)\big)u_i\big(\hat{x}(t)\big), \tag{3.19}$$

$$\dot{w}(t) = f\big(w(t)\big) + g_i\big(w(t)\big)u_i\big(w(t)\big), \tag{3.20}$$

where $w(T_i^b) = \hat{x}(T_i^b)$. The system of Eq. (3.20) is exactly the closed-loop system under full state feedback and has an asymptotically (and exponentially) stable equilibrium at the origin, for all initial conditions within Ω_i. The system of Eq. (3.19) is the closed-loop system under output feedback and (from Proposition 3.1) has an asymptotically (and locally exponentially) stable equilibrium at the origin, for all initial conditions within $\Omega_{b,i} \subset \Omega_i$ and for all $\varepsilon_i \le \varepsilon_i^*$. Since $x(t) = \hat{x}(t) + O(\varepsilon_i)$ for all $t \ge T_i^b$, we have that $x(T_i^b) = \hat{x}(T_b^i) + O(\varepsilon_i)$ and, when $\varepsilon_i = 0$, the two systems of Eqs. (3.19)–(3.20) become identical. Let $F_i(\cdot) = f(\cdot) + g_i(\cdot)u_i(\cdot)$, and $x(T_i^b) = \hat{x}(T_b^i) + O(\varepsilon_i) := \eta(\varepsilon_i)$, where η is a continuous function that depends smoothly on ε_i, then we can write

$$\begin{aligned} \dot{x}(t) &= F_i\big(x(t), \varepsilon_i\big), & x\big(T_i^b\big) &= \eta(\varepsilon_i), \\ \dot{w}(t) &= F_i\big(w(t)\big), & w\big(T_i^b\big) &= \eta(0). \end{aligned} \tag{3.21}$$

It is clear from the above representation that the state equations for both the filter system and the closed-loop system, as well as their initial conditions at T_i^b, are identical when $\varepsilon_i = 0$. Therefore, we can use the theory of regular perturbations (see Chap. 8 in [76]) to establish the closeness of solutions between the two systems over the infinite time interval. In particular, since $F_i(\cdot)$ is continuous and bounded on $\Omega_{b,i}$, and the w-system is exponentially stable, an application of the result of Theorem 8.2 in [76] yields that there exists $\varepsilon_i'' > 0$ such that for all $\varepsilon_i \in (0, \varepsilon_i'']$, $x(t) = w(t) + O(\varepsilon_i)$ for all $t \ge T_i^b$. We therefore have that, for $\varepsilon_i \in (0, \min\{\varepsilon_i^{**}, \varepsilon_i''\}]$, $r(t) = \|\hat{x}(t) - w(t)\| = \|\hat{x}(t) - x(t) + x(t) - w(t)\| \le \|\hat{x}(t) - x(t)\| + \|x(t) - w(t)\| \le (k_1 + k_2)\varepsilon_i$ for all $t \ge T_i^b$. This implies that given any positive real number $\delta_{m,i}$, there exists $\varepsilon_i' > 0$ such that $\|\hat{x}(t) - w(t)\| \le \delta_{m,i}$ for all $\varepsilon_i \in (0, \varepsilon_i']$, for all $t \ge T_i^b$, where $\varepsilon_i' = \min\{\varepsilon_i^{**}, \varepsilon_i'', \delta_{m,i}/(k_1 + k_2)\}$.

To summarize, we conclude that given the set of positive real numbers $\{\delta_{b,i}, \delta_{\zeta,i}, \delta_{m,i}, T_i^b\}$, there exists a positive real number, $\varepsilon_i' > 0$, such that if $\varepsilon_i \in (0, \varepsilon_i']$, $V_i(x(0)) \le \delta_{b,i}$, $\|\tilde{y}(0)\| \le \delta_{\zeta,i}$, $w(T_i^b) = \hat{x}(T_i^b)$, the residual satisfies a relation of the form $r(t) \le \delta_{m,i}$ for all $t \ge T_i^b$. This completes the proof of the proposition. □

Note that the bound $\delta_{m,i}$ can be chosen arbitrarily small by choosing the observer gain to be sufficiently large. Note also that, unlike the case of full state feedback, the fault-detection filter is initialized only after the passage of some short period of time, $[0, T_i^b]$ (which can be chosen arbitrarily small by increasing the observer gain), to ensure that the closed-loop state estimates have converged sufficiently close to the true closed-loop states and thus—by setting the filter state w at this time equal to the value of the state estimate—ensure that the filter state is initialized sufficiently close to the true values of the state. From this point onwards, the filter simply integrates a replica of the dynamics of the process in the absence of errors. In the absence of actuator faults, the difference between the filter states and the process states is a function of the initial error, which can be bounded from above by a value that can be made as small as desired by decreasing the initial error, which in turn can be done by appropriate choice of the observer parameters.

Having established a bound on the residual in the absence of faults, we are now ready to proceed with the design of the switching logic. To this end, consider the nonlinear system of Eq. (3.1) where, for each control configuration, an output feedback controller of the form of Eq. (3.17) is available and, given the desired output feedback stability regions $\Omega_{b,i} \subset \Omega_i$, $i = 1, \dots, N$, as well as the desired values for $\delta_{m,l}$, T_b^l, an appropriate observer gain has been determined (e.g., $\varepsilon_l \leq \min\{\varepsilon_l^*, \varepsilon_l', \varepsilon_l^{**}\}$ to guarantee both stability and satisfaction of the desired bound on the residual) and the sets $\Omega_{s,i}$ (see Proposition 3.2) have been computed. The implementation of the fault-detection filter and fault-tolerant controller is described in Theorem 3.2 below.

Theorem 3.2 *Let $k(0) = i$ for some $i \in \mathcal{K}$, $x(0) \in \Omega_{b,i}$, $w(T_i^b) = \hat{x}(T_i^b)$, and consider a fault for which $r(T_i^s) \geq \delta_{m,i}$, where $T_i^s > T_i^b$ is the earliest time for which $r(t) \geq \delta_{m,i}$. Then under the switching rule*

$$k(t) = \begin{cases} i, & 0 \leq t < T_i^s, \\ j \neq i, & t \geq T_i^s,\ \hat{x}(T_i^s) \in \Omega_j^s \end{cases} \tag{3.22}$$

the origin of the closed-loop system is asymptotically stable.

Proof Consider the nonlinear system of Eq. (3.1), under the output feedback controller of Eq. (3.17), and the system of Eq. (3.18), where $k(0) = i$ for some $i \in \mathcal{K}$, $x(0) \in \Omega_{b,l}$, $w(T_i^b) = \hat{x}(T_i^b)$, $\varepsilon_l \leq \min\{\varepsilon_l^*, \varepsilon_l', \varepsilon_l^{**}\}$, where ε_l^*, ε_l^{**} were defined in Proposition 3.1 and ε_l' was defined in Proposition 3.3. Since we consider only faults for which $r(T_i^s) \geq \delta_m^l$, where $T_i^s > T_i^b$ is the earliest time for which $r(t) \geq \delta_m^l$, it follows that

(a) In the absence of such faults, no switching takes place and configuration i is implemented for all times. Since $x(0) \in \Omega_{b,l}$ and $\varepsilon_l \leq \varepsilon_l^*$, asymptotic closed-loop stability of the origin follows directly from Proposition 3.1.
(b) In the case that such faults take place, the earliest time a fault is detected is $T_i^s > T_i^b$ and we have, from Eq. (3.22), that $k(t) = i$ for $0 \leq t < T_i^s$. From the stability of the ith closed-loop system established in Proposition 3.1, we have that the closed-loop trajectory stays bounded within $\Omega_{b,l}$ for $0 \leq t < T_i^s$. At time T_i^s, the supervisor switches to a control configuration j for which $\hat{x}(T_i^s) \in \Omega_{s,j}$. By design, $\hat{x}(t) \in \Omega_{s,j} \Rightarrow x(t) \in \Omega_{b,j}$ for all $t \geq T_i^s > T_i^b$. From this point onwards, configuration j is implemented in the closed-loop system for all future times and, since $x(T_i^s) \in \Omega_{b,j}$, asymptotic closed-loop stability of the origin follows from the result of Proposition 3.1. This completes the proof of Theorem 3.2. □

The design and implementation of the fault-detection filter and fault-tolerant controller proceed as follows:

1. Given the nonlinear process of Eq. (3.1), identify the available control configurations, $k = 1, \ldots, N$. For each configuration, design the output feedback controller of Eq. (3.17), and for a given choice of the output feedback stability region, $\Omega_{b,i}$, determine a stabilizing observer gain, ε_i^*.
2. Given any positive real numbers, $\delta_{m,l}$ and T_i^b, determine the observer gain, ε_i', for which the maximum possible difference between the filter states and the state estimates, in the absence of faults, is less than the threshold $\delta_{m,i}$ for all times greater than T_i^b.

3. Given the output feedback stability region, $\Omega_{b,i}$, determine the maximum error, $e^*_{m,i}$, and the set $\Omega_{s,i}$ such that if $\|x - \hat{x}\| \leq e_{m,i} \leq e^*_{m,i}$ (i.e., the error between the estimates and the true values of the states is less than $e_{m,i}$) and $\hat{x} \in \Omega_{s,i}$ (i.e., the state estimates belong to $\Omega_{s,i}$), then $x \in \Omega_{b,i}$ (i.e., the state belongs to the output feedback stability region).
4. For a choice of $e_{m,i} \in (0, e^*_{m,i}]$ and given T_i^b, determine the observer gain, ε_i^{**}, for which the maximum possible difference between the states and the state estimates, in the absence of faults, is less than the threshold $e_{m,i}$ for all times greater than T_i^b. Set $\varepsilon_i := \min\{\varepsilon_i^*, \varepsilon_i', \varepsilon_i^{**}\}$. Note that this choice guarantees that by time T_i^b: (i) the residual is within the desired threshold and (ii) the presence of $\hat{x}$ within $\Omega_{s,i}$ guarantees that x belongs to $\Omega_{b,i}$.
5. Initialize the closed-loop system such that $x(0) \in \Omega_{b,i}$, for some $i \in \mathcal{K}$, and start generating the state estimates $\hat{x}(t)$. At time T_i^b, initialize and start integrating the filter dynamics of Eq. (3.18) with $w(T_i^b) = \hat{x}(T_i^b)$, where $\hat{x}$ is the state estimate generated by the high-gain observer.
6. At the earliest time $T_i^s > T_i^b$ that $r(t) > \delta_{m,i}$ (implying that the difference between the expected evolution of the process states and the estimates of the process states is more than what can be accounted for by the error in the initialization of the filter states, implying that a fault has occurred), activate the backup configuration for which $\hat{x}(T_i^s) \in \Omega_{s,j}$ (note that since $t = T_i^s > T_i^b$, we have that $\|x(T_i^s) - \hat{x}(T_i^s)\| \leq e_{m,i}$; this, together with $\hat{x}(T_i^s) \in \Omega_{s,j}$, implies that $x(T_i^s) \in \Omega_{b,j}$, i.e., the state belongs to the stability region of configuration j). Implement the backup configuration j to achieve closed-loop stability.

Theorem 3.2 considers faults that are "observable" from the filter's residual in the sense that if the residual in Eq. (3.18) exceeds the allowable threshold $\delta_{m,i}$ at any time, then the supervisor can conclude with certainty that a fault has occurred. On the other hand, if the residual does not exceed the allowable threshold, it might still be possible that some "unobservable" fault—the effect of which is within the filter threshold—has taken place. Note that in contrast to the case of full state feedback, the states in this case are only known up to a certain degree of accuracy. Therefore, any fault that causes a difference in the closed-loop behavior that is within that margin of (i.e., indistinguishable from) the effect of the estimation error will, in principle, go undetected. This class of faults is not considered in Theorem 3.2 since its effect on closed-loop stability cannot be discerned from the behavior of the residual. This, however, is not a restriction since the observability threshold $\delta_{m,i}$ is a design parameter and can be chosen arbitrarily small, thus rendering the possibility of major (i.e., destabilizing) faults that cannot be detected quite small. Ultimately, the choice of $\delta_{m,i}$ reflects a fundamental tradeoff between the need to avoid false alarms that could be caused by estimation errors (this favors a relatively large threshold) and the need to minimize the possibility of some faults going undetected (this favors a relatively small threshold).

Note that for all times prior to T_i^b, the filter is inactive. Until this time, the state estimates have not yet converged close enough to the true values of the states, and no inference about the state of the system can be drawn by looking at the evolution of the state estimate, and therefore no inference about any possible faults can be drawn via the fault-detection filter. If a fault occurs within this time, the filter will detect its occurrence only after the time T_i^b. By choosing a larger value of the observer gain, however, the time T_i^b can be reduced further, if so desired. Note also that while we consider the problem of unavailability of some of the state variables as measurements, we do not consider the problem of sensor faults, i.e., we assume that the sensors do not malfunction both in the state and output feedback cases; these issues will be studied in Chaps. 8 and 9. In the event of availability of multiple measurements in a way that each of them can be used to estimate of the process states, the estimates of the states generated using the different measurements can be used to also detect sensor faults.

Remark 3.5 The central idea behind the model-based fault-detection filter design, that of comparing the evolution of the process to the expected evolution of the process in the absence of faults, can also be used to design a rule-based fault-detection filter. One example of a rule-based fault-detection filter is to declare a fault if the state estimates, after a time T_i^b, touch the boundary of $\Omega_{s,i}$, indicating that the closed-loop states themselves may be about to escape the output feedback stability region $\Omega_{b,i}$. The rule-based fault detection filter, however, would be able to detect the fault only when the state estimates hit the boundary of $\Omega_{s,i}$, as opposed to the model-based fault detection filter, which detects a fault as soon as the effect of the fault on the closed-loop evolution goes beyond a prescribed threshold. This delay in a rule-based approach could result in the state escaping the stability region of the available backup configurations (see the simulation for an example). Also, it may happen that the fault causes the closed-loop process states evolving within $\Omega_{s,i}$ to neither escape $\Omega_{s,i}$ nor converge to the origin. The rule based fault-detection filter would not be able to detect such a fault. In contrast, the model-based fault-detection filter of Theorem 3.2, is able to detect faults that have an effect, up to a desirable threshold, on the evolution of the closed-loop process. Note also that the model-based fault-detection filter of Theorem 3.2 and the rule-based fault-detection filter discussed above differ only in that the model-based filter of Theorem 3.2 uses a more quantitative knowledge of the closed-loop dynamics to predict the expected closed-loop trajectory, instead of using the qualitative knowledge that the fault-free closed-loop state trajectory does not escape the stability region.

3.5.3 Simulation Results

In this subsection, we first illustrate the implementation of the proposed fault-tolerant control methodology to the chemical reactor introduced as a motivating example to clearly explain the main ideas behind the application of the proposed fault-detection and fault-tolerant control method, and then demonstrate an application to a chemical reactor example, investigating issues such as uncertainty and measurement noise.

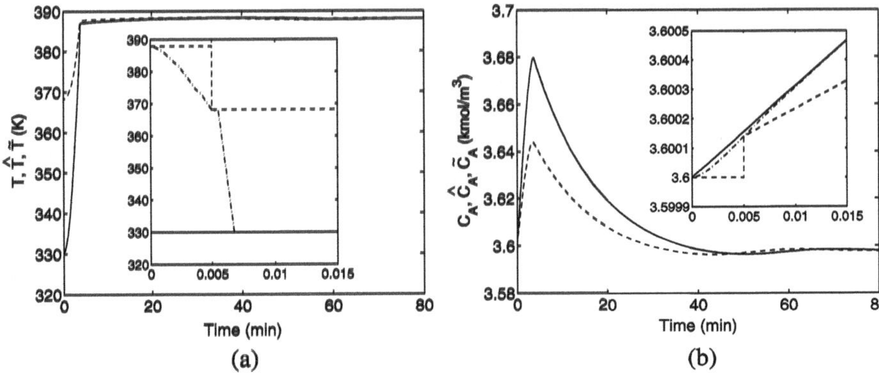

Fig. 3.8 Evolution of the closed-loop (**a**) temperature (*solid line*), estimate of temperature (*dash–dotted line*) and the temperature profile generated by the filter (*dashed line*) and (**b**) concentration (*solid line*), estimate of concentration (*dash-dotted line*) and the concentration profile generated by the filter (*dashed line*) under control configuration 1 when the fault detection filter is initialized at $t = 0.005$ minutes

For the chemical reactor of Sect. 3.3, Fig. 3.11 depicts the stability region, in the (T, C_{A}) space, for each configuration. The desired steady-state is depicted with an asterisk that lies in the intersection of the three stability regions. For the first two control configurations, a state estimator of the form of Eq. (3.17) is designed. For thresholds of $\delta_m = 0.0172$ and 0.00151 in the fault detection filters, the parameters in the observer of Eq. (3.17) are chosen as $L_1 = L_2 = 100$, $a_1^{(1)} = a_1^{(2)} = 10$, and $a_2^{(1)} = a_2^{(2)} = 20$. For the third configuration, the estimates, $\hat{T}$, $\hat{C}_{\mathrm{A}}$ are generated as follows:

$$\begin{aligned}
\frac{d\hat{T}}{dt} &= \frac{F}{V}(T_{\mathrm{A0}} - \hat{T}) + \sum_{i=1}^{3} \frac{(-\Delta H_i)}{\rho c_p} k_{i0} e^{\frac{-E_i}{R\hat{T}}} \hat{C}_{\mathrm{A}} + \alpha_1 (C_{\mathrm{A}} - \hat{C}_{\mathrm{A}}), \\
\frac{d\hat{C}_{\mathrm{A}}}{dt} &= \frac{F}{V}(C_{\mathrm{A0}} - \hat{C}_{\mathrm{A}}) - \sum_{i=1}^{3} k_{i0} e^{\frac{-E_i}{R\hat{T}}} \hat{C}_{\mathrm{A}} + \alpha_2 (C_{\mathrm{A}} - \hat{C}_{\mathrm{A}}),
\end{aligned} \tag{3.23}$$

where $\alpha_1 = -10^4$ and $\alpha_2 = 10$. The reactor is initialized at $T(0) = 330$ K, $C_{\mathrm{A}}(0) = 3.6$ kmol/m^3, $C_{\mathrm{B}}(0) = 0.0$ kmol/m^3, using the Q-control configuration, while the state estimates are initialized at $\hat{T}(0) = 390$ K, $\hat{C}_{\mathrm{A}}(0) = 3.6$ kmol/m^3 and the supervisor proceeds to monitor the evolution of the closed-loop estimates.

We first demonstrate the need to wait for a sufficient time before initializing the filter. To this end, consider the fault-detection filter initialized at $t = 0.005$ minutes $\equiv T_1^b$ at which time the state estimates (dash-dotted lines in Fig. 3.8) have not converged to the true values (solid lines in Fig. 3.8). As a result, the fault-detection filter shows a false alarm (see Fig. 3.9(a)) by crossing the threshold even when control configuration 1 is functioning properly (see Fig. 3.9(b)) and stabilizes the closed-loop system. Note that while the initialization of the filter at a time when

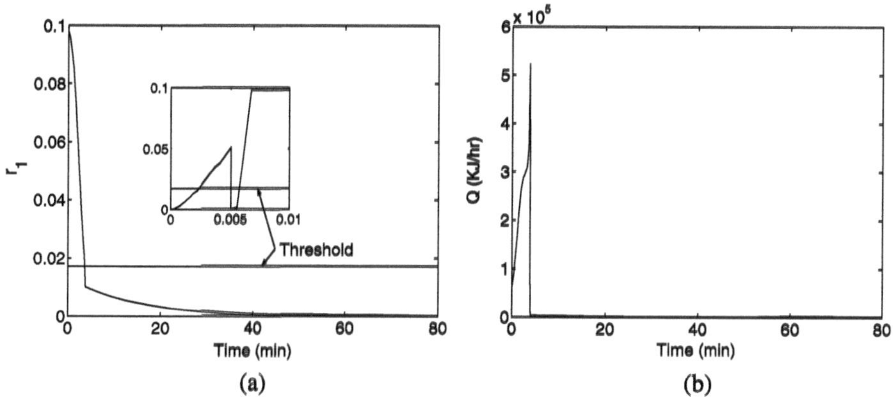

Fig. 3.9 Evolution of (**a**) the residual and (**b**) the manipulated input profile for the first control configuration when the fault detection filter is initialized at $t = 0.005$ minutes

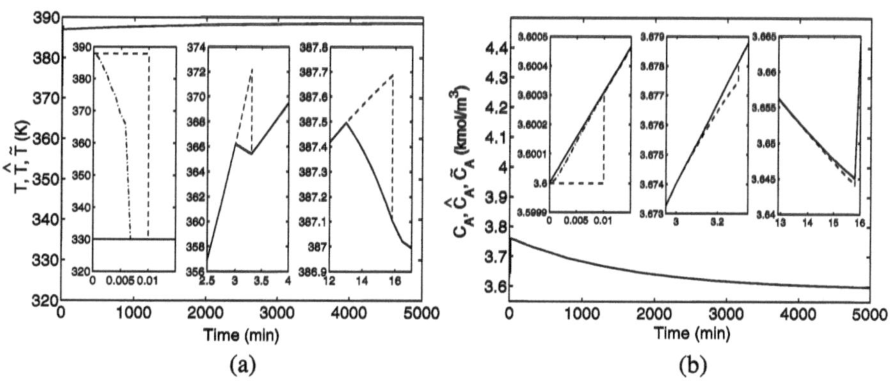

Fig. 3.10 Evolution of the closed-loop (**a**) temperature (*solid line*), estimate of temperature (*dash-dotted line*) and the temperature profile generated by the filter (*dashed line*) and (**b**) concentration (*solid line*), estimate of concentration (*dash-dotted line*) and the concentration profile generated by the filter (*dashed line*) under the switching rule of Eq. (3.22) subject to failures in control systems 1 and 2

the state estimates have not converged leads to the residual crossing the threshold, the residual eventually goes to zero as expected, since both the filter states and the closed-loop process states eventually stabilize and go to the same equilibrium point.

We now demonstrate the application of the fault-detection filter and fault-tolerant controller of Theorem 3.2. Starting from the same initial conditions, the estimates of T and C_{A} (dash-dotted lines in Fig. 3.10(a)–(b)) converge very quickly to the true values of the states (solid lines in Fig. 3.10(a)–(b)). The states in the fault-detection filter are initialized and set equal to the value of the state estimates at $t = 0.01$ minutes $\equiv T_1^b$; note that by this time the estimates have converged to the true values. By initializing the fault-detection filter appropriately, a false alarm is prevented (the value of $r_1(t)$ does not hit the threshold in the absence of a fault after a time $t = 0.01$

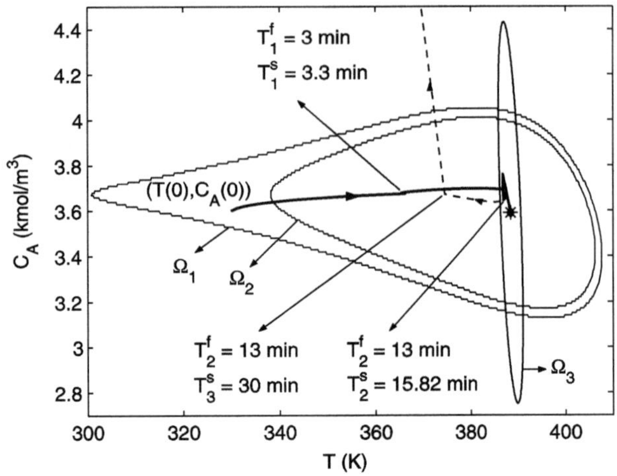

Fig. 3.11 Evolution of the closed-loop state trajectory under the switching rule of Eq. (3.22) subject to failures in control systems 1 and 2, using an appropriate fault-detection filter (*solid line*) and in the absence of a fault-detection filter (*dashed line*)

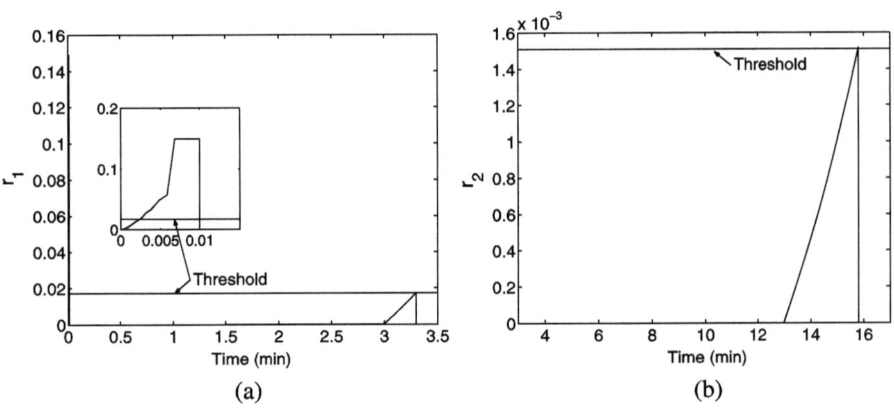

Fig. 3.12 Evolution of the residual for (**a**) the first control configuration and (**b**) the second control configuration

minutes, see Fig. 3.12(a)). As shown by the solid lines in Fig. 3.11, the controller proceeds to drive the closed-loop trajectory towards the desired steady-state, up until the Q-configuration fails after 3.0 minutes $\equiv T_1^f$ of reactor startup (see solid lines in Fig. 3.14(a)). Note that at this time, the value of $r_1(t)$ becomes non-zero and hits the threshold at $t = 3.3$ minutes $\equiv T_1^s$. From Fig. 3.11, it is clear that the failure of the primary control configuration occurs when the closed-loop trajectory is within the stability region of the second control configuration, and outside the stability region of the third control configuration. Therefore, on the basis of the switching logic of Eq. (3.22), the supervisor activates the second configuration (with T_{A0} as the manipulated input). The result is shown by the solid line in Fig. 3.11 where it is seen that upon switching to the T_{A0}-configuration, the corresponding controller continues to drive the state trajectory closer to the desired steady-state.

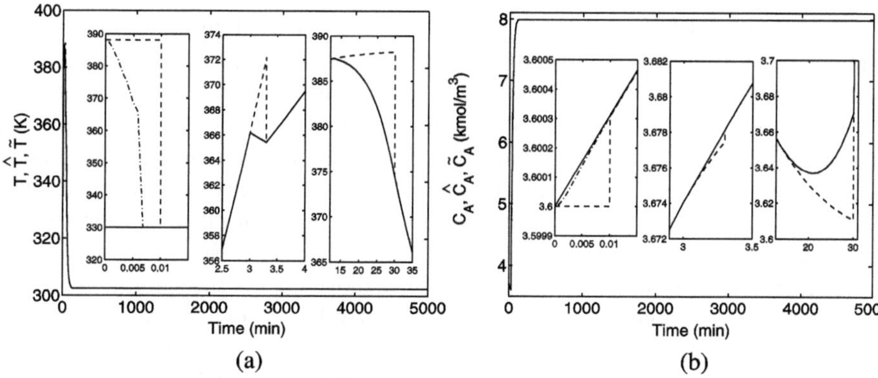

Fig. 3.13 Evolution of the closed-loop (**a**) temperature (*solid line*), estimate of temperature (*dash–dotted line*) and the temperature profile generated by the filter (*dashed line*) and (**b**) concentration (*solid line*), estimate of concentration (*dash-dotted line*) and the concentration profile generated by the filter (*dashed line*) under the switching rule of Eq. (3.22) subject to failures in control systems 1 and 2 in the absence of a fault-detection filter

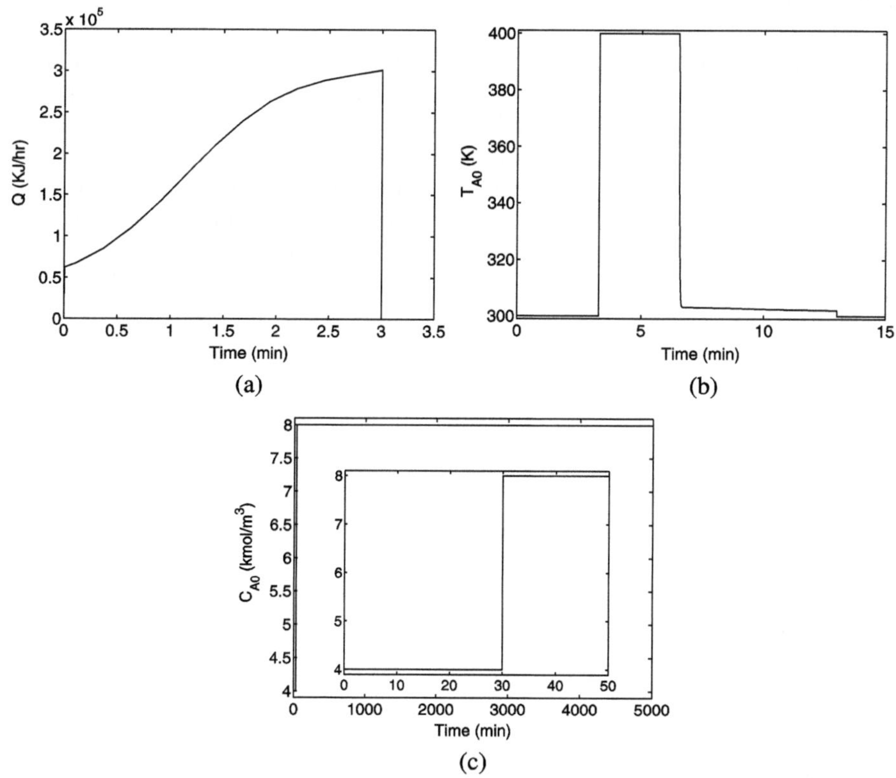

Fig. 3.14 Manipulated input profiles under (**a**) control configuration 1, (**b**) control configuration 2, and (**c**) control configuration 3 under the switching rule of Eq. (3.22) subject to failures in control systems 1 and 2 in the presence (*solid lines*) and absence (*dashed lines*) of a fault-detection filter

When a second failure occurs (this time in the T_{A0}-configuration) at $t = 13.0$ minutes $\equiv T_2^f$ (which is simulated by fixing T_{A0} for all $t \geq 13.0$ minutes, see solid lines in Fig. 3.14(b)) before the process has reached the steady state, the filter detects this failure via the value of $r_2(t)$ hitting the threshold (see Fig. 3.12(b)). From the solid line in Fig. 3.11, it is clear that the failure of the second control configuration occurs when the closed-loop trajectory is within the stability region of the third configuration. However, if the fault-detection filter is not in place and the backup configuration is implemented late in the closed-loop (at $t = 30$ minutes $\equiv T_3^s$), by this time the state of the closed-loop system has moved out of the stability region of the third control configuration, and closed-loop stability is not achieved (see dashed line in Fig. 3.11, see also Fig. 3.13 and dashed lines in Fig. 3.14). In contrast, when the fault-detection filter is in place, it detects a fault at $t = 15.82$ minutes $\equiv T_2^s$ and when the supervisor switches to configuration 3, closed-loop stability is achieved (see solid line in Fig. 3.11).

3.6 Conclusions

In this chapter, fault-detection and fault-tolerant control strategies for single-input nonlinear systems were presented. The problem was first studied in the case that the whole state feedback is available and then extended to the case of output feedback. The presented framework integrates fault-detection, feedback, and supervisory control together and provides effective fault-tolerant control for nonlinear systems. Simulation studies were presented to demonstrate the implementation and evaluate the effectiveness of the presented fault-tolerant control scheme.

Chapter 4
Integrated Fault-Detection and Isolation and Fault-Tolerant Control

4.1 Introduction

This chapter considers the problem of implementing fault tolerant control on a multi-input multi-output nonlinear system subject to multiple faults in the control actuators and constraints on the manipulated inputs. To illustrate some of the ideas behind the design of the fault-detection and isolation filter and subsequent reconfiguration, the case where all the states of the system are measured is first considered. The state measurements and the model are used to design filters that essentially capture the difference between the fault-free evolution and the evolution of the system to detect and isolate faults. Once a fault is detected and isolated, out of the available backup configurations, a configuration is chosen that (i) does not use the failed control actuator and (ii) guarantees the stability of the closed-loop system starting from the system state at the time of the failure. To be able to ascertain the second condition, Lyapunov-based controllers are used in designing the control laws for the individual control configurations which provide an explicit characterization of the set of initial conditions starting from where the closed-loop stability is guaranteed. The more complicated and realistic problem where all the system states are not measured is considered next. First, output-feedback controllers are designed that use a combination of state estimators and state-feedback controllers in a way that allows for an explicit characterization of the output-feedback stability region. The state estimates are employed in the design of the fault-detection and isolation filters, and also in devising the reconfiguration rule that determines which of the backup control configurations should be implemented in the closed-loop system. The implementation of the fault-detection and isolation filters and reconfiguration strategy is first illustrated via a chemical reactor example under state-feedback, and then issues such as uncertainty, measurement noise, and applicability in an output-feedback setting are investigated in further chemical reactor examples. Finally, the application of the integrated fault detection and isolation and fault-tolerant control framework to a reverse osmosis water desalination process is presented.

P. Mhaskar et al., *Fault-Tolerant Process Control*, DOI 10.1007/978-1-4471-4808-1_4,

4.2 Preliminaries

We consider nonlinear systems with input constraints, described by

$$\begin{aligned} \dot{x} &= f(x) + G_{k(t)}(x)\big(u_{k(t)}(x) + \tilde{u}_{k(t)}(t)\big), \qquad y(x) = h(x), \\ u_k &\in \mathbf{U}_k, k(t) \in \mathcal{K} = \{1, \dots, N\}, \quad N < \infty, \end{aligned} \tag{4.1}$$

where $x \in \mathbb{R}^n$ denotes the vector of state variables, $y \in \mathbb{R}^m$ denotes the vector of measured variables, and $u_{k(t)}(x) \in \mathbb{R}^m$ denotes the control action prescribed by the control law for the vector of constrained manipulated inputs under the kth configuration. $\tilde{u}_{k(t)}$ denotes the unknown fault vector with and $u_{k(t)}(x) + \tilde{u}_{k(t)}$ taking values in a nonempty convex subset $\mathbf{U}_k$ of $\mathbb{R}^m$, where $\mathbf{U}_k = \{u_k + \tilde{u}_k \in \mathbb{R}^m : \|u_k + \tilde{u}_k\| \le u_k^{\max}\}$, $u_k^{\max} > 0$ is the magnitude of input constraints and $f(0) = 0$. The vector function $f(x)$ and the matrices $G_k(x) = [g_{1,k}(x) \dots g_{m,k}(x)]$ are assumed to be sufficiently smooth on their domains of definition. The function $k(t)$, which takes values in the finite index set $\mathcal{K}$, represents a discrete state that indexes the matrix $G_k(\cdot)$ as well as the manipulated input $u_k(\cdot)$ and the possible faults in the manipulated inputs $\tilde{u}_k(\cdot)$. For each value that k assumes in $\mathcal{K}$, the process is controlled via a different set of manipulated inputs which defines a given control configuration. Throughout the chapter, we assume that for any $u_k \in \mathbf{U}_k$ the solution of the system of Eq. (4.1) exists and is continuous for all t.

To illustrate some of the ideas behind the fault detection and isolation filter design and reconfiguration strategy, we begin by assuming that all the states are available as measurements. We next review one example of a state-feedback controller that provides an explicit estimate of the stability region for the closed-loop system subject to constraints (for more details on the controller design and the proof, see [46] and [85]).

Theorem 4.1 (Cf. [46]) *Consider the switched nonlinear system of Eq.* (4.1) *for a configuration k for which a Control Lyapunov Function V_k exists, with $\tilde{u}_k(t) \equiv 0$, under state-feedback using the following bounded nonlinear feedback controller*:

$$u_k = -w_k\big(x, u_k^{\max}\big)\big(L_{G_k} V_k(x)\big)^T, \tag{4.2}$$

where

$$w_k\big(x, u_k^{\max}\big) = \begin{cases} \dfrac{\alpha_k(x) + \sqrt{\alpha_k^2(x) + (u_k^{\max}\|b_k^T(x)\|)^4}}{\|b_k^T(x)\|^2[1 + \sqrt{1 + (u_k^{\max}\|b_k^T(x)\|)^2}]}, & b_k^T(x) \ne 0, \\ 0, & b_k^T(x) = 0, \end{cases} \tag{4.3}$$

with $\alpha_k(x) = L_{f_k} V_k(x) + \rho_k V_k(x)$, $\rho_k > 0$, and $b_k(x) = L_{G_k} V_k(x)$. Assume that the set $\Phi_k(u_k^{\max})$ of x satisfying

$$L_{f_k} V_k(x) + \rho_k V_k(x) \le u_k^{\max}\big\|\big(L_{G_k} V_k(x)\big)^T\big\| \tag{4.4}$$

contains the origin and a neighborhood of the origin. Also, let $\Omega_k(u_k^{\max}) := \{x \in \mathbb{R}^n : V_k(x) \le c_k^{\max}\}$ *be a level set of* V_k*, completely contained in* Φ_k*, for some* $c_k^{\max} > 0$*. Then for all* $x(0) \in \Omega_k(u_k^{\max})$ *the control law guarantees that the origin of the closed-loop system is asymptotically stable.*

4.3 State-Feedback Fault-Tolerant Control

In this section, we first consider the problem under state-feedback to illustrate the main idea behind the fault detection and isolation filter and fault-tolerant controller design.

4.3.1 State-Feedback Fault Detection and Isolation Filter

To be able to detect the occurrence of a fault in a control actuator via observing the state evolution, it is necessary that the control actuator influences the evolution of at least some of the states. To be able to isolate the occurrence of a fault, it becomes further necessary that the control actuator in question be the only one influencing at least some state. To understand this better, consider the following single state, two input example: $\dot{x} = x + u_1(x) + \tilde{u}_1 + u_2(x) + \tilde{u}_2$. As is clear from the equation, the faults in the manipulated inputs u_1 and u_2 effect the evolution of the state additively, i.e., as the sum $(\tilde{u}_1 + \tilde{u}_2)$. While it may be possible to detect that a fault has occurred in either u_1 or u_2 (if the faults do not cancel out each other, i.e., if $\tilde{u}_1 + \tilde{u}_2 \neq 0$), it is not possible, in this case, to determine by observing the evolution of the system state (and finding it to be different when compared to the expected evolution with $\tilde{u}_1 = \tilde{u}_2 = 0$) whether $\tilde{u}_1 \neq 0$ or $\tilde{u}_2 \neq 0$, or both. In other words, while it may be possible to detect the occurrence of a fault, it is not possible to isolate it. Below we formulate a verifiable assumption on the structure of the system of Eq. (4.1) that allows for fault detection and isolation.

Assumption 4.1 Consider the system of Eq. (4.1) in configuration k under state-feedback. Then for every input $u_{j,k}$, $j = 1, \dots, m$, there exists a unique state $x_{i,k}$, $i \in \{1, \dots, n\}$ such that with $x_{i,k}$ as output, the relative degree of $x_{i,k}$ with respect to $u_{j,k}$ and only with respect to $u_{j,k}$ is equal to 1.

Consider now the system of Eq. (4.1) in configuration k for which Assumption 4.1 holds. Theorem 4.2 below formulates the fault detection and isolation filter and outlines its fault detection and isolation properties.

Theorem 4.2 *Consider the system of Eq.* (4.1) *in configuration k under the control law of Eq.* (4.2). *Let the fault detection and isolation filter for the* j*th manipulated* *input in the kth configuration be described by*

$$\begin{aligned}
\dot{\tilde{x}}_{i,k} &= f_i(x_1, \ldots, \tilde{x}_{i,k}, \ldots, x_n) \\
&\quad + g_{j,k}[i](x_1, \ldots, \tilde{x}_{i,k}, \ldots, x_n)u_{j,k}(x_1, \ldots, \tilde{x}_{i,k}, \ldots, x_n), \qquad (4.5)\\
e_{i,k} &= \tilde{x}_{i,k} - x_i,
\end{aligned}$$

where $g_{j,k}[i]$ denotes the ith element of the vector $g_{j,k}$, $\tilde{x}_{i,k}(0) = x_i(0)$, and the subscripts i, k refer to the ith state under the kth control configuration. Let $T^f_{j,k}$ be the earliest time for which $\tilde{u}_{j,k} \neq 0$, then the fault detection and isolation filter of Eq. (4.5) *ensures that $e_{i,k}(T^{f^+}_{j,k}) \neq 0$. Also, $e_{i,k}(t) \neq 0$ only if $\tilde{u}_{j,k}(s) \neq 0$ for some $0 \leq s < t$.*

Proof Part 1: We first show the *only if* part of the theorem by contradiction. To this end, consider the equation describing the evolution of the ith state, x_i, described by

$$\dot{x}_i = f_i(x) + g_{j,k}[i](x)\big(u_{j,k}(x) + \tilde{u}_{j,k}(t)\big) \tag{4.6}$$

and let us assume that $\tilde{u}_{j,k}(s) = 0$, for all $0 \leq s < t$. Then for all $0 \leq s < t$ Eq. (4.6) reduces to

$$\dot{x}_i = f_i(x) + g_{j,k}[i](x)u_{j,k}(x). \tag{4.7}$$

Since $x_i(0) = \tilde{x}_{i,k}(0)$, we therefore have that $\dot{x}_i(s) = \dot{\tilde{x}}_{i,k}(s)$ for $s = 0$ and subsequently for all $0 \leq s < t$. Therefore, $e_{i,k}(s) = 0$ for all $0 \leq s < t$, which leads to a contradiction. This means that the assumption that $\tilde{u}_{j,k}(s) = 0$, for all $0 \leq s < t$ does not hold, i.e., $\tilde{u}_{j,k}(s) \neq 0$ for some $0 \leq s < t$. This completes the proof of the first part of the theorem.

Part 2: To prove the *if* part of the theorem, consider once again Eq. (4.5) and Eq. (4.6) with $\tilde{u}^k_j(t) = 0$ for all $t \leq T^f_k$. Then following the line of reasoning as in Part 1, we get that $x_i(T^f_{j,k}) = \tilde{x}_{i,k}(T^f_{j,k})$. Since $\tilde{u}_{j,k}(T^f_{j,k}) \neq 0$, we get that $\dot{x}_i(T^f_{j,k}) \neq \dot{\tilde{x}}_{i,k}(T^f_{j,k})$, and therefore, that $x_i(T^{f^+}_{j,k}) \neq \tilde{x}_{i,k}(T^{f^+}_{j,k})$, i.e., $e_{i,k}(T^{f^+}_{j,k}) \neq 0$. This completes the proof of Theorem 4.2. □

Remark 4.1 As stated in Theorem 4.2 above, the fault detection and isolation filter performs the task of detection as well as isolation. Specifically, the *if* part of the theorem characterizes the detection capabilities where the residual for a manipulated input becomes nonzero if a fault occurs in the given manipulated input. The *only if* part of the theorem allows isolation since a residual is non-zero only if a fault has occurred at some previous time in the given manipulated input. Note that in general it is possible that a fault occurs for some time and disappears, and also the fault profile is such that after some time the evolution of the system becomes identical again to the fault-free system, in which case the residual would once again go back to zero. The immediate detection capability of the filter above, however, precludes the possibility that such a fault goes undetected.

Remark 4.2 Note that Assumption 4.1 is a sufficient condition that allows fault detection and isolation filter design, and can be readily relaxed. For instance, if the inputs influence the evolution of the states in an 'upper triangular' or 'lower triangular' form, fault detection and isolation is possible using the same idea as in Theorem 4.2 above. As an illustration, consider a two state two input system of the form

$$\begin{aligned} \dot{x}_1 &= f_1(x) + g_1[1](x)\big(u_1(x) + \tilde{u}_1(t)\big), \\ \dot{x}_2 &= f_2(x) + g_1[2](x)\big(u_1(x) + \tilde{u}_1(t)\big) + g_2[2](x)\big(u_2(x) + \tilde{u}_2(t)\big), \end{aligned} \tag{4.8}$$

where $f_i(\cdot)$ denotes the ith elements of the vector function $f(\cdot)$ and $g_i[j]$ denotes the jth element of the vector g_i. While this system does not satisfy Assumption 4.1, fault detection and isolation can still be achieved. Specifically, a filter design of the form of Eq. (4.5) can be used to build a detection filter for the first manipulated input. The dynamics of the second filter can then be designed as

$$\begin{aligned} \dot{\tilde{x}}_2 &= f_2(x_1, \tilde{x}_2) + g_1[2](x_1, \tilde{x}_2)\big(u_1(x_1, \tilde{x}_2)\big) + g_2[2](x_1, \tilde{x}_2)\big(u_2(x_1, \tilde{x}_2)\big), \\ e_2 &= \tilde{x}_2 - x_2. \end{aligned} \tag{4.9}$$

In this setup, faults in u_1 will be captured in both e_1 and e_2, while faults in u_2 will only effect e_2. The task of fault detection and isolation can therefore be carried out via a simple process of elimination.

Remark 4.3 Even in cases where the structure of the process dynamic model does not allow for complete isolation of a fault (i.e., more than one manipulated input has a relative degree one with respect to a given state), the proposed method can still isolate the failure to a subset of the entire group of active manipulated inputs. This would be especially useful in the case of high-dimensional process systems with a large number of states and inputs where several redundant inputs are used simultaneously. However, once a subset of control actuators including the failed ones has been identified by the filter, nothing can be said about which actuator(s) of the ones in this subset has actually failed. Therefore, in order to guarantee stability in the controller reconfiguration phase, the worst case scenario, where all the actuators in this subset have failed, must be assumed and the supervisor must then switch to a fall-back configuration that does not implement any of the control actuators included in this subset.

4.3.2 State-Feedback Fault-Tolerant Controller

Given that a fault is detected and isolated using the filters designed in the previous section, the problem that we address in this section is that of determining an appropriate backup configuration. The first requirement for an appropriate backup control configuration is that it does not use the faulty control actuator. Secondly, the limitations imposed by the presence of input constraints must be accounted for, and in particular, a backup configuration should be implemented for which the state of the closed-loop system resides in its stability region. This idea is formalized in Theorem 4.3 below.

Theorem 4.3 *Consider the closed-loop system of Eqs.* (4.1)–(4.2) *under state-feedback and let* $x(0) := x_0 \in \Omega_{k_0}$ *for some* $k_0 \in \mathcal{K}$. *Let* T_{j,k_0} *be the earliest time such that* $e_{i,k_0} \neq 0$ *for some* i *corresponding to a manipulated input* u_{j,k_0} *in Eq.* (4.5). *Then the following switching rule*:

$$k(t) = \begin{cases} k_0, & 0 \leq t < T_{j,k_0}, \\ q \neq k_0, & t \geq T_{j,k_0},\ x(T_{j,k_0}) \in \Omega_q,\ u_{j,k_0} \notin u_q \end{cases} \tag{4.10}$$

guarantees asymptotic stability of the origin of the closed-loop system.

Proof We consider the two cases:

1. $e_{i,k_0}(t) = 0$ for all $t \geq 0$ for all $i \in \{1, \dots, n\}$ and
2. $e_{i,k_0}(t) \neq 0$ for some T_{j,k_0} for some $j \in \{1, \dots, m\}$.

Case 1: $e_{i,k_0}(t) = 0\ \forall t \geq 0$ for all $j \in \{1, \dots, m\}$ implies (using Theorem 4.2) that $\tilde{u}_{j,k}(t) = 0$ for all $t \geq 0$ and for all $j \in \{1, \dots, m\}$. The switching rule of Eq. (4.10) then dictates that $k(t) = k_0\ \forall t \geq 0$. Since $x(0) \in \Omega_{k_0}$, asymptotic stability of the origin of the closed-loop system follows from Theorem 4.1.

Case 2: If $e_{i,k_0}(t) \neq 0$ for some T_{j,k_0} for some $j \in \{1, \dots, m\}$, the switching rule dictates switching to configuration q such that $x(T_{j,k_0}) \in \Omega_q$. Stability of the origin of the closed-loop system again follows from Theorem 4.1. This completes the proof of Theorem 4.3. □

Remark 4.4 Early detection of a fault enhances the chances that corrective action can be taken in time to achieve fault-tolerant control. Specifically, it may happen that a fault occurs when the closed-loop state resides in the stability region of one of the backup configurations, but the destabilizing effect of the fault may drive the state outside the stability region of the backup configuration by the time the fault is detected. Theorem 4.2 guarantees that a fault is detected as soon as it occurs. Note also that in the presence of plant–model mismatch or unknown disturbances, the value of $e_{i,k}(t)$ will be nonzero even in the absence of faults. The presence of time-varying disturbances $\theta(t)$ with known bounds θ_b on the disturbances can be accounted for in the filter design as well as reconfiguration. Specifically, the filter can be redesigned to declare a fault only if the value of $e_{i,k}(t)$ increases beyond some threshold, $\delta(\theta_b)$, where $\delta(\theta_b)$ accounts for the deviation of the plant dynamics from the nominal dynamics in the absence of faults. Further robust controllers can be utilized and the robust stability regions can be used as criteria for deciding which backup configuration should be implemented in the closed-loop system.

Remark 4.5 In the event that the process state at the time of the failure of the primary control configuration lies in the stability region of more than one backup control configurations, additional performance considerations such as ease and/or cost of implementing one control configuration over another can be used in choosing the backup control configuration to be implemented [111]. Note that the set of initial conditions starting from where a given control configuration can stabilize a steady state—the so-called null-controllable region—is fundamentally limited by the constraints on the available control action, and that different control laws typically provide estimates of the stability region which are subsets of the null-controllable region. If the state at the time of a failure lies outside the stability region of all the backup configurations, then this indicates that the backup configurations do not have enough control action available and calls for increasing the allowable control action.

4.4 Output-Feedback Fault-Tolerant Control

In the previous section, we assumed the availability of all the state measurements to illustrate the design of the fault detection and isolation filters and the controller reconfiguration strategy. In this section, we consider the case where only some of the process states are available for measurement. The unavailability of some states as measurements necessitates estimating the states from the measurements for the purposes of fault detection and isolation, feedback control and controller reconfiguration. To this end, we next review an output-feedback controller design [46] that provides estimates of the states (for other examples of nonlinear observer and output-feedback controller designs, see [74, 77]) along with an explicit characterization of the output feedback stability region.

4.4.1 Output Feedback Controller

To design the output feedback controllers for the individual configurations, we will use the following assumption:

Assumption 4.2 Consider the system of Eq. (4.1) in configuration k with $\tilde{u}_k \equiv 0$. There exists a set of integers $r_{1,k}, r_{2,k}, \dots, r_{m,k}$ (with $r_{1,k} + r_{2,k} + \cdots + r_{m,k} = n$ for each k) and a coordinate transformation $\zeta_k = \chi_k(x)$ such that the representation of the system of Eq. (4.1), in the ζ_k coordinates, takes the form:

$$\begin{aligned} \dot{\zeta}_{1,k}^{(i)} &= \zeta_{2,k}^{(i)}, \\ &\vdots \\ \dot{\zeta}_{r_{i,k}-1}^{(i)} &= \zeta_{r_{i,k}}^{(i)}, \\ \dot{\zeta}_{r_{i,k}}^{(i)} &= L_f^{r_{i,k}} h_i(x) + L_{g_{i,k}} L_f^{r_{i,k}-1} h_i(x) u_{i,k}, \end{aligned} \tag{4.11}$$

where $x = \chi_k^{-1}(\zeta_k)$ and $\zeta_k = [\zeta_k^{(1)^T} \dots \zeta_k^{(m)^T}]^T$.

Theorem 4.4 (Cf. [46]) *Consider the constrained nonlinear process of Eq.* (4.1) *with* $\tilde{u}_k(t) \equiv 0$ *for which Assumption* 4.2 *holds, under the output feedback controller using the kth control configuration*:

$$\dot{\tilde{y}}_{i,k} = \begin{bmatrix} -L_{i,k}a_{i,k}^{(1)} & 1 & 0 & \dots & 0 \\ -L_{i,k}^2 a_{i,k}^{(2)} & 0 & 1 & \dots & 0 \\ \vdots & \vdots & \vdots & \ddots & \vdots \\ -L_{i,k}^{r_i} a_{i,k}^{(r_i)} & 0 & 0 & \dots & 0 \end{bmatrix} \tilde{y}_{i,k} + \begin{bmatrix} L_{i,k}a_{i,k}^{(1)} \\ L_{i,k}^2 a_{i,k}^{(2)} \\ \vdots \\ L_{i,k}^n a_{i,k}^{(r_i)} \end{bmatrix} y_{i,k}, \tag{4.12}$$

$$u_k = -w_k\big(\hat{x}, u_k^{\max}\big)\big(L_{G_k}V_k(\hat{x})\big)^T,$$

where $\hat{x} = \chi_k^{-1}(\text{sat}(\tilde{y}_k))$, $\tilde{y}_k = [\tilde{y}_{(1,k)}^T \cdots \tilde{y}_{(m,k)}^T]^T$, $i = 1, \dots, m$, *and where the parameters* $a_{i,k}^{(1)}, \dots, a_{i,k}^{(r_i)}$ *are chosen such that the polynomial* $s^{r_i} + a_{i,k}^{(1)} s^{n-1} + a_{i,k}^{(2)} s^{r_i-2} + \dots + a_{i,k}^{(r_i)} = 0$ *is Hurwitz*, $\hat{x} = \chi_k^{-1}(\text{sat}(\tilde{y}))$, $\text{sat}(\cdot) = \min\{1, \zeta_{\max,k}/\|\cdot\|\}(\cdot)$, *with* $\zeta_{\max,k} = \beta_\zeta(\delta_{\zeta,k}, 0)$ *where* β_ζ *is a class* $\mathcal{KL}$ *function and* $\delta_{\zeta,k}$ *is the maximum value of the vector* $[l_1^T(x) l_2^T(x) \dots l_m^T(x)]^T$ *for* $V_k(x) \le \delta_{b,k}$, *where* $l_i(x) = [h_i(x) L_f h_i(x) \dots L_f^{r_i-1} h_i(x)]^T$, *and let* $\varepsilon_k = \max_i 1/L_{i,k}$. *Then, given* $\Omega_{b,k} := \{x \in \mathbb{R}^n | V_k(x) \le \delta_{b,k}\}$ *and positive real numbers* $e_{m,k}$, $\tilde{u}_k^*$, *and* d_k, *there exist* $\varepsilon_k^* > 0$, $T_k^b > 0$ *such that if* $\varepsilon_k \in (0, \varepsilon_k^*]$, $x(0) \in \Omega_{b,k}$, *and* $\|\tilde{y}(0)\| \le \delta_{\zeta,k}$, *the origin of the closed-loop system is asymptotically (and locally exponentially) stable, and if* $\|\tilde{u}_k(t)\| \le \tilde{u}_k^*$ *then* $\|x(t) - \hat{x}(t)\| \le e_{m,k}$ *for all* $t \ge T_k^b$ *and* $\limsup_{t\to\infty} x(t) = d_k$.

Remark 4.6 Theorem 4.4 above provides the estimation and controller design that guarantees asymptotic stability in the case of fault-free system as well as practical stability in the presence of 'small' faults (that preserve stability). The result relies on closeness of the state estimates to the true states over the infinite time interval. In fault detection and isolation, the closeness of solution would be required to hold even in the presence of large, possibly destabilizing faults, at least up-to some finite time to be able to detect and isolate the faults. This requirement is formalized in Assumption 4.3 below.

Assumption 4.3 Consider the system of Eq. (4.1) in configuration k under the output feedback controller of Theorem 4.4. There exist positive real numbers $T_{\text{close}} > T_k^b$ and δ_k such that if $\|\tilde{u}_k(t)\| > \tilde{u}_k^*$ for some $T_{\text{fault}} > T_k^b$ where $\tilde{u}_k^*$ was defined in Theorem 4.4, then $\|x(t) - \hat{x}(t)\| \le e_{m,k}$ for all $t \in [T_k^b, T_{\text{fault}} + T_k^{\text{close}}]$ and $\| \int_{T_k^b}^t g_{j,k}[i](x(\tau))\tilde{u}_{j,k}(\tau)\,d\tau \| > \delta_k$ for some $t \in [T_{\text{fault}}, T_{\text{fault}} + T_k^{\text{close}}]$.

Due to the lack of full state measurements, the reconfiguration decision needs to be done based only on the available state estimates. It is therefore necessary to be able to make reliable inferences regarding the states using the state estimates. Proposition 4.1 below establishes the existence of a set, $\Omega_{s,k} := \{x \in \mathbb{R}^n : V_k(x) \le \delta_{s,k}\}$, such that once the state estimation error has fallen below a certain value (note that the decay rate can be controlled by adjusting L_k), the presence of the state within the output feedback stability region, $\Omega_{b,k}$, can be guaranteed by verifying the presence of the state estimates in the set $\Omega_{s,k}$. A similar approach was employed in the construction of the output feedback stability regions $\Omega_{b,k}$ and the regions for the state estimates $\Omega_{s,k}$ in the context of output feedback control of linear systems in [107], and for nonlinear systems in [51]. For a proof of the proposition, see [51]; see also Chap. 3.

Proposition 4.1 *Given any positive real number $\delta_{b,k}$, there exist positive real numbers $e^*_{m,k}$ and $\delta_{s,k}$ such that if $\|x - \hat{x}\| \le e_{m,k}$, where $e_{m,k} \in (0, e^*_{m,k}]$, and $V_k(\hat{x}) \le \delta_{s,k}$, then $V_k(x) \le \delta_{b,k}$.*

4.4.2 Output-Feedback Fault Detection and Isolation Filter

The output feedback fault detection and isolation filter uses the same principle as the state feedback fault detection and isolation filter while using the state estimates to implement the filter. For the system of Eq. (4.1), the fault detection and isolation filter for the jth manipulated input in the kth configuration is designed as:

$$\begin{aligned}\dot{\tilde{x}}_{i,k} &= f_i(\hat{x}_{1,k}, \ldots, \tilde{x}_{i,k}, \ldots, \hat{x}_{n,k}) \\ &\quad + g_{j,k}[i](\hat{x}^k_1, \ldots, \tilde{x}_{i,k}, \ldots, \hat{x}_{n,k}) u_{j,k}(\hat{x}_{1,k}, \ldots, \tilde{x}_{i,k}, \ldots, \hat{x}_{n,k}), \\ e_{i,k} &= \hat{x}_{i,k} - \tilde{x}_{i,k},\end{aligned} \tag{4.13}$$

where $g_{j,k}[i]$ denotes the ith element of the vector $g_{j,k}$, and $\tilde{x}_{i,k}(T^b_k) = \hat{x}_{i,k}(T^b_k)$, where T^b_k was defined in Theorem 4.4

Proposition 4.2 *Consider the nonlinear system of Eq.* (4.1), *for a fixed mode under the output feedback controller of Eq.* (4.12) *and the filter of Eq.* (4.13). *Given $\tilde{u}^*_{j,k}$, δ_k and T^{close}_k there exist positive real numbers $\delta_{j,k}$ and ε^{**}_k such that if $|\tilde{u}_{j,k}(t)| \ge \tilde{u}^*_{j,k}$ for some $T^{\text{fault}}_k \ge T_{b,k}$ and $\varepsilon_k \le \min\{\varepsilon^*_k, \varepsilon^{**}_k\}$ then $e_{i,k}(t) > \delta_{j,k}$ for some $t \in [T^{\text{fault}}_k, T^{\text{fault}}_k + T^{\text{close}}_k]$.*

Proof Consider the filter of Eq. (4.13) and the evolution of x_i for $t \in [T^b_k, T^{\text{fault}}_k + T^{\text{close}}_k]$, i.e., consider the systems

$$\begin{aligned}\dot{\tilde{x}}_{i,k} &= f_i(x) + g_{j,k}[i](x)\big(u_{j,k}(x)\big) + \big(f_i(\hat{x}_{1,k}, \ldots, \tilde{x}_{i,k}, \ldots, \hat{x}_{n,k}) - f_i(x)\big) \\ &\quad + \big(g_{j,k}[i](\hat{x}_{1,k}, \ldots, \tilde{x}_{i,k}, \ldots, \hat{x}_{n,k}) u_{j,k}(\hat{x}_{1,k}, \ldots, \tilde{x}_{i,k}, \ldots, x_{n,k} + \hat{x}_{n,k}) \\ &\quad - g_{j,k}[i](x) u_{j,k}(x)\big)\end{aligned} \tag{4.14}$$

and

$$\dot{x}_{i,k} = f_i(x) + g_{j,k}[i](x)\big(u_{j,k}(x) + \tilde{u}_{j,k}(t)\big). \tag{4.15}$$

Therefore,

$$\begin{aligned}\dot{x}_{i,k} - \dot{\tilde{x}}_{i,k} &= g_{j,k}[i](x)\tilde{u}_{j,k}(t) + \big(f_i(x) - f_i(\hat{x}_{1,k}, \ldots, \tilde{x}_{i,k}, \ldots, \hat{x}_{n,k})\big) \\ &\quad + \big(g_{j,k}[i](x)u_{j,k}(x)\big) \\ &\quad - g_{j,k}[i](\hat{x}_{1,k}, \ldots, \tilde{x}_{i,k}, \ldots, \hat{x}_{n,k})u_{j,k}(\hat{x}_{1,k}, \ldots, \tilde{x}_{i,k}, \ldots, x_{n,k} + \hat{x}_{n,k}).\end{aligned} \tag{4.16}$$

Note that $\hat{x}(T_b) - x(T_b)$ can be made as small as desired by choosing a sufficiently small ε. From the continuity of $f_i(\cdot)$ and $g_{j,k}[i](\cdot)$, this implies that the last two terms in Eq. (4.16) can be made as small as desired. The difference between $\dot{x}_{i,k}$ and $\dot{\tilde{x}}_{i,k}$ can therefore be made as close as desired to $g_{j,k}[i](x)(\tilde{u}_{j,k}(t))$. Using Assumption 4.3, therefore, given a time $T^{\text{close}} > T_k^b$, there exists a positive real number $\delta_{j,k}^* = \delta_k^*$ such that if $|\tilde{u}_{j,k}(t)| > \tilde{u}_{j,k}^*$ for some $T_k^{\text{fault}} \geq T_k^b$ then $\|x_{i,k}(t) - \tilde{x}_{i,k}(t)\| \geq \delta_{j,k}^*$ for some $t \in [T_k^{\text{fault}}, T_k^{\text{fault}} + T_k^{\text{close}}]$. Finally, once again since $\hat{x}(t) - x(t)$ can be made as close as desired (up until T_k^{close}), then given that $\|x_{i,k}(t) - \tilde{x}_{i,k}(t)\| \geq \delta_{j,k}^*$, there exists a positive real number $\delta_{j,k}$ such that $e_{i,k} = \|\hat{x}_{i,k}(t) - \tilde{x}_{i,k}(t)\| \geq \delta_{j,k}$ for some $t \in [T_k^{\text{fault}}, T_k^{\text{fault}} + T_k^{\text{close}}]$. In summary, there exists a positive real number ε_k^{**} such that if $\varepsilon_k \leq \min\{\varepsilon_k^*, \varepsilon_k^{**}\}$ and $|\tilde{u}_{j,k}(t)| \geq \tilde{u}_{j,k}^*$ for some $T_k^{\text{fault}} \geq T_{b,k}$ then $e_{i,k}(t) > \delta_{j,k}$ for some $t \in [T_k^{\text{fault}}, T_k^{\text{fault}} + T_k^{\text{close}}]$. □

Remark 4.7 Note that unlike the case of full state-feedback, the fault detection filter is initialized only after the passage of some short period of time, T_k^b (which can be chosen arbitrarily small by increasing the observer gain), to ensure that the closed-loop state estimates have converged sufficiently close to the true closed-loop states and thus—by setting the filter state $\tilde{x}_{i,k}$ at this time equal to the value of the state estimate—ensure that the filter state is initialized sufficiently close to the true values of the state. Note also that unlike the case of full state availability, where the filter is able to immediately detect and isolate the occurrence of fault, the lack of measurements which induces the error in the initialization of the filter states allows detection of only such faults that impact the states of the closed-loop system above a certain threshold. The key is to ensure that only such faults go undetected which do not impact undesirably on the stability of the closed-loop system. In the subsequent section, we design an output-feedback fault detection and isolation and fault-tolerant control structure that ensures detection and isolation of destabilizing faults.

4.4.3 *Output-Feedback Fault Detection and Isolation and Fault Tolerant Control*

Having designed the state estimators and controllers and output feedback fault detection and isolation filters, in this section we present an integrated output-feedback fault detection and isolation and fault-tolerant controller structure. To this end, consider the nonlinear system of Eq. (4.1), for which the output feedback controller of Eq. (4.12) and the filters of Eq. (4.13) have been designed for each manipulated input under the primary configuration, $k(0) = k_0$ under possible faults in only one control actuator. The theorem below formalizes the integrated output-feedback fault detection and isolation and fault-tolerant control structure.

Theorem 4.5 *Let $k(0) = k_0$ for some $k_0 \in \mathcal{K}$, $x(0) \in \Omega_{b,k_0}$, $\tilde{x}_{i,k}(T^b_{i,k}) = \hat{x}(T^b_{i,k})$. Given a positive real number d_{k_0} there exist positive real numbers $\delta_{i,k}$ and ε^{***}_k such that if $\varepsilon_k \in (0, \varepsilon^{***}_k]$ then under the switching rule*

$$k(t) = \begin{cases} k_0, & 0 \le t < T_{\text{detect}}, \\ q \neq k_0, & t \ge T_{\text{detect}},\ \hat{x}(T_{\text{detect}}) \in \Omega_{s,q}, u_{j,k_0} \notin u_q, \end{cases} \tag{4.17}$$

where T_{detect} is the earliest time for which $e_{i,k} > \delta_{i,k}$ for some $i \in \{0, \ldots, n\}$, we have that $\limsup_{t\to\infty} \|x(t)\| \le d_{k_0}$.

Proof We consider the two cases:

1. $e_{i,k}(t) \le \delta_{i,k}$ $\forall t$ and
2. $e_{i,k}(t) > \delta_{i,k}$ for some $t = T_{\text{detect}}$.

Case 1: From Theorem 4.4, we have that given a positive real number d_k, there exist positive real numbers ε^{**}_k and $\tilde{u}^*_k$ such that if $\|\tilde{u}_{j,k}(t)\| \le \tilde{u}^*_k$, then $\limsup_{t\to\infty} x(t) = d_{k_0}$. For such choices of ε^{**}_k and $\tilde{u}^*_k$, we have from Proposition 4.2 that there exists a positive real number $\delta_{i,k}$ such that if $\varepsilon_k \in (0, \min\{\varepsilon^*_k, \varepsilon^{**}_k\} = \varepsilon^{***}_k]$ then $e_{i,k} \le \delta_{i,k} \Rightarrow \|\tilde{u}_{j,k}(t)\| \le \tilde{u}^*_k$. Therefore, for the above choices of $\tilde{u}^*_k$, ε^{***}_k, and $\delta_{j,k}$, we have that $e_{i,k}(t) \le \delta_{i,k}$ implies $\|\tilde{u}_{i,k_0}(t)\| \le \tilde{u}^*_{i,k_0}$, yielding $\limsup_{t\to\infty} \|x(t)\| = d_{k_0}$.

Case 2: The switching rule of Eq. (4.17) ensures that at $t = T_{\text{detect}}$, $\hat{x}(t) \in \Omega_{s,q}$, which in turn implies that $x(t) \in \Omega_{b,q}$ (Proposition 4.1). This, together with the switching to the qth control configuration, ensures asymptotic stability of the origin of the closed-loop system (Theorem 4.4). In either case, we get that $\limsup_{t\to\infty} \|x(t)\| \le d_{k_0}$. This completes the proof of the theorem. □

The design of the output feedback fault detection and isolation filter and controller reconfiguration is best understood through the following algorithm

1. Given the system of the form of Eq. (4.1), design the output feedback controller of Eq. (4.12) that also yields estimates of the states, and estimate the output feedback stability regions of the control configurations, $\Omega_{b,k}$, and the sets $\Omega_{s,k}$, defined in Proposition 4.1, and compute the values of T^b_k. For an initial condition in the stability region of the k_0th control configuration, initialize the state estimator and the output feedback controller as described in Theorem 4.4.
2. After a time $T^b_{k_0}$, initialize the fault detection and isolation filters of the form of Eq. (4.13) using the values of the state estimates at time $T^b_{k_0}$.

3. Monitor the evolution of the residuals (e_{i,k_0}). If any of the residuals go above the threshold, it implies that a possibly destabilizing fault has occurred.
4. Switch to a configuration q for which the closed-loop state estimates at the time of fault detection lie in $\Omega_{s,q}$, where $\Omega_{s,q}$ was defined in Proposition 4.1 (this ensures that the states are in the output feedback stability region of the qth configuration) and one which does not involve the failed control actuator.
5. Implement this control configuration to achieve closed-loop stability and fault-tolerant control.

Remark 4.8 Note that while the above switching rule provides a sufficient condition for practical stability, it is not a necessary condition. In other words, the value of the residual going above the threshold does not imply that a destabilizing fault has occurred. However, the value of the residual being less than the threshold does ensure that no destabilizing fault has occurred. So while the above switching logic may trigger a switching where simply continuing with the primary control configuration could have preserved stability (i.e., it allows for false alarms), it is designed to preclude the possibility that a destabilizing fault takes place and reconfiguration is not executed. This, however, is not a limitation of the proposed filter, but stems simply from the fundamental problem of differentiating between the error introduced in the filtering system due to the presence of estimation errors and those due to the faults.

Remark 4.9 Note that while the algorithm above is written for the case of a single fault, generalization to multiple faults, whether simultaneous or otherwise, is straightforward. The current fault detection filter design can detect and isolate multiple faults, while the reconfiguration rule can be 're-initialized' after the first backup control configuration is activated to handle subsequent faults (see the simulation section for a demonstration).

4.5 Simulation Examples

We demonstrate the application of the proposed fault detection and isolation and reconfiguration strategy to two chemical reactors configured to operate in series. To this end, consider two well mixed, non-isothermal continuous stirred tank reactors (see Fig. 4.1), where three parallel irreversible elementary exothermic reactions of the form $\mathrm{A} \xrightarrow{k_1} \mathrm{B}$, $\mathrm{A} \xrightarrow{k_2} \mathrm{U}$, and $\mathrm{A} \xrightarrow{k_3} \mathrm{R}$ take place. A is the reactant species, B is the desired product, and U and R are undesired byproducts. The feed to the first reactor consists of pure A at a flow rate F_0, molar concentration $C_{\mathrm{A}0}$ and temperature T_0. The output from the first reactor is fed to the second reactor along with a fresh feed that consists of pure A at a flow rate F_3, molar concentration $C_{\mathrm{A}03}$, and temperature T_{03}. Due to the non-isothermal nature of the reactions, jackets are used to remove or provide heat to the reactors. Under standard modeling assumptions, a mathematical model of the process can be derived from material and energy

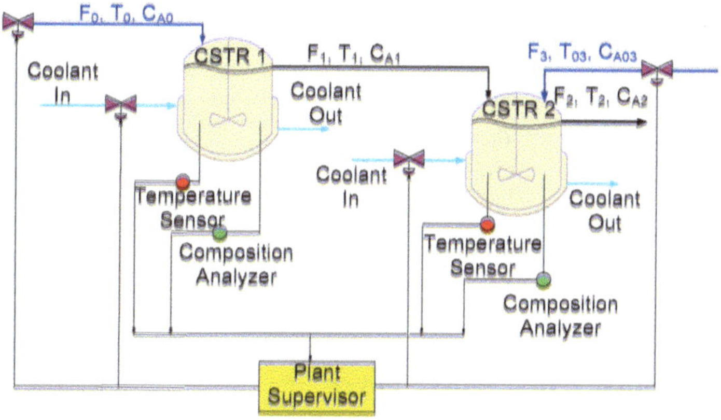

Fig. 4.1 A schematic of two CSTRs operating in series

balances and takes the following form:

$$\begin{aligned}
\frac{dT_1}{dt} &= \frac{F_0}{V_1}(T_0 - T_1) + \sum_{i=1}^{3} \frac{(-\Delta H_i)}{\rho c_p} R_i(C_{A1}, T_1) + \frac{Q_1}{\rho c_p V_1},\\
\frac{dC_{A1}}{dt} &= \frac{F_0}{V_1}(C_{A0} - C_{A1}) - \sum_{i=1}^{3} R_i(C_{A1}, T_1),\\
\frac{dT_2}{dt} &= \frac{F_0}{V_2}(T_1 - T_2) + \frac{F_3}{V_2}(T_{03} - T_2) + \sum_{i=1}^{3} \frac{(-\Delta H_i)}{\rho c_p} R_i(C_{A2}, T_2) + \frac{Q_2}{\rho c_p V_2},\\
\frac{dC_{A2}}{dt} &= \frac{F_0}{V_2}(C_{A1} - C_{A2}) + \frac{F_3}{V_2}(C_{A03} - C_{A2}) - \sum_{i=1}^{3} R_i(C_{A2}, T_2),
\end{aligned} \tag{4.18}$$

where $R_i(C_{Aj}, T_j) = k_{i0} \exp(\frac{-E_i}{RT_j}) C_{Aj}$, for $j = 1, 2$. T, C_A, Q, and V denote the temperature of the reactor, the concentration of species A, the rate of heat input/removal from the reactor, and the volume of reactor, respectively, with subscript 1 denoting CSTR 1 and subscript 2 denoting CSTR 2. ΔH_i, k_i, E_i, $i = 1, 2, 3$, denote the enthalpies, pre-exponential constants and activation energies of the three reactions, respectively, c_p and ρ denote the heat capacity and density of the fluid. The values of the process parameters can be found in Table 4.1. CSTR 1, with $Q_1 = 0$, has three steady-states: two locally asymptotically stable and one unstable at $(T_1^s, C_{A1}^s) = (388.57\text{ K}, 3.59\text{ kmol/m}^3)$. The unstable steady-state of CSTR 1 corresponds to three steady-states for CSTR 2 (with $Q_2 = 0$), one of which is unstable at $(T_2^s, C_{A2}^s) = (429.24\text{ K}, 2.55\text{ kmol/m}^3)$.

The control objective is to stabilize the reactors at the (open-loop) unstable steady-state. Operation at this point is typically sought to avoid high temperatures while simultaneously achieving reasonable reactant conversion. To accomplish this

Table 4.1 Process parameters and steady-state values for the chemical reactors of Eq. (4.18)

$F_0 = 4.998\ \text{m}^3/\text{hr}$	$F_1 = 4.998\ \text{m}^3/\text{hr}$
$F_3 = 30.0\ \text{m}^3/\text{hr}$	$V_1 = 1.0\ \text{m}^3$
$V_2 = 3.0\ \text{m}^3$	$R = 8.314\ \text{kJ/kmol K}$
$T_0 = 300.0\ \text{K}$	$T_{03} = 300.0\ \text{K}$
$C_{A0} = 4.0\ \text{kmol/m}^3$	$C_{A03}^s = 3.0\ \text{kmol/m}^3$
$\Delta H_1 = -5.0 \times 10^4\ \text{kJ/kmol}$	$\Delta H_2 = -5.2 \times 10^4\ \text{kJ/kmol}$
$\Delta H_3 = -5.4 \times 10^4\ \text{kJ/kmol}$	$k_{10} = 3.0 \times 10^6\ \text{hr}^{-1}$
$k_{20} = 3.0 \times 10^5\ \text{hr}^{-1}$	$k_{30} = 3.0 \times 10^5\ \text{hr}^{-1}$
$E_1 = 5.0 \times 10^4\ \text{kJ/kmol}$	$E_2 = 7.53 \times 10^4\ \text{kJ/kmol}$
$E_3 = 7.53 \times 10^4\ \text{kJ/kmol}$	$\rho = 1000.0\ \text{kg/m}^3$
$c_p = 0.231\ \text{kJ/kg K}$	$T_1^s = 388.57\ \text{K}$
$C_{A1}^s = 3.59\ \text{kmol/m}^3$	$T_2^s = 429.24\ \text{K}$
$C_{A2}^s = 2.55\ \text{kmol/m}^3$	

objective in the presence of actuator failures, we consider the following manipulated input candidates:

1. Rate of heat input into reactor one, Q_1, subject to the constraint $|Q_1| \le 1.4 \times 10^7\ \text{kJ/hr}$.
2. Reactor one inlet stream temperature, $T_0 - T_0^s$, subject to the constraint $|T_0 - T_0^s| \le 60\ \text{K}$.
3. Reactor one inlet reactant concentration, $C_{A0} - C_{A0}^s$, subject to the constraint $|C_{A0} - C_{A0}^s| \le 4.0\ \text{kmol/m}^3$.
4. Rate of heat input into reactor two, Q_2, subject to the constraint $|Q_2| \le 4.2 \times 10^7\ \text{kJ/hr}$.
5. Reactor two inlet stream temperature, $T_{03} - T_{03}^s$, subject to the constraint $|T_{03} - T_{03}^s| \le 60\ \text{K}$.
6. Reactor two inlet reactant concentration, $C_{A03} - C_{A03}^s$, subject to the constraint $|C_{A03} - C_{A03}^s| \le 3.0\ \text{kmol/m}^3$.

The above manipulated inputs can be used in various combinations to stabilize the reactors using measurements of the reactor temperatures and reactant concentrations provided by the sensors (full state-feedback) and to employ reconfiguration. The primary control configuration ($k = 1$) involves four inputs consisting of the two heating jackets and the two inlet stream concentrations (Q_1, Q_2, C_{A0}, and C_{A03}). In the event of a partial failure in this configuration, the supervisor needs to detect and isolate the fault and activate a fall-back configuration in order to maintain closed-loop stability.

We first illustrate the application of the fault detection and isolation and fault-tolerant control under state-feedback control. A quadratic Lyapunov function of the form $V_k = x^T P_k x$, where P_k is a positive-definite symmetric matrix that satisfies the Riccati inequality $A^T P_k + P_k A - P_k b_k b_k^T P_k < 0$, is used in controller design with A and b based on the linearized system around the desired steady-state.

1. For the primary control configuration, the manipulated inputs are scaled to give

$$b_1^* = \begin{bmatrix} 1 & 0 & 0 & 0 \\ 0 & 0.0198 & 0 & 0 \\ 0 & 0 & 1 & 0 \\ 0 & 0 & 0 & 0.0297 \end{bmatrix}$$

and

$$P_1 = \begin{bmatrix} 1.2290 & 2.2195 & 0.0203 & 0.1733 \\ 2.2195 & 28.4462 & 0.1396 & 8.8183 \\ 0.0203 & 0.1396 & 1.6150 & 9.8728 \\ 0.1733 & 8.8183 & 9.8728 & 145.7245 \end{bmatrix}.$$

2. The fall-back control configuration involves four manipulated inputs given by $u_2 = [T_0 - T_0^s \; C_{A0} - C_{A0}^s \; T_{03} - T_{03}^s \; C_{A03} - C_{A03}^s]'$. Scaling the manipulated input yields

$$b_2^* = \begin{bmatrix} 1 & 0 & 0 & 0 \\ 0 & 0.1333 & 0 & 0 \\ 0 & 0 & 2 & 0 \\ 0 & 0 & 0 & 0.2 \end{bmatrix}$$

and

$$P_2 = \begin{bmatrix} 1.1991 & 1.8730 & .00051 & 0.0236 \\ 1.8730 & 12.6725 & 0.0093 & 0.4141 \\ 0.0051 & 0.0093 & 0.6150 & 1.9055 \\ 0.0236 & 0.4141 & 1.9055 & 17.9826 \end{bmatrix}.$$

The state-feedback controller of Eq. (4.2) is subsequently designed for both the control configurations, and their stability region characterization, yielding $c_1^{\max}$ and $c_2^{\max}$ equal to 7.2 and 1.9, respectively. The fault detection filters are designed using Eq. (4.5) and the reactors as well as the filter states for the first control configuration are initialized at $T_1(0) = 386.8$ K, $C_{A1}(0) = 3.6$ kmol/m^3, $T_2(0) = 430.5$ K, $C_{A2}(0) = 2.56$ kmol/m^3. This initial condition is within the stability region of the primary control configuration ($V_1(x) = 6.64 \le c_1^{\max} = 7.2$). As shown by the solid lines in Figs. 4.2, 4.3, 4.4 and 4.5, the controller proceeds to drive the closed-loop trajectory toward the desired steady-state until the heating jackets fail simultaneously 0.1 minutes after reactor startup. As can be seen in Figs. 4.6 and 4.7, the values of only the residuals $e_{1,1}(t)$ and $e_{3,1}(t)$ become nonzero, thereby detecting as well as isolating the faults in the control actuators. If the supervisor does not perform any switching at this point, closed-loop stability is not achieved (dashed lines in Figs. 4.2–4.5). Note that this occurs because the actuators heating/cooling jackets have failed, but the controller still tries to use the heat supplied to/removed from the reactors as manipulated inputs. Having identified that the faults occurred in the actuators changing Q_1 and Q_2, the supervisor can implement the fall-back configuration (using T_0, C_{A0}, T_{03}, and C_{A03} as the manipulated inputs, $k = 2$) since the fall-back configuration does not use the failed actuators. Furthermore, at the time when the

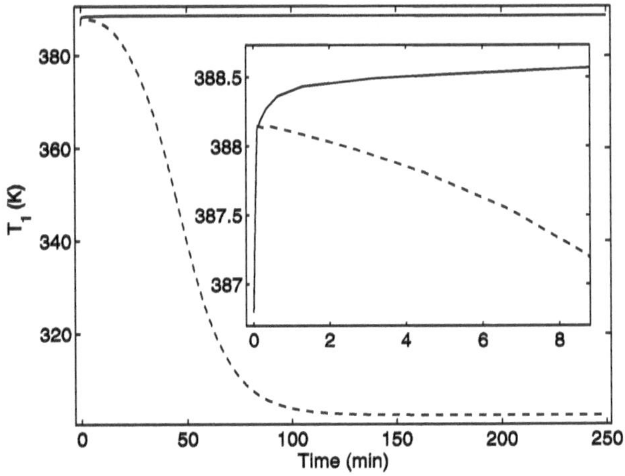

Fig. 4.2 Evolution of reactor one closed-loop temperature profile under the switching rule of Theorem 4.3 (*solid line*) and in the absence of fault-tolerant control (*dashed line*) subject to simultaneous failures in both the heating jackets

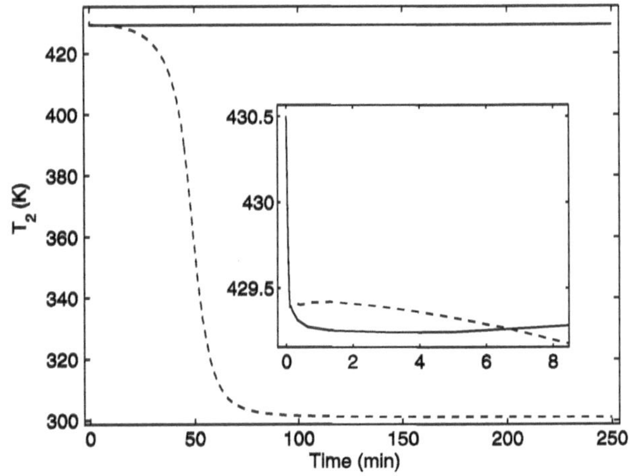

Fig. 4.3 Evolution of reactor two closed-loop temperature profile under the switching rule of Theorem 4.3 (*solid line*) and in the absence of fault-tolerant control (*dashed line*) subject to simultaneous failures in both the heating jackets

fault is detected, the state of the closed loop system is within the stability region of the backup control configuration ($V_2(x(t = 0.162)) = 0.221 < c_2^{\max} = 1.9$). The supervisor therefore activates the fall-back configuration (solid lines in Figs. 4.2–4.5) which stabilizes the closed-loop system and achieves fault-tolerant control.

The next simulation illustrates the application of fault detection and isolation and fault-tolerant control when not all of the process states are available for measurement. In this case, the output-feedback methodology is implemented on the same two-reactor system used for the previous simulation study with changes to the parameters $F_3 = 4.998\ \mathrm{m}^3/\mathrm{hr}$ and $V_2 = 0.5\ \mathrm{m}^3$. This changes the unstable steady state of the second reactor to $T_2^s = 433.96$ K and $C_{A2}^s = 2.88\ \mathrm{kmol/m}^3$. The dynamics for the controller are designed using the same state-feedback methodologies as in the previous simulation study. However, the controller utilizes the state estimates to compute a control action. The fault detection and isolation filter is designed based on Eq. (4.13).

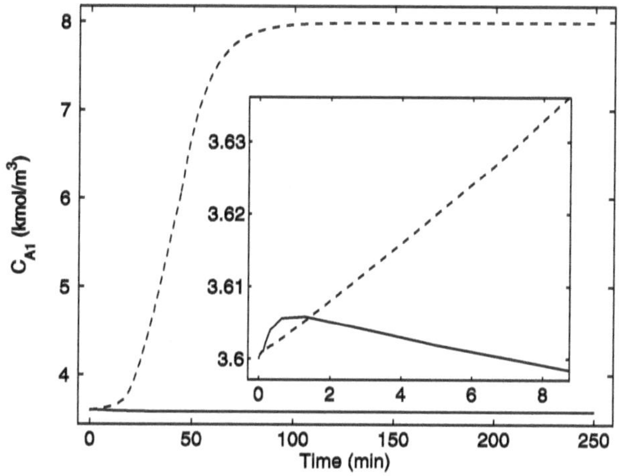

Fig. 4.4 Evolution of reactor one closed-loop reactant concentration profile under the switching rule of Theorem 4.3 (*solid line*) and in the absence of fault-tolerant control (*dashed line*) subject to simultaneous failures in both the heating jackets

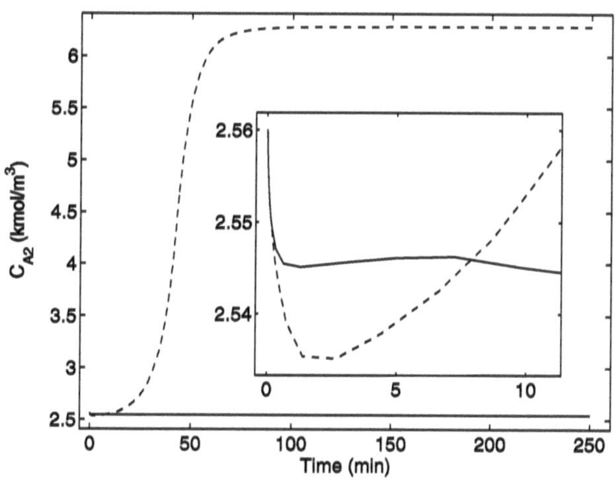

Fig. 4.5 Evolution of reactor two closed-loop reactant concentration profile under the switching rule of Theorem 4.3 (*solid line*) and in the absence of fault-tolerant control (*dashed line*) subject to simultaneous failures in both the heating jackets

The control objective is to stabilize the reactor at the open-loop unstable steady-state using measurements of C_{A1} and C_{A2}. The available manipulated inputs include the rate of heat input into reactor one, Q_1, subject to the constraint $|Q_1| \leq 2.333 \times 10^6$ kJ/hr, the rate of heat input into reactor two, Q_2, subject to the constraint $|Q_2| \leq 1.167 \times 10^6$ kJ/hr, and a duplicate backup heating configuration for reactor one, Q_3, subject to the constraint $|Q_3| \leq 2.333 \times 10^6$ kJ/hr.

The primary control configuration ($k = 1$) consists of the manipulated inputs Q_1 and Q_2, while the backup configuration ($k = 2$) consists of manipulated inputs Q_2 and Q_3. In order to implement the state-feedback Lyapunov-based controllers, estimates of T_1 and T_2 are generated using a state estimator of the form of Eq. (4.12) with $L_{i,k} = 10000$, $a_{i,k}^{(1)} = 5$, and $a_{i,k}^{(2)} = 1$ for $i = 1, 2$ and $k = 1, 2$. The reactors are initialized at $T_1(0) = 386.97$ K, $C_{A1}(0) = 3.59$ kmol/m^3, $T_2(0) = 432.36$ K, and $C_{A2}(0) = 2.88$ kmol/m^3. The state estimator is initialized at the steady-state values

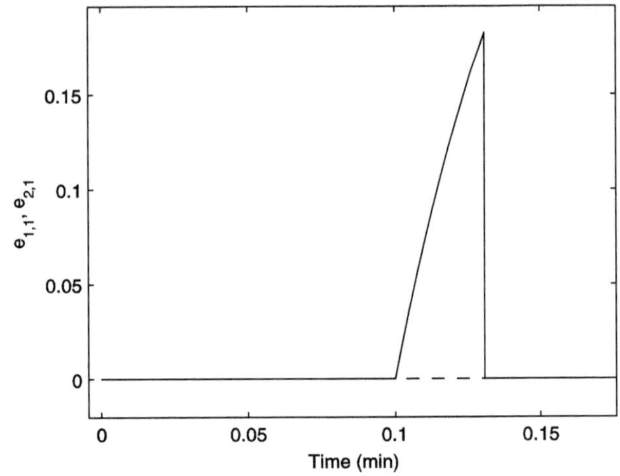

Fig. 4.6 Evolution of residuals $e_{1,1}$ (*solid line*) and $e_{2,1}$ (*dashed line*) corresponding to the manipulated inputs in the first reactor

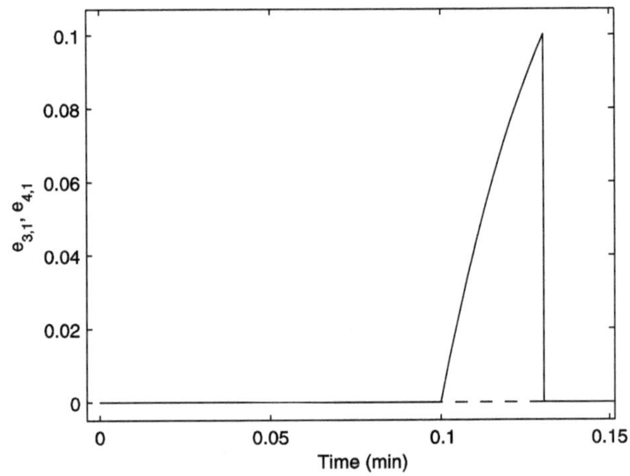

Fig. 4.7 Evolution of residuals $e_{3,1}$ (*solid line*) and $e_{4,1}$ (*dashed line*) corresponding to the manipulated inputs in the second reactor

for this system ($\tilde{T}_1(0) = 388.57$ K, $\tilde{C}_{A1}(0) = 3.59$ kmol/m^3, $\tilde{T}_2(0) = 433.96$ K, and $\tilde{C}_{A2}(0) = 2.88$ kmol/m^3). The fault detection filter states are initialized with the value of the state estimates at $t = 0.0022\,\text{min} \equiv T_1^b$. Note that by this time the estimates have converged sufficiently close to the true values as can be seen as the dash-dotted lines in Fig. 4.8.

As shown by the solid line in Fig. 4.8, the controller drives the closed-loop system to the desired steady-state (for the sake of brevity, only T_1 is shown). A complete failure occurs in Q_1 early on at $T_f = 0.01$ min while the system is still moving toward the desired steady-state. If the fault is not detected and no switching takes place the value of T_1 moves away from the desired operating temperature shown as the dotted line in Fig. 4.8. However, when the fault detection and isolation filter is utilized we can see the filter value $\hat{T}_1$, dashed line in Fig. 4.8, diverges from the estimated value $\tilde{T}_1$. This discrepancy causes the residual $e_{1,1}(t)$ corresponding to Q_1 to rise to the threshold value of 0.01 K (chosen to ensure that all destabilizing

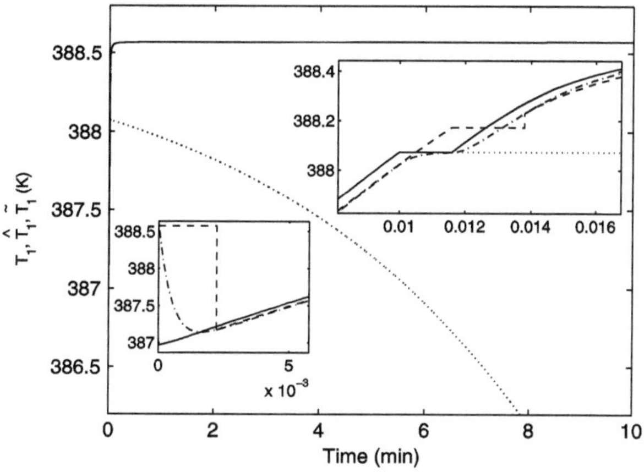

Fig. 4.8 Evolution of the closed-loop temperature (*solid line*), estimate of temperature (*dash-dotted line*), and the temperature profile generated by the FDI filter (*dashed line*) with fault-tolerant control in place. Evolution of the temperature (*dotted line*) without fault-tolerant control in place

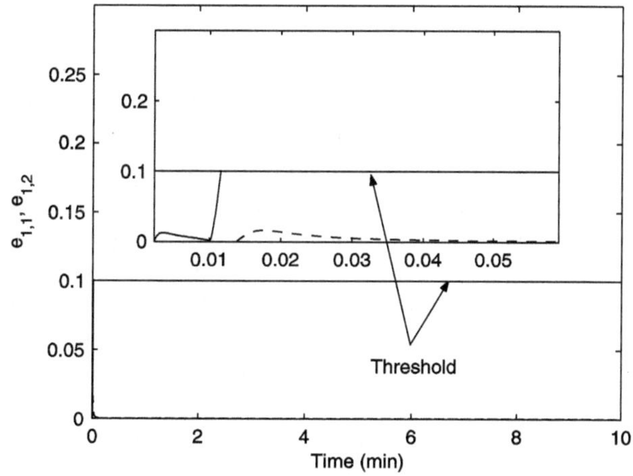

Fig. 4.9 Evolution of the residual corresponding to Q_1 before switching ($k=1$, *solid line*), and Q_3 after switching ($k=2$, *dashed line*). A fault is declared when $e_{1,1}$ reaches the threshold at 0.1

faults are detected) at time $t = 0.0116$ min, as shown in Fig. 4.9. A fault in Q_1 is declared at this time, and the supervisor checks the value of the Lyapunov function for $k=2$. Since $V_2(0.0116) = 0.38 < c_2^{\max} = 9.4$ the supervisor activates the fall-back configuration to achieve closed-loop stability despite actuator failure in Q_1. The fault detection and isolation filter is restarted 0.0022 minutes later at $T_2^b = 0.0138$ min. As expected, no fault is declared at any time in Q_2 as can be seen in Fig. 4.10. In summary, the output-feedback fault detection and isolation and fault-tolerant control system is able to detect and isolate the fault to allow reconfiguration and drive the system to the desired steady state (solid line in Fig. 4.8).

The application and effectiveness of the proposed fault-detection and isolation and fault-tolerant control method has been illustrated in the case of both state and output feedback. Next, this method is applied in the presence of uncertainty and measurement noise. To this end consider the two reactor system used in the previous example with full-state feedback.

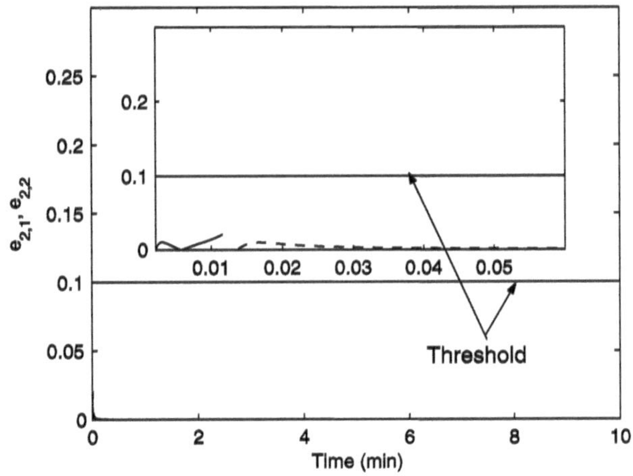

Fig. 4.10 Evolution of the residual corresponding to Q_2 before switching ($k = 1$, *solid line*), and after switching ($k = 2$, *dashed line*). No fault is declared

The control objective is to stabilize the reactor at the open-loop unstable steady-state where $(T_1^s, C_{A1}^s) = (388.57\ \text{K},\ 3.59\ \text{kmol/m}^3)$ and $(T_2^s, C_{A2}^s) = (433.96\ \text{K},\ 2.88\ \text{kmol/m}^3)$. The measurements of temperature and concentration are assumed to contain a noise of magnitude 1 K and 0.1 kmol/m^3, respectively. Also, the concentrations of A in the inlet streams C_{A0} and C_{A03} used in the process model are 10 % smaller than the values used in the filter equations and the controller. The available manipulated inputs include the rate of heat input into reactor one, Q_1, subject to the constraint $|Q_1| \leq 2.333 \times 10^6$ kJ/hr, the rate of heat input into reactor two, Q_2, subject to the constraint $|Q_2| \leq 1.167 \times 10^6$ kJ/hr, and a duplicate backup heating configuration for reactor two, Q_3, subject to the constraint $|Q_3| \leq 1.167 \times 10^6$ kJ/hr.

The primary control configuration consists of the manipulated inputs Q_1 and Q_2, while the backup configuration comprises manipulated inputs Q_1 and Q_3. As before, quadratic Lyapunov functions of the form $V_k = x^T P_k x$ are used for controller design. the controller design yields a stability region estimate with $c_1^{\max}$ and $c_2^{\max}$ both approximately equal to 9.4. Note that all the information about the stability region is completely contained in the values of $c_1^{\max}$ and $c_2^{\max}$. Specifically, the presence of the closed-loop state in the stability region can be ascertained by simply evaluating the value of the Lyapunov-function and checking against the value of $c^{\max}$.

In the first scenario, the ability to detect a fault in the presence of multiple disturbances and noise is demonstrated. The reactors, as well as the fault detection filter for the first control configuration are initialized at the desired unstable steady-state. For the sake of brevity, only the evolution of T_2 and of the residuals are shown. As can be seen in Fig. 4.11(a), the controller maintains the closed-loop trajectory near the desired steady-state until heating jacket two (Q_2) fails 40 min after reactor startup. If a fault-detection and isolation filter is not in place, and the fault is not detected, closed-loop stability is not achieved (dotted lines in Fig. 4.11(a)). The fault-detection and isolation filter designed using the proposed methodology, however, detects this fault when the value of residual $e_{2,1}(t)$ becomes greater than the threshold value of 2.0 K at $t = 40.79$ min (see Fig. 4.11(c)) while $e_{1,1}(t)$ (Fig. 4.11(b))

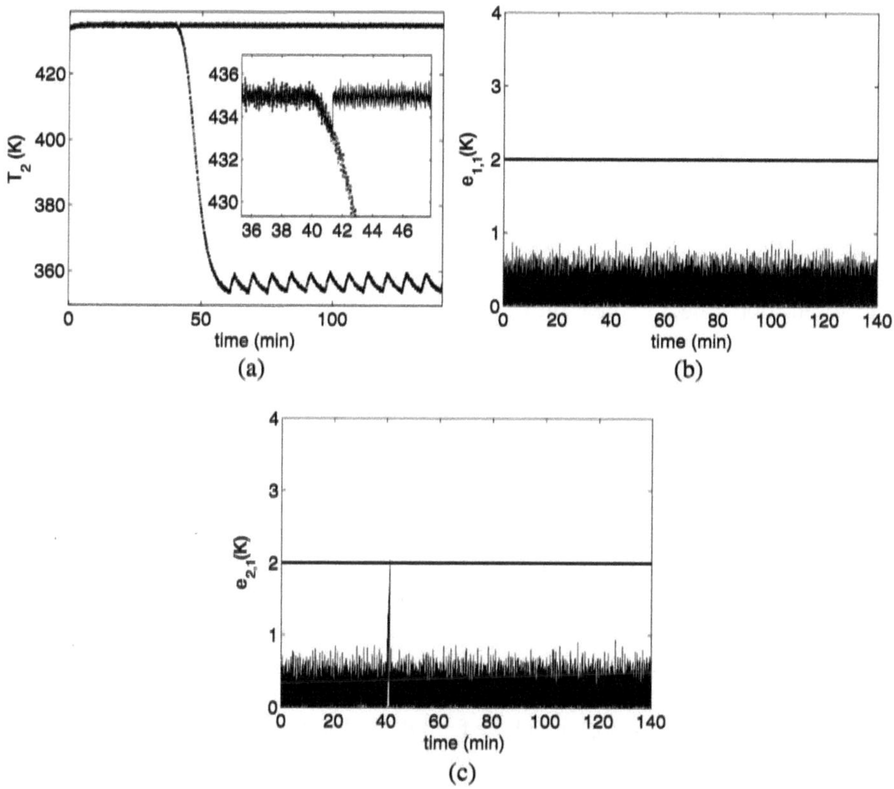

Fig. 4.11 (**a**) Temperature profile of reactor two with reconfiguration (*solid line*) and without reconfiguration (*dotted line*), (**b**) Q_1 residual profile, and (**c**) Q_2 residual profile (note fault detection at time $t = 40.79$ min)

remains below the threshold of 2.0, allowing the detection and isolation of the fault. While at the time of the failure ($t = 40$ min), the state of the closed-loop system is within the stability region of the backup-configuration, but the time that the failure is detected at $t = 40.79$ min, operation of reactor two in an open-loop fashion for 0.79 min results in the state moving out of the stability region of the backup configuration ($V_2(40.79) = 73.17 > c_2^{\max} = 9.4$) and stability is not guaranteed after switching. However, it is possible that stability may still be achieved by using the fall-back configuration. In particular, having been alerted by the fault-detection and isolation filter of the occurrence of the fault, the supervisor activates the fall-back configuration (with Q_1 and Q_3 as the manipulated inputs, solid lines in Fig. 4.11(a)) and is able to drive the system to the desired steady-state and enforce closed-loop stability.

Detection of faults in the presence of process disturbances and noise is clearly possible using the methodology above. In order to guarantee stability after switching, however, the disturbances acting on the system should be reduced or the constraints on the control action should be relaxed to enlarge the estimate of the closed-loop stability region. In the second scenario, the ability to detect a fault in the presence of noise and single disturbance (in contrast to two disturbances in the first scenario), then switch to a fall-back configuration with guaranteed stability is demonstrated. In this case, the measurements of temperature and concentration are again assumed to contain noise of magnitude 1 K and 0.1 kmol/m^3, respectively. Also, the concentration of A in the inlet stream C_{A03} used in the process model is 10 % smaller than the values used in the filter equations and the controller. The reactors as well as the fault detection filter for the first control configuration are initialized at the desired steady state. As can be seen in Fig. 4.12(a), the controller maintains the closed-loop trajectory near the desired steady-state until heating jacket two (Q_2) fails 40 min after reactor startup. If a fault-detection filter is not in place and the fault is not detected, closed-loop stability is not achieved (dotted lines in Fig. 4.12(a)). The implemented fault-detection and isolation filter detects this fault when the value of the residual $e_{2,1}(t)$ becomes greater than the threshold value of 2.0 at 41.33 min (see Fig. 4.12(c)) while $e_{1,1}(t)$ (Fig. 4.12(b)) remains below the threshold of 2.0, allowing the detection and isolation of the fault. In this scenario, by the time that the fault is detected, the state of the closed-loop system resides within the stability region of configuration two ($V_2 = 8.03 < c_2^{\max} = 9.4$). Therefore, the supervisor activates the fall-back configuration with Q_1 and Q_3 as the manipulated inputs (solid lines in Fig. 4.12(a)) and the control system is able to drive the process to the desired steady-state and enforce closed-loop stability.

Remark 4.10 In order to implement the fault detection and isolation filter on process systems accounting for noise, disturbances, and/or output feedback considerations, one needs to decide on a value for the detection threshold for each individual residual. Given the complexity of the closed-loop system, there is no simple and explicit way (formula) to directly calculate this threshold; a trial-and-error procedure needs to be followed. However, there are several things to consider when choosing an appropriate threshold value. The threshold should be chosen large enough so that noisy data, system disturbances, or discrepancies due to estimation error do not cause frequent false alarms. The threshold must also be chosen small enough so that at the time of detection the state of the system is within the stability region of a fall-back configuration. These two considerations will give a reasonable range of threshold values to implement on the fault detection and isolation filter.

4.6 Application to a Reverse Osmosis Desalination Process

In this section, we focus on FTC of a reverse osmosis (RO) process. First, a detailed mathematical model that adequately describes the process evolution is derived. A family of candidate control configurations are identified, and Lyapunov-based feedback control laws are constructed for each configuration such that closed-loop stability is guaranteed within an associated constrained stability region. Subsequently, an FDI filter that observes the deviation of the process states from the

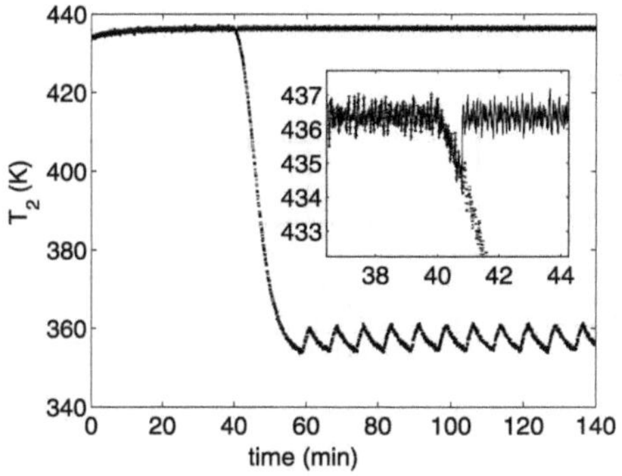

(a)

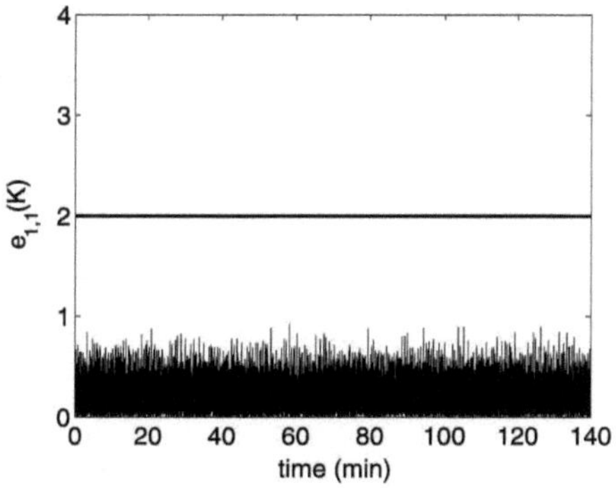

(b)

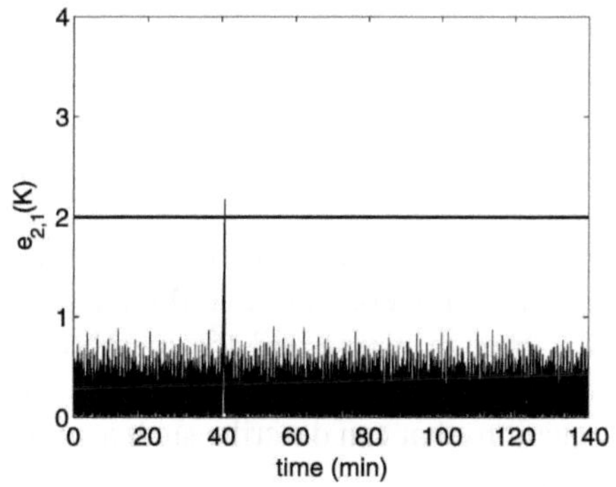

(c)

Fig. 4.12 (**a**) Temperature profile of reactor two with reconfiguration (*solid line*) and without reconfiguration (*dotted line*), (**b**) Q_1 residual profile, and (**c**) Q_2 residual profile (note fault detection at time $t = 41.33$ min)

expected closed-loop behavior is developed to detect and isolate actuator failures. A supervisory switching logic is then derived, on the basis of stability regions and FDI filter information, to orchestrate switching between the available control configurations in a way that guarantees closed-loop stability in the event of actuator faults. The effectiveness of the proposed FDIFTC structure is demonstrated through simulation. For more results on FTC of RO processes, please refer to [102].

4.6.1 Process Description and Modeling

Figure 4.13 shows a schematic of an elementary RO desalination process. This is a single-unit RO system with no pre-treatment or post-treatment units. Feed brackish or seawater enter the system through the high pressure pump. This high-pressure water then flows across an RO membrane, and low salinity product water permeates. Concentrated brine then continues through a throttling valve and is discharged

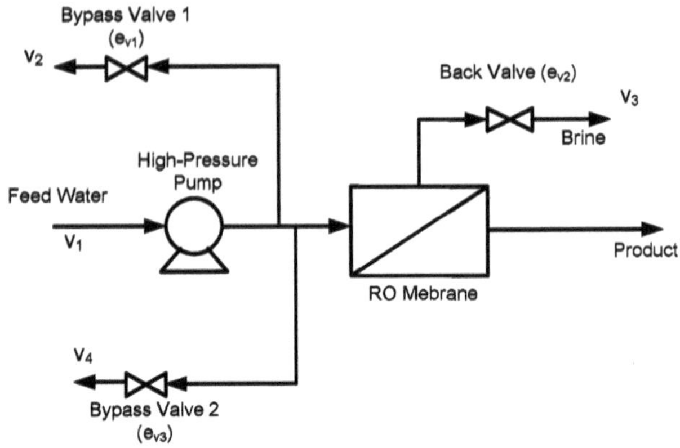

Fig. 4.13 Single membrane unit reverse osmosis desalination process

at atmospheric pressure. The RO plant consists of a high pressure pump, three automated valves, membrane unit, and required plumbing and tanks. The valve settings can be manipulated in real-time based on measurement information which includes the flow velocities.

The first principles model of this system is based on a macroscopic kinetic energy balance. This model assumes an incompressible fluid and constant internal volume and mass. Skin friction through piping and the membrane system are negligible relative to hydraulic losses in the throttling valves and across the membrane. Three ordinary differential equations that can describe such a system are derived and they have the following form:

$$
\begin{aligned}
\frac{dv_2}{dt} &= \frac{1}{\rho V}\left(\frac{W_p}{v_1(v_2, v_3, v_4)} - \frac{1}{2}e_{v1}v_2\right),\\
\frac{dv_3}{dt} &= \frac{1}{\rho V}\left(\frac{W_p}{v_1(v_2, v_3, v_4)} - \frac{1}{2}e_{v2}v_3\right),\\
\frac{dv_4}{dt} &= \frac{1}{\rho V}\left(\frac{W_p}{v_1(v_2, v_3, v_4)} - \frac{1}{2}e_{v3}v_4\right),\\
v_1 &= -\frac{1}{2}b + \frac{1}{2}\sqrt{b^2 + 4c},\\
b &= -\left(v_2 + v_3 + v_4 - \frac{A_m K_m \Delta\pi}{\rho A_p}\right),\\
c &= \frac{A_m K_m W_p}{\rho A_p^2},
\end{aligned}
\tag{4.19}
$$

where v_1, feed velocity, is a nonlinear function of v_2, v_3, and v_4. v_2, v_3, and v_4 are the velocities of bypass discharge one, brine discharge, and bypass discharge two, respectively. ρ is the fluid density, V is the internal volume, W_p is the power delivered by the pump, A_p is the pipe cross-sectional area. e_{v1}, e_{v2}, and e_{v3} are the frictional valve constants. A_m is the membrane area, K_m is a membrane mass

Table 4.2 Process parameters and steady-state values for the desalination process

$\rho = 1000$ kg/m^3	$V = 10$ L
$W_p = 104.4$ W	$A_p = 0.25$ in^2
$A_m = 5$ m^2	$K_m = 9.218 \times 10^{-9}$ s/m
$\Delta\pi = 200$ psi	$e_{v1}^{s1} = 100$
$e_{v2}^{s1} = 230$	$e_{v3}^{s1} = 10^{-8}$
$v_2^{s1} = 1.0547$ m/s	$v_3^{s1} = 0.4625$ m/s
$v_4^{s1} = 1.07 \times 10^{-6}$ m/s	$P^{s1} = 243.7$ psi
$e_{v1}^{s2} = 150$	$e_{v2}^{s2} = 230$
$e_{v3}^{s2} = 300$	$v_2^{s2} = 0.7092$ m/s
$v_3^{s2} = 0.4625$ m/s	$v_4^{s2} = 0.3546$ m/s
$P^{s2} = 243.7$ psi	

transfer coefficient, and $\Delta\pi$ is the osmotic pressure. The potential manipulated inputs of the model are the valve constants (e_{v1}, e_{v2}, and e_{v3}) which can be manipulated in practice by an automated electric motor that partially opens or closes the valves. The measured outputs are the velocities of the fluid in the bypass lines, and brine velocity (v_2, v_3, and v_4). Internal pressure, P can be related to feed velocity by $P = \frac{W_p}{v_1 A_p}$. The product velocity, v_5, can be related to the system pressure by $v_5 = \frac{A_m K_m}{\rho A_p}(P - \Delta\pi)$. Table 4.2 shows the parameter values used for this example.

The control objective is to stabilize the process at the desired steady-state. There are at least two unique configurations that will give simultaneous independent control of transmembrane pressure and brine flow-rate. Configuration one, u_1, uses the back valve and the first bypass valve (e_{v1}, e_{v2}) as manipulated inputs. The valves are subjected to input constraints of the form $0 < e_{v1} < 200$ and $130 < e_{v2} < 330$. Configuration two, u_2, uses the back valve with the second bypass valve (e_{v2}, e_{v3}) as manipulated inputs. These valves are subjected to input constraints of the form $130 < e_{v2} < 330$ and $200 < e_{v3} < 400$. The first control configuration, u_1, will be considered as the primary configuration. However, in the event of a failure the plant supervisor may need to implement the fall-back configuration, u_2, to maintain closed-loop stability. By observing the evolution of the plant the FDI filters can detect and isolate an actuator fault. If there is a fall-back control configuration available that is able to stabilize the RO plant, then the supervisor will initiate a mode transition to the fall-back configuration. These issues are addressed in detail in the next section.

4.6.2 Fault-Detection and Isolation and Fault-Tolerant Control

Given the properties of the dynamic model, Eq. (4.19), it can be shown that both configurations, u_1 and u_2, satisfy the requirements of achieving fault-detection and isolation of actuator faults. This section discusses the four steps to implement FDIFTC on the RO process. The first step is to synthesize stabilizing feed-back controllers for each configuration. The second step is to explicitly characterize the constrained stability region associated with each configuration. The third step is to design FDI filters for each manipulated input. The final step is to design the switching law that orchestrates the reconfiguration of the control system in a way that guarantees closed-loop stability in the event of faults in the active control configuration.

To present results in a convenient form, the model of Eq. (4.19) is written in deviation variable form around the desired steady state. This is defined as $x = [x_1\ x_2\ x_3]^T$ where $x_1 = v_2 - v_{2_s}$, $x_2 = v_2 - v_{2_s}$, and $x_3 = v_4 - v_{4_s}$. The plant can then be described by the following nonlinear continuous-time system:

$$\begin{aligned} \dot{x}(t) &= f_{k(t)}\big(x(t)\big) + g_{k(t)}\big(x(t)\big)u_{k(t)}, \\ |u_{k(t),i}| &\le u_{k,i}^{\max}, \\ k(t) &\in K = \{1, 2\}, \end{aligned} \tag{4.20}$$

where $x(t) \in \mathbb{R}^3$ denotes the vector of process state variables and $u_{k(t)}$ is a vector of inputs where $u_{k,i}(t) \in [-u_{k,i}^{\max}, {u_{k,i}}^{\max}] \subset \mathbb{R}$ denotes the ith constrained manipulated input associated with the kth control configuration. The function $k(t)$, which takes values in the finite set K, represents a discrete state that indexes the vector fields $f_k(\cdot)$, $g_k(\cdot)$ and the manipulated inputs $u_k(\cdot)$. The explicit form of the vector fields can be obtained by comparing Eqs. (4.19) and (4.20) and is omitted for brevity. For each value that k assumes in K, the process is controlled via a different set of manipulated inputs which define a given control configuration. Switching between the two available configurations is handled by the high-level supervisor. The control objective is to stabilize the process in the presence of actuator constraints and possible faults. The state feedback problem where measurements of all process states are available for all times is considered to simplify presentation of the results.

4.6.2.1 Constrained Feedback Controller Synthesis

In this step, we synthesize for each control configuration a feedback controller that enforces asymptotic closed-loop stability in the presence of actuator constraints. To accomplish this task, first a quadratic Lyapunov function of the form $V_k = x^T P_k x$ is defined, where P_k is a positive-definite symmetric matrix that satisfies the Riccati inequality. This Lyapunov function is used to synthesize a bounded nonlinear feedback control law for each control-loop (see [85] and [46]) of the form

$$u_k = -r\big(x, u_k^{\max}\big) L_{\bar{g}_k} V_k, \tag{4.21}$$

where

$$r = \frac{L_{\bar{f}_k}^{*} V_k + \sqrt{(L_{\bar{f}_k}^{*} V_k)^2 + (u_k^{\max} |L_{\bar{g}_k} V_k|)^4}}{(|L_{\bar{g}_k} V_k|)^2 (1 + \sqrt{1 + (u_k^{\max} |L_{\bar{g}_k} V_k|)^2})} \tag{4.22}$$

and $L^*_{\bar{f}_k} V_k = L_{\bar{f}_k} V_k + \alpha V_k$, $\alpha > 0$. The scalar function $r(\cdot)$ in Eqs. (4.21) and (4.22) can be considered as a nonlinear controller gain. It can be shown that each control configuration asymptotically stabilizes the states in each mode. This controller gain, which depends on both the size of actuator constraints, $u_k^{\max}$, and the particular configuration used is shaped in a way that guarantees constraint satisfaction and asymptotic stability within a well-characterized region in the state space. The characterization of this region is discussed in the next step.

Actuator constraints place fundamental limitations on the initial conditions from which the closed-loop system is asymptotically stable. It is important for the control system designer to explicitly characterize these limitations by identifying, for each control configuration, the set of initial conditions for which the constrained closed-loop system is asymptotically stable. This is necessary for the design of an appropriate switching policy that ensures the fault-tolerance of the closed-loop system. The feedback controller of Eq. (4.21) that is synthesized for each configuration provides such a characterization. Specifically, using a Lyapunov argument, one can show that the set

$$\Theta\left(u_k^{\max}\right) = \left\{x \in \mathbb{R}^3 : L^*_{\bar{f}_k} V_k \le u_k^{\max} |L_{\bar{g}_k} V_k|\right\} \tag{4.23}$$

describes a region in the state-space where the control action satisfies the constraints and the time-derivative of the corresponding Lyapunov function is negative-definite along the trajectories of the closed-loop system (see [28]). Note that the size of the set depends on the magnitude of the constraints. The set becomes smaller as the constraints become tighter (smaller $u_{k,i}^{\max}$). For a given control configuration, the above inequality can be used to estimate the associated stability region. This can be done by constructing the largest invariant subset of Θ, which is denoted by $\Omega(u_k^{\max})$. Initial conditions within the set $\Omega(u_k^{\max})$ ensure that the closed-loop trajectory stays within the region defined by $\Theta(u_k^{\max})$, and thereby V_k continues to decay monotonically, for all times that the kth control configuration is active (see [45] for further discussion on this issue). An estimate of $\Omega(u_k^{\max})$ is obtained by defining a composite Lyapunov function of the form $V_{C_k} = x^T P_C x$, where P_C is a positive definite matrix, and choosing a level set of V_{C_k}, Ω_{C_k}, for which $\dot{V}_{C_k} < 0$ for all x in Ω_{C_k}. The value $c_k^{\max}$ represents a level set on V_{C_k} where $\dot{V}_{C_k} < 0$.

The third step in implementing FDIFTC is that of designing appropriate fault-detection filters. The filters should detect and isolate the occurrence of a fault in an actuator by observing the behavior of the closed-loop process. The FDI filter design for the primary control configuration takes the form:

$$\begin{aligned}
\frac{d\tilde{v}_2}{dt} &= \frac{1}{\rho V}\left(\frac{W_p}{v_1(\tilde{v}_2, v_3, v_4)} - \frac{1}{2} e_{v1}(\tilde{v}_2, v_3, v_4)\tilde{v}_2\right), \\
\frac{d\tilde{v}_3}{dt} &= \frac{1}{\rho V}\left(\frac{W_p}{v_1(v_2, \tilde{v}_3, v_4)} - \frac{1}{2} e_{v2}(v_2, \tilde{v}_3, v_4)\tilde{v}_3\right), \\
r_{1,1} &= |v_2 - \tilde{v}_2|, \\
r_{1,2} &= |v_3 - \tilde{v}_3|,
\end{aligned} \tag{4.24}$$

where $\tilde{v}_2$ and $\tilde{v}_3$ are the filter states for valve one and two, respectively. $r_{k,i}$ is the residual associated with the ith input of the kth configuration. The filter states are initialized at the same value as the process states ($\tilde{x}(0) = x(0)$) and essentially predict the evolution of the process in the absence of actuator faults. The residual associated with each manipulated input captures the difference between the predicted evolution of the states in the absence of a fault on that actuator and the evolution of the measured process state. If a given residual becomes nonzero, a fault is declared on the associated input.

The final step is to design a switching logic that the plant supervisor will use to decide what fall-back control configuration to implement given an actuator failure. The supervisor should only implement those configurations that will guarantee closed-loop stability and do not utilize a failed actuator. This requires that the supervisor only activates fall-back control configurations for which the state is within the associated stability region at the time of fault-detection. Let the initial actuator configuration be $k(0) = 1$, T_{fault} be the time of an actuator failure, and T_{detect} be the earliest time at which the value of $r_{1,i}(t) > \delta_{r_{1,i}} > 0$ (for the ith input where $\delta_{r_{1,i}}$ is the ith detection threshold). The switching rule given by

$$k(t \geq T_{\text{detect}}) = 2 \quad \text{if } x(T_{\text{detect}}) \in \Omega_{C_2}\left(u_2^{\max}\right) \tag{4.25}$$

guarantees asymptotic closed-loop stability if u_2 does not include any faulty actuators. The switching law requires monitoring of FDI filters and process state location with respect to fall-back stability regions.

4.6.3 Simulation Results

A simulation has been performed to demonstrate the implementation of the proposed FDIFTC strategy on the RO plant of Fig. 4.13. The states in the mathematical model given in Eq. (4.19) may not be the system parameters of interest for the operator because bypass flows (v_2 and v_4) do not interact with the membrane unit. Pressure and brine flow, P and v_3, are useful parameters to regulate because they directly effect the membrane unit. Hence, two steady-states have been considered, each one of them has the same system pressure and brine flow rate (v_3), but different bypass flows (v_2 and v_4). The first steady-state corresponds to bypass valve two being closed. The parameters and steady-state values can be seen in Table 4.2. Under these operating conditions the open-loop system behaves in a stable fashion at each steady-state.

First, nonlinear feedback control under the primary configuration, u_1, was considered. The bounded nonlinear controller was synthesized using Eq. (4.21) and (4.22), with $\alpha = 0.1$. The stability region for the primary configuration was estimated using the Lyapunov function, $V_1 = x^T P_1 x$, yielding a $c_1^{\max} = 1$ (note that this value of $c_1^{\max}$ represents a sufficiently large region of the state space for this simulation, in general much higher values can be considered). Figure 4.14 shows

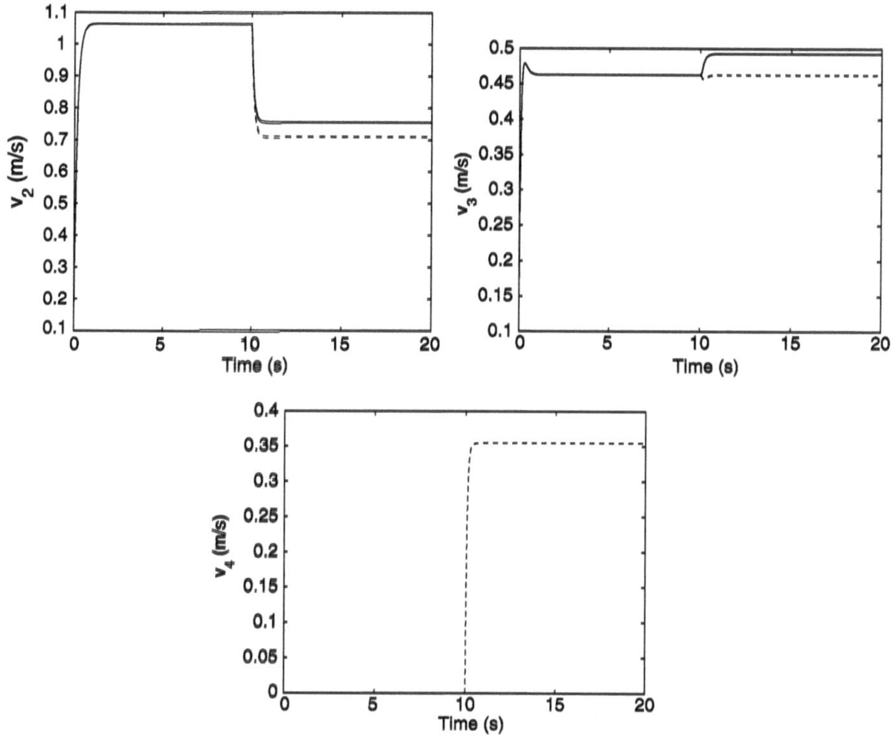

Fig. 4.14 Evolution of the closed-loop state profiles under fault-tolerant control (*dashed line*) and without fault tolerant-control (*solid line*). FTC recovers the desired brine flow, v_3

the evolution of the closed-loop state profiles starting from the initial condition $v_2 = v_3 = 0.1$ m/s and $v_4 = 0.001$ m/s for which $V_1(x(0)) = 0.0263$. Evolution of the system pressure is shown in Fig. 4.15. Since the initial state was within the stability region of the primary control configuration, $V_1(x(0)) = 0.0263 \leq c_1^{\max} = 1$, the primary control configuration was able to stabilize the system at the desired steady-state.

Next, a fault in the primary configuration (in e_{v1} specifically) at a time $T_{\text{fault}} = 10$ s was considered. In this case, the fall-back configuration, u_2, was available with valve three, e_{v3}, as one of the manipulated inputs. The quadratic Lyapunov function $V_2 = x^T P_2 x$ and $\alpha = 0.1$ was used to design the controller. The stability region was also estimated using V_2 yielding a $c_2^{\max} = 1$.

To demonstrate the advantage of operating under the FDIFTC structure consider the case where no control system reconfiguration takes place after T_{fault}. The system is initialized at $v_2 = v_3 = 0.1$ m/s and $v_4 = 0.001$ m/s, and the primary control configuration operates normally until the time $T_{\text{fault}} = 10$ s. At this time, valve one stops operating and is partially closed, $e_{v1} = 150$. As shown by the solid lines in Figs. 4.14 and 4.15, the states move away from the desired values, and settle at a new, undesired, steady-state.

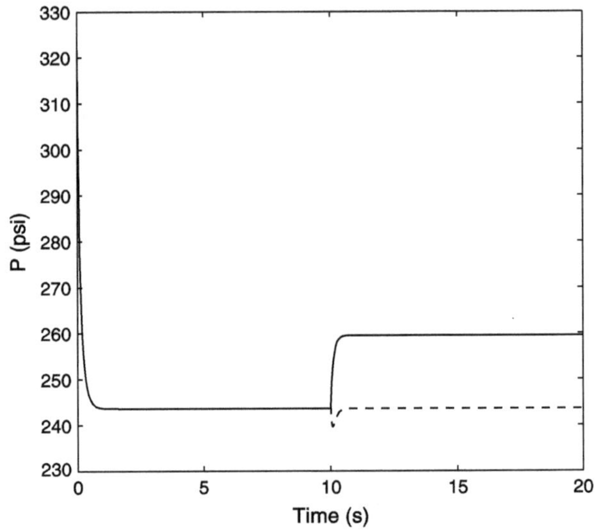

Fig. 4.15 Evolution of the closed-loop pressure profile under fault tolerant control (*dashed line*) and without fault tolerant control (*solid line*). FTC recovers the desired operating pressure

However, by implementing the FDIFTC structure the fault can be mitigated. The residual value associated with valve one, $r_{1,1}$, becomes nonzero and reaches the detection threshold, $\delta_{r_{1,1}} = 0.01$, at $T_{\text{detect}} = 10.004$ s when the fault is declared. The residual value associated with valve two, $r_{1,2}$, remains at zero, indicating that the fault is effecting only valve one. At time T_{detect}, the value of the fall-back Lyapunov function is checked against the fall-back stability region to see if switching would guarantee stability. The value of $V_2(x(T_{\text{detect}})) = 0.0119 < c_2^{\max} = 1$, so reconfiguration to the fall-back controller, $k = 2$, does guarantee closed-loop stability. The evolution of the system states and pressure under the proposed FDIFTC structure can be seen in Figs. 4.14 and 4.15 (solid lines). This automated reconfiguration allowed the closed-loop system to maintain pressure and brine flow at the desired values.

4.7 Conclusions

In this chapter, we extended the results of Chap. 3 to include multi-input multi-output nonlinear systems subject to multiple faults in the control actuators and constraints on the manipulated inputs. A fault-tolerant control framework integrating fault detection, fault isolation, and feedback control configurations was discussed. In order to illustrate the ideas, we considered the case that the state feedback is available first and then considered the case of output feedback. Applications of the methods to a chemical reactor and a reverse osmosis water desalination process were presented to demonstrate the applicability and effectiveness of the methods.

Chapter 5
Safe-Parking

5.1 Introduction

In Chaps. 3 and 4, we presented fault handling methods that assume availability of sufficient residual control effort or redundant control configurations to preserve operation at the nominal equilibrium point. In particular, if redundant control configurations are available, control-loop reconfiguration (activating an appropriately chosen fall-back configuration) can be implemented to preserve closed-loop stability at the nominal equilibrium point. In this chapter, we consider the scenario where a fault results in temporary loss of stability that cannot be handled by redundant control loops. In other words, we consider faults for which there simply does not exist a fall-back configuration that allows continuation of operation at the nominal, desired equilibrium point. Handling such faults requires the design of a mechanism that achieves the transition of the plant to an appropriately chosen temporary operating point in such a way that nominal operation can be resumed safely and smoothly. In the absence of a framework for handling such faults, ad-hoc approaches could result in performance degradation or even result in process shutdowns.

Motivated by the above considerations, this chapter considers the problem of control of nonlinear systems subject to input constraints and destabilizing faults in the control actuators. Specifically, faults are considered that cannot be handled via robust control approaches or activation of redundant control configurations, and necessitate fault-rectification. A safe-parking framework is developed to address the problem of determining how to run the process during fault-rectification to enable smooth resumption of nominal operation. The rest of the chapter is organized as follows: we first present, in Sect. 5.2.1, the class of processes considered, followed by a styrene polymerization process in Sect. 5.2.2 and review a Lyapunov-based predictive controller in Sect. 5.2.3. The safe-parking problem is formulated in Sect. 5.3.1, and safe-parking designs that address stability and performance objectives are presented in Sects. 5.3.2 and 5.3.3, respectively. A chemical reactor example is used to illustrate the details of the safe-parking framework in Sect. 5.3.4 while application to the styrene polymerization process, subject to parametric uncertainty and disturbances, is demonstrated in Sect. 5.4. Finally, in Sect. 5.5 we summarize our results.

P. Mhaskar et al., *Fault-Tolerant Process Control*, DOI 10.1007/978-1-4471-4808-1_5,

5.2 System Description

In this section, we describe the class of processes considered, present a polystyrene process example to motivate the proposed framework, and review a Lyapunov-based model predictive control design.

5.2.1 Process Description

We consider nonlinear process systems subject to input constraints and failures described by

$$\dot{x}(t) = f\big(x(t)\big) + G\big(x(t)\big)u_\sigma(t), \quad u_\sigma(\cdot) \in \mathbf{U}, \tag{5.1}$$

where $x \in \mathbb{R}^n$ denotes the vector of state variables, $u_\sigma(t) \in \mathbb{R}^m$ denotes the vector of constrained manipulated inputs, taking values in a nonempty convex subset $\mathbf{U}$ of $\mathbb{R}^m$, where $\mathbf{U} = \{u \in \mathbb{R}^m : u_{\min} \le u \le u_{\max}\}$, where $u_{\min}, u_{\max} \in \mathbb{R}^m$ denote the constraints on the manipulated inputs, and $u^{\text{norm}} > 0$ is such that $|u| \le u^{\text{norm}} \Longrightarrow u \in \mathbf{U}$, $f(0) = 0$ and $\sigma \in \{1, 2\}$ is a discrete variable that indexes the fault-free and faulty operation ($\sigma = 1$ denotes fault-free operation and $\sigma = 2$ denotes faulty operation). The vector function $f(x)$ and the matrix $G(x) = [g^1(x) \ldots g^m(x)]$ where $g^i(x) \in \mathbb{R}^n$, $i = 1, \ldots, m$ are assumed to be sufficiently smooth on their domains of definition. Throughout the chapter, we assume that for any $u \in \mathbf{U}$ the solution of the system of Eq. (5.1) exists and is continuous for all t, and we focus on the state feedback problem where $x(t)$ is assumed to be available for all t.

5.2.2 Motivating Example

To motivate the safe-parking framework and to demonstrate an application of our results, we introduce in this section a polystyrene polymerization process. To this end, consider the following model for a polystyrene polymerization process given in [69] (also studied in, e.g., [136] and [93]).

$$\begin{aligned}
\dot{C}_I &= \frac{(F_i C_{If} - F_t C_I)}{V_{pr}} - k_d C_I, \\
\dot{C}_M &= \frac{(F_m C_{Mf} - F_t C_M)}{V_{pr}} - k_p C_M C_P, \\
\dot{T} &= \frac{F_t(T_f - T)}{V_{pr}} + \frac{(-\Delta H)}{\rho c_p} k_p C_M C_P - \frac{hA}{\rho c_p V}(T - T_c), \\
\dot{T}_c &= \frac{F_c(T_{cf} - T_c)}{V_c} + \frac{hA}{\rho_c C_{pc} V_c}(T - T_c),
\end{aligned}$$

Table 5.1 Styrene polymerization parameter values and units

Parameter	Value	Unit	Parameter	Value	Unit
F_i	0.3	L/s	F_m	1.05	L/s
F_s	1.275	L/s	F_t	2.625	L/s
F_c	1.31	L/s	$C_{If,n}$	0.5888	$\mathrm{kmol/m^3}$
C_I	0.067	$\mathrm{kmol/m^3}$	$C_{Mf,n}$	9.975	$\mathrm{kmol/m^3}$
C_M	3.968	$\mathrm{kmol/m^3}$	$T_{f,n}$	306.71	K
T	303.55	K	$T_{cf,n}$	294.85	K
T_c	297.95	K	A_d	5.95×10^{14}	$\mathrm{s^{-1}}$
A_t	1.25×10^{10}	$\mathrm{s^{-1}}$	A_p	1.06×10^{8}	$\mathrm{kmol/(m^3\,s)}$
E_d/R	14.897×10^3	K	E_t/R	8.43×10^2	K
E_p/R	3.557×10^3	K	f	0.6	
ΔH	-1.67×10^4	kJ/kmol	ρc_p	360	$\mathrm{kJ/(m^3\,K)}$
hA	700	J/(K s)	$\rho_c c_{pc}$	966.3	$\mathrm{kJ/(m^3\,K)}$
V_{pr}	3.0	$\mathrm{m^3}$	V_c	3.312	$\mathrm{m^3}$

$$C_P = \left[\frac{2fk_dC_I}{k_t}\right]^{\frac{1}{2}}, \tag{5.2}$$

$$k_d = A_d e^{\frac{-E_d}{RT}},$$

$$k_p = A_p e^{\frac{-E_p}{RT}},$$

$$k_t = A_t e^{\frac{-E_t}{RT}},$$

where C_I, C_{If}, C_M, C_{Mf} refer to the concentrations of the initiator and monomer in the reactor and inlet stream, respectively, T and T_f refer to the reactor and inlet stream temperatures, and T_c and T_{cf} refer to the coolant jacket and inlet temperatures, respectively. The manipulated inputs are the monomer (F_m) and coolant (F_c) flow rates. As is the practice with the operation of the polystyrene polymerization process [69], the solvent flow rate is also changed in proportion to the monomer flow rate. The values of the process parameters are given in Table 5.1.

The control objective is to stabilize the reactor at the equilibrium point ($C_I = 0.067\ \mathrm{kmol/m^3}$, $C_M = 3.97\ \mathrm{kmol/m^3}$, $T = 303.55$ K, $T_c = 297.95$ K), corresponding to the nominal values of the manipulated inputs of $F_c = 1.31$ L/s and $F_m = 1.05$ L/s. The manipulated inputs are constrained as $0 \le F_m \le 31.05$ L/s and $0 \le F_c \le 31$ L/s.

Consider the scenario where the valve manipulating the coolant flow rate fails and reverts to the fail-safe position (fully open). With the coolant flow rate set to the maximum, there simply does not exist an admissible value of the functioning manipulated input F_m such that the nominal equilibrium point remains an equilibrium point for the process, precluding the possibility of continued operation at the nominal equilibrium point. The key problem is to determine how to operate the process under failure condition so that upon fault-recovery, nominal operation can be resumed efficiently. We will demonstrate the application as well as investigate the robustness of the proposed safe-parking framework via the styrene polymerization process in Sect. 5.4, while illustrating the details of the proposed framework using an illustrative chemical reactor in Sect. 5.3.4.

5.2.3 Lyapunov-Based Model Predictive Control

In this section, we briefly review a recent result on the design of a Lyapunov-based predictive controller that possesses an explicitly characterized set of initial conditions from where it is guaranteed to be feasible, and hence stabilizing in the presence of input constraints. Consider the system of Eq. (5.1), for $\sigma(t) = 1$ (i.e., under no fault, where all the manipulated inputs can be changed via a feedback law), under the predictive controller [108] of the form:

$$u_1(\cdot) = \text{argmin}\{J(x, t, u(\cdot)) | u(\cdot) \in S\}, \tag{5.3}$$

$$\text{s.t.} \quad \dot{x} = f(x) + G(x)u(t), \tag{5.4}$$

$$\dot{V}(x(\tau)) \le -\varepsilon^* \quad \forall \tau \in [t, t + \Delta) \quad \text{if } V(x(t)) > \delta', \tag{5.5}$$

$$V(x(\tau)) \le \delta' \quad \forall \tau \in [t, t + \Delta) \quad \text{if } V(x(t)) \le \delta', \tag{5.6}$$

where $S = S(t, T)$ is the family of piecewise continuous functions (functions continuous from the right), with period Δ, mapping $[t, t + T]$ into U, and T is the horizon. Equation (5.4) is the nonlinear model describing the time evolution of the state x, V is a control Lyapunov function and δ', ε^* are parameters to be determined. A control $u(\cdot)$ in S is characterized by the sequence $\{u[j]\}$ where $u[j] := u(j\Delta)$ and satisfies $u(t + \tau) = u[j]$ for all $\tau \in [t + j\Delta, t + (j + 1)\Delta)$. The performance index is given by

$$J(x, t, u(\cdot)) = \int_t^{t+T} \left[\|x^u(s; x, t)\|_{Q_w} + \|u(s)\|_{R_w}\right] ds, \tag{5.7}$$

where Q_w is a positive semi-definite symmetric matrix and R_w is a strictly positive definite symmetric matrix. $x^u(s; x, t)$ denotes the solution of Eq. (5.1), due to control u, with initial state x at time t. The minimizing control $u[1] \in S$ is then applied to the plant over the interval $[t, t + \Delta)$ and the procedure is repeated indefinitely.

The stability properties of the predictive controller are characterized using a bounded controller of the form (e.g., see [46, 85]):

$$u(x) = -k(x)(L_G V)'(x), \tag{5.8}$$

$$k(x) = \frac{L_f V(x) + \sqrt{(L_f V(x))^2 + (u_{\max}\|(L_G V)'(x)\|)^4}}{\|(L_G V)'(x)\|^2[1 + \sqrt{1 + (u_{\max}\|(L_G V)'(x)\|)^2}]}, \tag{5.9}$$

when $L_G V(x) \neq 0$ and $k(x) = 0$ when $L_G V(x) = 0$ where $L_G V(x) = [L_{g_1} V(x) \dots L_{g_m} V(x)]'$ and $g_i(x)$ is the ith column of the matrix $G(x)$. For the controller of Eq. (5.8)–(5.9), one can show, using a standard Lyapunov argument, that whenever the closed-loop state, x, evolves within the region described by the set

$$\Pi = \{x \in \mathbb{R}^n : L_f V(x) \leq u^{\text{norm}} \|(L_G V)'(x)\|\}, \tag{5.10}$$

where $u^{\text{norm}} > 0$ is such that $\|u\| \leq u^{\text{norm}}$ implies $u \in \mathbf{U}$, then the control law satisfies the input constraints, and the time-derivative of the Lyapunov function is negative-definite. An estimate of the stability region can be constructed using a level set of V, i.e.,

$$\Omega = \{x \in \mathbb{R}^n : V(x) \leq c^{\max}\}, \tag{5.11}$$

where $c^{\max} > 0$ is the largest number for which $\Omega \subseteq \Pi$. Closed-loop stability and feasibility properties of the closed-loop system under the Lyapunov-based predictive controller are inherited from the bounded controller under discrete implementation and are formalized in Theorem 1 below (for a proof, see [108]).

Theorem 5.1 [108] *Consider the constrained system of Eq.* (5.1) *under the MPC law of Eqs.* (5.3)–(5.7). *Then, given any* $d \geq 0$, $x_0 \in \Omega$, *where* Ω *was defined in Eq.* (5.11), *there exist positive real numbers* δ', ε^*, *and* Δ^*, *such that if* $\Delta \in (0, \Delta^*]$, *then the optimization problem of Eqs.* (5.3)–(5.7) *is feasible for all times*, $x(t) \in \Omega$ *for all* $t \geq 0$, *and* $\limsup_{t\to\infty} \|x(t)\| \leq d$.

Remark 5.1 The predictive controller formulation of Eqs. (5.3)–(5.7) requires that the value of the Lyapunov function decrease during the first step only. Practical stability of the closed-loop system is achieved since only the first move of the set of calculated moves is implemented and the problem is re-solved at the next time step. If the optimization problem is initially feasible and continues to be feasible, then every control move that is implemented enforces a decay in the value of the Lyapunov function, leading to stability. Furthermore, the constraint of Eq. (5.5) is guaranteed to be satisfied since the control action computed by the bounded controller design provides a feasible initial guess to the optimization problem. Finally, since the state is initialized in Ω, which is a level set of V, the closed-loop system evolves so as to stay within Ω, thereby guaranteeing feasibility at future times. The key idea in the predictive control design is to identify stability constraints that can (a) be shown to be feasible and (b) upon being feasible can guarantee stability. Note that the model predictive controller of Eqs. (5.3)–(5.7) is only used to illustrate the safe-parking framework, and any other controller that provides an explicit characterization of the closed-loop stability region can be used within the proposed framework.

5.3 Safe-Parking of Nonlinear Process Systems

We first formalize the problem in Sect. 5.3.1, and present a safe-parking algorithm focusing on closed-loop stability in Sect. 5.3.2. We then incorporate performance considerations in the safe-parking framework in Sect. 5.3.3, where we also summarize results on implementation of the safe-parking approach subject to limited availability of measurements and uncertainty, as well as application to chemical processes described by distributed parameter systems.

5.3.1 Problem Definition

We consider faults where one of the control actuators fails and reverts to the fail-safe value. Examples of fail-safe positions include fully open for a valve controlling a coolant flow rate, fully closed for a valve controlling a steam flow, etc. (generalization to the case where multiple actuators fail and get 'stuck' at non-nominal values is discussed in Remark 5.4). Specifically, we characterize the fault occurring w.l.o.g., in the first control actuator at a time T^{fault}, subsequently rectified at a time T^{recovery} (i.e., for $t \leq T^{\text{fault}}$ and $t > T^{\text{recovery}}$, $\sigma(t) = 1$, and $\sigma(t) = 2$ for $T^{\text{fault}} < t \leq T^{\text{recovery}}$), as $u_2^1(t) = u_{\text{failed}}^1$, with $u_{\min}^1 \leq u_{\text{failed}}^1 \leq u_{\max}^1$, where u^i denotes the ith component of a vector u, for all $T^{\text{fault}} < t \leq T^{\text{recovery}}$, leaving only u_2^i, $i = 2, \dots, m$ available for feedback control. With $u_2^1(t) = u_{\text{failed}}^1$, there exists a (possibly connected) manifold of equilibrium points where the process can be stabilized, which we denote as the candidate safe-park set $X_c := \{x_c \in \mathbb{R}^n : f(x_c) + g^1(x_c)u_{\text{failed}}^1 + \sum_{i=2}^m g^i(x_c)u_2^i = 0,\ u_{\min}^i \leq u_2^i \leq u_{\max}^i, i = 2, \dots, m\}$. The safe-park candidates therefore represent equilibrium points that the system can be stabilized at, subject to the failed actuator, and with the other manipulated inputs within the allowable ranges. Note that if $u_{\text{failed}}^1 \neq 0$, then it may happen that $0 \notin X_c$, i.e., if the failed actuator is frozen at a non-nominal value, then it is possible that the process simply cannot be stabilized at the nominal equilibrium point using the functioning control actuators. In other words, if one of the manipulated inputs fails and reverts to a fail-safe position, it may happen that no admissible combination of the functioning inputs exists for which the nominal equilibrium point continues to be an equilibrium point. Maintaining the functioning actuators at the nominal values may drive the process state to a point from where it may not be possible to resume nominal operation upon fault-recovery, or even if possible, may not be 'optimal'. We define the safe-parking problem as the one of identifying safe-park points $x_s \in X_c$ that allow efficient resumption of nominal operation upon fault-recovery.

5.3.2 Safe-Parking to Resume Nominal Operation

In this section, we present a safe-parking framework and a controller that executes safe-parking as well as resumption of nominal operation. To account for the presence of constraints on the manipulated inputs, the key requirements for a safe-park point include that the process state at the time of the failure resides in the stability region for the safe-park point (so the process can be driven to the candidate safe-park point), and that the safe-park point should reside in the stability region under nominal operation (so the process can be returned to nominal operation). These requirements are formalized in Theorem 5.2 below. To this end, consider the system of Eq. (5.1) for which the first control actuator fails at a time T^{fault} and is reactivated at time T^{recovery}, and for which the stability region under nominal operation, denoted by Ω_n, has been characterized using the predictive controller formulation of Eqs. (5.3)–(5.7). Similarly, for a candidate safe-park point x_c, we denote Ω_c as the stability region (computed a priori) under the predictive controller of Eqs. (5.3)–(5.7), and u_{2,x_c} as the control law designed to stabilize at the candidate safe-park with $u_{1,n}$ being the nominal control law.

Theorem 5.2 *Consider the constrained system of Eq.* (5.1) *under the MPC law of Eqs.* (5.3)–(5.7). *If* $x(T^{\text{fault}}) \in \Omega_c$ *and* $\Omega_c \subset \Omega_n$, *then the switching rule*

$$u(t) = \begin{cases} u_{1,n}, & 0 \le t < T^{\text{fault}}, \\ u_{2,x_c}, & T^{\text{fault}} \le t < T^{\text{recovery}}, \\ u_{1,n}, & T^{\text{recovery}} \le t \end{cases} \tag{5.12}$$

guarantees that $x(t) \in \Omega_n \ \forall t \ge 0$ *and* $\limsup_{t\to\infty} \|x(t)\| \le d$.

Proof We consider the two possible cases: first, if no fault occurs ($T^{\text{fault}} = T^{\text{recovery}} = \infty$), and second, if a fault occurs at a time $T^{\text{fault}} < \infty$ and is recovered at a time $T^{\text{fault}} \le T^{\text{recovery}} < \infty$.

Case 1: The absence of a fault implies $u(t) = u_{1,n} \ \forall t \ge 0$. Since $x(0) \in \Omega_n$, and the nominal control configuration is implemented for all times, we have from Theorem 5.1 that $x(t) \in \Omega_n \ \forall t \ge 0$ and $\limsup_{t\to\infty} \|x(t)\| \le d$.

Case 2: At time T^{fault}, the control law designed to stabilize the process at x_c is activated and implemented till T^{recovery}. Since $x(T^{\text{fault}}) \in \Omega_c \subset \Omega_n$, we have that $x(t) \in \Omega_n \ \forall T^{\text{fault}} \le t \le T^{\text{recovery}}$. At a time T^{recovery}, we therefore also have that $x(T^{\text{recovery}}) \in \Omega_n$. Subsequently, as with Case 1, the nominal control configuration is implemented for all time thereafter, we have that $x(t) \in \Omega_n \ \forall t \ge T^{\text{recovery}}$. In conclusion, we have that $x(t) \in \Omega_n \ \forall t \ge 0$ and $\limsup_{t\to\infty} \|x(t)\| \le d$. This completes the proof of Theorem 5.2. □

Remark 5.2 The statement of Theorem 5.2 requires that for a safe-park point, the stability (and invariant) region be such that the process state at the time of the failure resides in the stability region for the safe-park point so the process can be driven to the point of safe-park with the depleted control action available. Note that this characterization can be done off-line. Specifically, for a fail-safe position of an actuator, the entire set of candidate safe-park points X_c can be computed off-line, and also, for any given point in this set, the stability region subject to depleted control action can also be computed off-line (as is done for the nominal equilibrium point). The statement of the theorem also requires that the stability (and invariant) region for a safe-park point be completely contained in the stability region under nominal operation, so the state trajectory always stays within the stability region under nominal operation. This requirement can be readily relaxed to only require that the state at the time of the failure resides in the stability region of the safe-park point. This will allow for the state trajectory to leave the stability region under nominal operation, and it may happen that at the time of fault-recovery, the closed-loop state trajectory does not reside in the stability region under nominal operation. However, to preserve closed-loop stability upon fault-recovery, the control law utilizing depleted control action may be continued up until the time that the state trajectory enters the stability region under nominal operation (this is guaranteed to happen because $x_c \in \Omega_n$), after which the control law utilizing all the manipulated inputs can be implemented to achieve closed-loop stability.

Remark 5.3 The key motivation, from a resumption of nominal operation stand point, for safe-parking is as follows: In the absence of a safe-park framework, if the control law still tries to utilize the available control action to preserve operation at the nominal operating point, the active actuators may saturate and drive the process state to a point starting from where resumption of nominal operation, even after fault-recovery, may not be achievable. Note that if continued operation at nominal operating point was possible either via the depleted control configuration or via control-loop reconfiguration (as developed in Chaps. 3 and 4), then reconfiguration-based fault-tolerant control approaches could be utilized. However, Theorem 5.2 addresses the problem where a fault occurs that precludes operation at nominal operating point, and provides an appropriately characterized safe-park point where the process can be temporarily 'parked' until nominal operation can be resumed.

Remark 5.4 Note that the assumption that the failed actuator reverts to the fail-safe position allows enumerating the possible fault situations for any given set of manipulated inputs a-priori to determine the safe-park candidates and then pick the appropriate safe-park point online (the condition $x_s \in \Omega_n$ can be verified off-line; however, $x(T^{\text{fault}}) \in \Omega_{x_s}$ can only be verified online, upon fault-occurrence; for further discussion on this point, see Remark 5.5). The assumption reflects the practice wherein actuators have a built-in fail-safe position that they revert to upon failure. The fail-safe positions are typically determined to minimize possibilities of excursions to dangerous conditions such as high temperatures and pressures (setting a coolant valve to fail to a fully open position, while setting a steam valve to fail to a shut position). In the event that the actuators experience a mechanical failure and are not able to revert to a fail-safe position, one can work with simplified (albeit without guarantees) estimates of the stability regions that can be generated much faster (and therefore computed online, upon fault-occurrence), to implement the proposed safe-parking mechanism. Specifically, instead of stability regions estimated by constructing invariant sets Ω within the set of initial conditions Π for which the Lyapunov-function can be made to decay, one can use the set Π (which is much easier to compute) to implement the proposed safe-park mechanism (see Sect. 5.4 for a demonstration). Note also that while the statement of Theorem 5.2

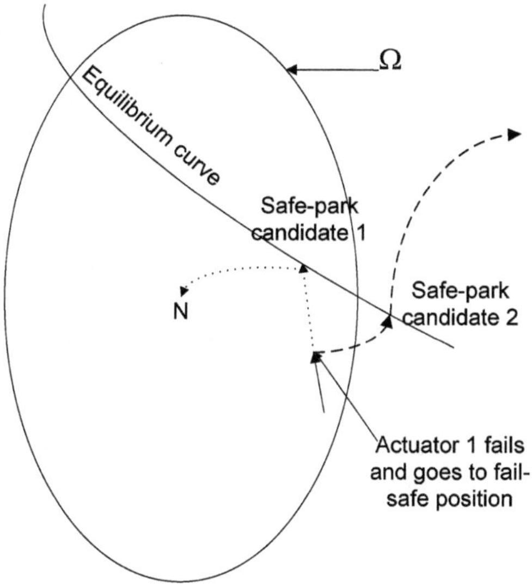

Fig. 5.1 A schematic illustrating the safe-parking framework for a process with two actuators. Ω denotes the stability region under nominal operation. Safe-parking candidates lie on the equilibrium curve corresponding to the fail-safe value of the first actuator, and admissible values of the second actuator. Arbitrarily choosing a safe-park candidate (e.g., safe-parking candidate 2) does not guarantee resumption of nominal operation upon fault-recovery, while choosing safe-park candidate 1 guarantees resumption of nominal operation upon fault-recovery

considers faults in one of the actuators, generalizations to multiple faults (simultaneous or otherwise) are possible, albeit involving the expected increase in off-line computational cost (due to the necessity of determining the safe-park points for all combinations of the faults in the control actuators).

Remark 5.5 The presence of constraints on the manipulated inputs limits the set of initial conditions from where the process can be driven to a desired equilibrium point. Control designs that allow an explicit characterization of their stability regions, and their use in deciding the safe-park point is therefore critical in determining the viability of a candidate safe-park point. Note also that while the schematic in Fig. 5.1 shows two dimensional representations of the stability region to enable visual verification of the presence of a candidate safe-park point in the stability region, the visual representation is *not* necessary. Specifically, the presence of a point in the stability region can be verified by evaluating the Lyapunov function value. Note that while the proposed safe-parking framework assumes a priori knowledge of the fail-safe positions of the actuators, it does not require a priori knowledge of the fault and recovery times, and only provides appropriate switching logic that is executed when and if a fault takes place and is subsequently rectified.

Remark 5.6 While the results in the present chapter are presented using the Lyapunov-based MPC of Eqs. (5.3)–(5.7), the use of this controller is not critical to the implementation of the proposed safe-parking design. Any other control law that provides an explicit characterization of the stability region subject to constraints can be used instead to implement the proposed safe-parking framework. With respect to the design of the Lyapunov-based predictive controller of Eqs. (5.3)–(5.7), we also note that while the use of a control Lyapunov function provides a better estimate of the stability region, even a quadratic Lyapunov function (chosen such that it is locally a control Lyapunov function) can be used to generate (possibly conservative) estimates of the stability region. For further discussion on this issue, see [109].

Remark 5.7 Implicit in the implementation of the proposed safe-parking mechanism is the assumption of the presence of fault-detection and isolation filters such as those presented in Chaps. 3 and 4. The proposed safe-parking framework determines the necessary course of action after a fault has been detected and isolated and can be readily integrated with any of the existing fault-detection and isolation structures.

5.3.3 Incorporating Performance Considerations in Safe-Parking

In the previous section, the requirements for an equilibrium point to be denoted a safe-park point was provided. A large set of equilibrium points may qualify as safe-park points and satisfy the requirements in Theorem 5.2. In this section, we introduce performance considerations in the eventual choice of the 'optimal' safe-park point. To this end, consider again the system of Eq. (5.1) for which the first control actuator fails at a time T^{fault} and is reactivated at time T^{recovery}, and for which the set of safe-park points, $x_s \in X_s$, have been characterized. For a given safe-park point (one that satisfies the requirements of Theorem 5.2), define the followings costs:

$$J_{tr} = \int_{T^{\text{fault}}}^{T^{\text{fault}}+T_s} \left[\left\| x^u(s; x, t) \right\|_{Q_{tr}^2} + \left\| u(s) \right\|_{R_{tr}^2} \right] ds, \tag{5.13}$$

where Q_{tr} and R_{tr} are positive definite matrices, the subscript tr signifying that this value captures the 'cost' associated with transitioning to the safe-park point, with T_s being the time required to go to a sufficiently close neighborhood of the safe-park point. This cost can be estimated online, upon fault-occurrence, by running fast simulations of the closed-loop system under the auxiliary controller of Eq. (5.8) (for further discussion on this issue, see Remark 5.8). Similarly, define

$$J_s = f_s(x_s, u_s), \tag{5.14}$$

where $f_s(x_s, u_s)$ is an appropriately defined cost function and the subscript s denotes that this value captures the 'cost' associated with operating at the safe-park point. Unlike the cost in Eq. (5.13), this cost does not involve an integration over time, and can be determined off-line. The framework allows for inclusion of (possibly nonlinear) costs associated with further unit operations that may have to be performed to recover useful products from the process operating at the safe-park point (for further discussion on this issue, see Remark 5.9). Finally, define

$$J_r = \int_0^{T_r} \left[\left\| x^u(s; x, t) \right\|_{Q_r^2} + \left\| u(s) \right\|_{R_r^2} \right] ds, \tag{5.15}$$

where Q_r and R_r are positive definite matrices, with the subscript r signifying that this value captures the 'cost' associated with resuming nominal operation, with T_r being the time required to return to a sufficiently close neighborhood of the nominal operating point, and the integration performed with the safe-park point as the initial condition. Again, this cost can be estimated off-line by running simulations of the closed-loop system under the auxiliary controller of Eq. (5.8). Consider now the safe-park points $x_{s,i} \in X_s$, $i = 1, \ldots, N_s$ where N_s is the number of safe-park points to be evaluated for optimality and let $J_{x_{s,i}} = J_{tr,i} + J_{s,i} + J_{r,i}$, $i = 1, \ldots, N_s$.

Theorem 5.3 *Consider the constrained system of Eq.* (5.1) *under the MPC law of Eqs.* (5.3)–(5.7) *and the switching rule*

$$u(t) = \begin{cases} u_{1,n}, & 0 \leq t < T^{\text{fault}}, \\ u_{2,x_{s,o}}, & T^{\text{fault}} \leq t < T^{\text{recovery}}, \\ u_{1,n}, & T^{\text{recovery}} \leq t, \end{cases} \tag{5.16}$$

where $o \in \{1, \ldots, N_s\} = \operatorname{argmin}_{i=1,N_s} J_{x_{s,i}}$ *guarantees that* $x(t) \in \Omega_n$ $\forall t \geq 0$ *and* $\limsup_{t\to\infty} \|x(t)\| \leq d$.

Proof Any $x_{s,o}$ chosen according to Theorem 5.3 satisfies the requirements of Theorem 5.2. $x(t) \in \Omega_n$ $\forall t \geq 0$ and $\limsup_{t\to\infty} \|x(t)\| \leq d$ follow from Theorem 5.2. □

Remark 5.8 Note that the cost of transitioning to the safe-park point J_{tr} can be estimated using the auxiliary controller of Eq. (5.8) since the auxiliary controller achieves decay of the same Lyapunov function as that used in the predictive controller design. This cost has to be estimated online because it depends on the process state at which the failure occurs (in the special case that faults occur after the process has been stabilized at the nominal operating points, this cost can also be computed off-line; see Sect. 5.4 for a demonstration). In contrast, the cost incurred in resuming nominal operation from the safe-park point can be computed off-line. Such a computation can be done by running simulations under the predictive controller to get a more accurate estimate of the 'cost'. Additional terms in J_{tr} and J_s can be readily included to cater to the specific process under consideration. Furthermore, the contribution of the cost J_s to the total cost can be appropriately scaled utilizing reasonable estimates of fault-rectification times. Specifically, if the malfunctioned actuator is known to require significant time to be rectified, then this cost can be 'weighed' more to recognize the fact that the process will deliver substantial amount of product corresponding to the safe-park point under consideration. If, on the other hand, it is known that the fault can be rectified soon, then the cost involving the resumption to nominal operation J_r, or alternatively, the time required to resume nominal operation can be given increased weight. Finally, if a significant number of safe-park points are to be evaluated for optimality (leading to possible computational issues), one safe-park point can be initially chosen to ensure stability and subsequently revised once the optimality computations are complete.

Remark 5.9 For the 'product' being generated during safe-parking, further unit operations may be required, ranging from simple separations to further processing, all of which may have associated costs. Possible loss of revenue during safe-park can be incorporated in the estimate J_s. If the process is connected to further units downstream, then increased utility costs associated with downstream processing can also be accounted for in this cost. Finally, we note that the costs outlined here are only some of the representative costs, and the framework allows for incorporating costs/revenues that may be specific to the process under consideration.

Remark 5.10 Note that while the set of safe-parking points (satisfying the requirements of Theorem 5.2) could be a continuous manifold of equilibrium points, safe-parking points to be evaluated for optimality can be picked by discretizing the manifold. The minimization in determining the optimal safe-park point can then be carried out by a simple procedure of comparison of the cost estimates associated with the finite number of safe-parking candidates. Choosing a finer discretization in evaluating the safe-parking candidates could possibly yield improved closed-loop costs, however, the approximations involved in the cost estimation (the cost of going to and returning from the safe-parking points are only approximately estimated using the auxiliary controller of Eq. (5.8)) could offset the benefits out of the finer discretization. Therefore, a balance has to be struck in picking the number of safe-parking points that will be evaluated for optimality that trades off the increased computational complexity, the approximations in cost estimation, and the improved performance derived out of the finer discretization.

Remark 5.11 Note that the proposed approach, with appropriate modifications, can also be used to handle faults in situations where not all states are available as measurements and in the presence on uncertainty. The key is to design the safe-park points (and their associated stability regions) accounting for uncertainty, and to only make the switching decision after the state estimator has converged (see [95] for details). In processes requiring a description by distributed parameter system models (e.g., diffusion–reaction processes), the approach can be implemented using controllers designed using reduced order models that ensure stability of the infinite dimensional system (see [95] for details).

Remark 5.12 The proposed approach can also be used to handle faults in a network of multiple units characterized by complex interconnections such as parallel and recycle streams. Furthermore, instead of 'safe-parking' the entire network, 'subsections' of the network can be identified (for faults in specific units) such that the rest of the network can continue nominal operation even during the fault (see [40] for details and simulation results).

Table 5.2 Chemical reactor parameters and steady-state values

$V = 0.1\ \text{m}^3$
$R = 8.314\ \text{kJ/(kmol K)}$
$C_{A0_s} = 0.73\ \text{kmol/m}^3$
$T_{0_s} = 310.0\ \text{K}$
$Q_s = 10.0\ \text{kJ/s}$
$\Delta H = -4.78 \times 10^4\ \text{kJ/kmol}$
$k_0 = 72 \times 10^9\ \text{min}^{-1}$
$E = 8.314 \times 10^4\ \text{kJ/kmol}$
$c_p = 0.239\ \text{kJ/(kg K)}$
$\rho = 1000.0\ \text{kg/m}^3$
$F = 100 \times 10^{-3}\ \text{m}^3/\text{min}$
$T_{Rs} = 393\ \text{K}$
$C_{As} = 0.447\ \text{kmol/m}^3$

5.3.4 Illustrative Simulation Example

We illustrate in this section the proposed safe-park framework via a continuous stirred tank reactor (CSTR). To this end, consider a CSTR where an irreversible, first-order exothermic reaction of the form $\text{A} \xrightarrow{k} \text{B}$ takes place. The mathematical model for the process takes the form:

$$\begin{aligned} \dot{C}_{\text{A}} &= \frac{F}{V}(C_{\text{A}0} - C_{\text{A}}) - k_0 e^{\frac{-E}{RT_R}} C_{\text{A}}, \\ \dot{T}_R &= \frac{F}{V}(T_0 - T_R) + \frac{(-\Delta H)}{\rho c_p} k_0 e^{\frac{-E}{RT_R}} C_{\text{A}} + \frac{Q}{\rho c_p V}, \end{aligned} \tag{5.17}$$

where C_{A} denotes the concentration of the species A, T_R denotes the temperature of the reactor, Q is the heat added to/removed from the reactor, V is the volume of the reactor, k_0, E, ΔH are the pre-exponential constant, the activation energy, and the enthalpy of the reaction, and c_p and ρ are the heat capacity and fluid density in the reactor. The values of all process parameters can be found in Table 5.2.

The control objective is to stabilize the reactor at the unstable equilibrium point $(C_{\text{A}}^s, T_R^s) = (0.447\ \text{kmol/m}^3, 393\ \text{K})$. Manipulated variables are the rate of heat input/removal, Q, and change in inlet concentration of species A, $\Delta C_{\text{A}0} = C_{\text{A}0} - C_{\text{A}0_s}$, with constraints: $|Q| \leq 32\ \text{kJ/s}$ and $0 \leq C_{\text{A}0} \leq 2\ \text{kmol/m}^3$. The heat input/removal Q consists of heating stream Q_1 and cooling stream Q_2 with the constraints on each as, $0\ \text{kJ/s} \leq Q_1 \leq 32\ \text{kJ/s}$ and $-32\ \text{kJ/s} \leq Q_2 \leq 0\ \text{kJ/s}$. The nominal operating point (N) corresponds to steady state values of the inputs $C_{\text{A}0} = 0.73\ \text{kmol/m}^3$ and $Q = 10\ \text{kJ/s}$.

For stabilizing the process at the nominal equilibrium point, the Lyapunov based MPC of Sect. 5.2.3 is designed using a quadratic Lyapunov function of the form $V = x^T P x$ with $P = \begin{bmatrix} 4.32 & 0 \\ 0 & 0.004 \end{bmatrix}$. The stability region is estimated and denoted by Ω in Fig. 5.2. Note that here stability region (Ω) is not a complete ellipse because of

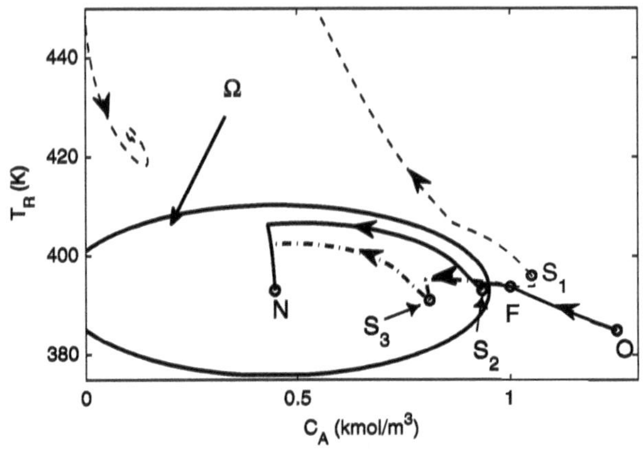

Fig. 5.2 Evolution of closed–loop states for the CSTR example. *Dashed line* (- -) indicates the case when a safe-park point S_1 is arbitrarily chosen (resulting in the inability to resume nominal operation upon fault-recovery) while the *solid line* (—) indicates the case when S_2 is chosen according to Theorem 5.2, guaranteeing resumption of nominal operation upon fault-recovery. The *dash-dotted lines* show the closed-loop response when optimality considerations are included in the choice of the safe-park point and S_3 is chosen

naturally invariant boundary of positive concentrations. We consider the problem of designing a safe-parking framework to handle temporary faults in the heating valve (resulting in a fail-safe value of $Q_1 = 0$). The nominal operating point corresponds to $Q_s = 10$ kJ/s, and no value of the functioning manipulated inputs -32 kJ/s $\leq Q_2 < 0$ kJ/s and $0 \leq C_{A0} \leq 2$ kmol/m^3 exists such that the nominal equilibrium point continues to be an equilibrium point of the process subject to the fault. For $Q_2 = -14.7$ kJ/s, $C_{A0} = 1.33$ kmol/m^3 and $Q_2 = -4$ kJ/s, $C_{A0} = 1.27$ kmol/m^3, the corresponding equilibrium points are $S_1 = (1.05\ \text{kmol/m}^3, 396\ \text{K})$ and $S_2 = (0.93\ \text{kmol/m}^3, 393\ \text{K})$, which we denote as safe-park candidates. For each of these safe-park candidates, we also design Lyapunov based MPC of Sect. 5.2.3 using $P = \begin{bmatrix} 12.56 & 0 \\ 0 & 0.049 \end{bmatrix}$ for S_1 and $P = \begin{bmatrix} 12.32 & 0 \\ 0 & 0.026 \end{bmatrix}$ for S_2. The matrices in the objective function (Eq. (5.7)), are chosen as $Q_w = \begin{bmatrix} 72.72 & 0 \\ 0 & 1 \end{bmatrix}$ and $R_w = \begin{bmatrix} 640 & 0 \\ 0 & 0.67 \end{bmatrix}$. Prediction and control horizons of 0.10 min and 0.02 min, respectively, are used in implementing the predictive controller. The discretized version of the stability constraint of the form $V(x(t + \Delta)) \leq 0.99V(x(t))$ is incorporated in the optimization problem.

Consider a scenario where the process starts from $O = (1.25\ \text{kmol/m}^3,\ 385\ \text{K})$ and the predictive controller drives the process toward the nominal operating point, $N = (0.447\ \text{kmol/m}^3,\ 393\ \text{K})$. At $t = 0.16$ min, when the process state is at $F = (0.9975\ \text{kmol/m}^3,\ 394.02\ \text{K})$, the heating valve fails, and reverts to the fail-safe position (completely shut), resulting in $Q_1 = 0$ kJ/s. This restricts the heat input/removal to -32 kJ/s $\leq Q < 0$ kJ/s instead of -32 kJ/s $\leq Q < 32$ kJ/s. We first consider the case where the safe-park candidate S_1 is arbitrarily chosen as the safe-park point, and the process is stabilized at S_1 until the fault is rectified. At

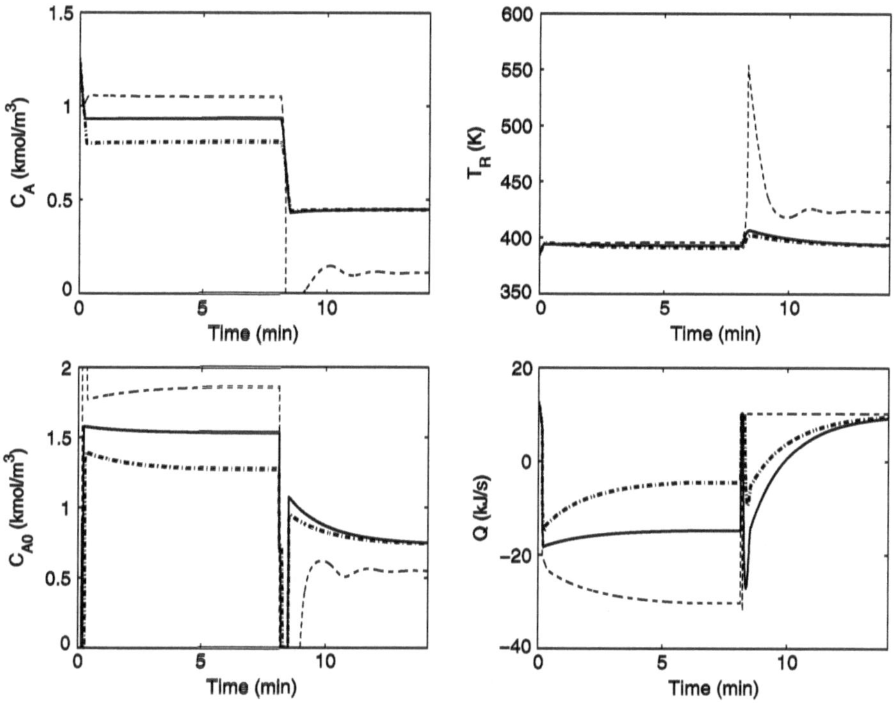

Fig. 5.3 Evolution of the closed-loop state (**a**)–(**b**) and input (**c**)–(**d**) profiles for the CSTR example. *Dashed lines* (- -) indicate the case when a safe-park point S_1 is arbitrarily chosen (resulting in the inability to resume nominal operation upon fault-recovery) while the *solid lines* (—) show the case when S_2 is chosen according to Theorem 5.2, guaranteeing resumption of nominal operation upon fault-recovery. The *dash-dotted lines* show the closed-loop response when optimality considerations are included in the choice of the safe-park point and S_3 is chosen

$t = 8.0$ min, the fault is rectified, however, we see that even after fault-recovery, nominal operation cannot be resumed (see dashed lines in Fig. 5.2). This happens because S_1 lies outside the stability region under nominal operation. In contrast, if S_2 is chosen as the safe-park point, we see that the process can be successfully driven to S_2 with limited control action as well as it can be successfully driven back to N after fault-recovery (see solid lines in Fig. 5.2). The state and input profiles are shown in Fig. 5.3. In summary, the simulation scenario illustrates the necessity to account for the presence of input constraints (characterized via the stability region) in the choice of the safe-park point.

Next, we demonstrate the incorporation of performance criterion in selecting the safe-park point. To this end, we consider another point S_3 (corresponding to $Q_2 = -14.6$ kJ/s, $C_{A0} = 1.53$ kmol/m^3), which is also inside the stability region of N, and is thereby also a viable safe-park point (i.e., either of S_2 or S_3 can be chosen as safe-park point from stability perspective). Using the approach in Sect. 5.3.3, the cost associated with operating at the two safe-park points is calculated utilizing

Table 5.3 Safe-parking cost estimates for the illustrative CSTR example of Sect. 5.3.4

	C_A	T	Objective function	
			Estimated using the bounded controller	Closed–loop process cost
S_2	0.9346	393	2406	4072
S_3	0.8107	391	1209	1105

$f(x_s, u_s) = \|x^u_{ss}\|_{Q_s^2} + \|u_{ss}\|_{R_s^2}$ and the weighting matrices in Eqs. (5.13)–(5.15) are chosen as $Q_{tr} = Q_r = Q_s = \begin{bmatrix} 727 & 0 \\ 0 & 10 \end{bmatrix}$ and $R_{tr} = R_r = R_s = \begin{bmatrix} 0.64 & 0 \\ 0 & 0.04 \end{bmatrix}$. At the time of the failure, the auxiliary controller of Eq. (5.8) is used to estimate J_{tr} and J_r, which are divided by T_s and T_r, to determine $J_{\text{safe-parking}} = \frac{J_{tr}}{T_s} + J_s + \frac{J_r}{T_r}$. Note that the computation of $J_{\text{safe-parking}}$ does not require prior information about the time of fault recovery. We also note that while in this illustrative simulation example, we only use two safe-park points for the purpose of illustration, the cost comparison can be carried out over a larger number of safe-park points (see the styrene process in Sect. 5.4). Table 5.3 shows the objective function value for the safe-park points calculated using the auxiliary controller. As can be seen from the table, the cost estimate for S_3 is significantly lower than for S_2, indicating that S_3 is a better choice for safe-parking the process. Subsequently, if S_3 is chosen as the safe-park point, it yields a closed-loop cost significantly lower than the closed-loop cost achieved when safe-parking the process at S_2 (the corresponding closed-loop state and input profiles are shown by the dash-dotted lines in Figs. 5.2–5.3).

5.4 Application to the Styrene Polymerization Process

In this section, we implement the proposed safe-parking framework on the styrene polymerization process described in Sect. 5.2.2. To evaluate the robustness of the proposed framework, we consider errors in the values of the parameters A_p, hA, and V_c of magnitude 1 %, 2 %, and 10 %, respectively, as well as sinusoidal disturbances in the initiator flowrate F_i of magnitude 10 % around the nominal values. The control objective is to stabilize the process at the nominal equilibrium point ($C_I = 0.067$ kmol/m^3, $C_M = 3.968$ kmol/m^3, $T = 303.55$ K, $T_c = 297.95$ K), corresponding to the nominal values of the manipulated inputs of $F_c = 1.31$ L/s and $F_m = 1.05$ L/s, while handling a fault in the valve manipulating the coolant flow rate.

For nominal operation, the predictive controller of Eqs. (5.3)–(5.7) is designed using a quadratic Lyapunov function of the form $V(x) = x'Px$ with

$$P = \begin{bmatrix} 3662.2 & 89.43 & -18.59 & -25.02 \\ 89.430 & 2.953 & -0.628 & -0.845 \\ -18.592 & -0.628 & 0.682 & -0.036 \\ -25.023 & -0.845 & -0.036 & 2.002 \end{bmatrix}$$

Table 5.4 Safe-parking cost estimates for the styrene polymerization process of Sect. 5.4

	C_I	C_M	T	T_c	Objective function	
					Bounded controller	Closed-loop process cost
N	0.0673	3.9685	303.55	297.95	−11.272	–
S_1	0.4298	1.165	297.36	294.91	−2.079	−2.144
S_2	0.2068	2.8708	299.25	294.95	−8.282	–
S_3	0.1362	3.4256	300.35	294.97	−9.407	–
S_4	0.1015	3.6998	301.14	294.99	−9.692	–
S_5	0.0809	3.8631	301.74	295.01	−9.734	−9.732
S_6	0.0673	3.9716	302.22	295.02	−9.655	–
S_7	0.0576	4.0488	302.62	295.03	−9.530	–
S_8	0.0503	4.1065	302.95	295.03	−9.383	–
S_9	0.0447	4.1513	303.22	295.04	−9.227	–
S_{10}	0.0402	4.1871	303.46	295.04	−9.069	–
S_{11}	0.0365	4.2163	303.67	295.05	−8.912	–

In Sect. 5.3.4, we demonstrated the implementation of the safe-parking framework where the fault occurs before the process is stabilized at the nominal equilibrium point. In this section, we consider faults that occur after the process has been stabilized at the nominal equilibrium point. Determination of the safe-park points and evaluation of the cost estimates for safe-park points can therefore be carried out off-line. The nominal operating point for the process is a stable operating point, and several safe-park points satisfy the requirements of Theorem 5.2 (guaranteeing resumption of nominal operation upon fault-recovery). Ten safe-park points are chosen to be evaluated for optimality and using the approach in Sect. 5.3.3, the cost associated with each safe-park point is estimated using the cost function, $f(x_s, u_s) = \|u_{ss}\|_{R_s^2} - q_s M_{\text{used}}$, where the first term represents the cost of the utilities, while the second term represents the value of the product formed (via computing the rate of consumption of the monomer). With such a formulation of the steady-state cost, the safe-park points where the rate of product formation is more are preferred. The weighting factors are chosen as $R_s = \left[\begin{smallmatrix} 0.25 & 0 \\ 0 & 0 \end{smallmatrix}\right]$ and $q_s = 0.5$. The weighting matrices in Eq. (5.13)–(5.15) are chosen as diagonal matrices with the elements on the diagonal as $Q_{tr} = Q_r = \text{diag}(1000,\ 1000,\ 10,\ 10)$ and $R_{tr} = R_r = \text{diag}(1,\ 1)$.

For the safe-park points, the costs are tabulated in Table 5.4. Note that the cost is the minimum for the nominal operating point, and out of the ten safe-park points, point S_5 ($C_I = 0.081$ kmol/m^3, $C_M = 3.863$ kmol/m^3, $T = 301.75$ K, $T_c = 295.01$ K) yields the lowest cost and is therefore picked as the optimal safe-park point. Closed-loop simulations are shown for the case where a fault occurs at 33.3 minutes and is rectified at 300 minutes. For the sake of comparison, closed-loop simulations are also shown when safe-park point S_1 ($C_I = 0.430$ kmol/m^3, $C_M = 1.165$ kmol/m^3, $T = 297.37$ K, $T_c = 294.91$ K) is picked. The equilibrium

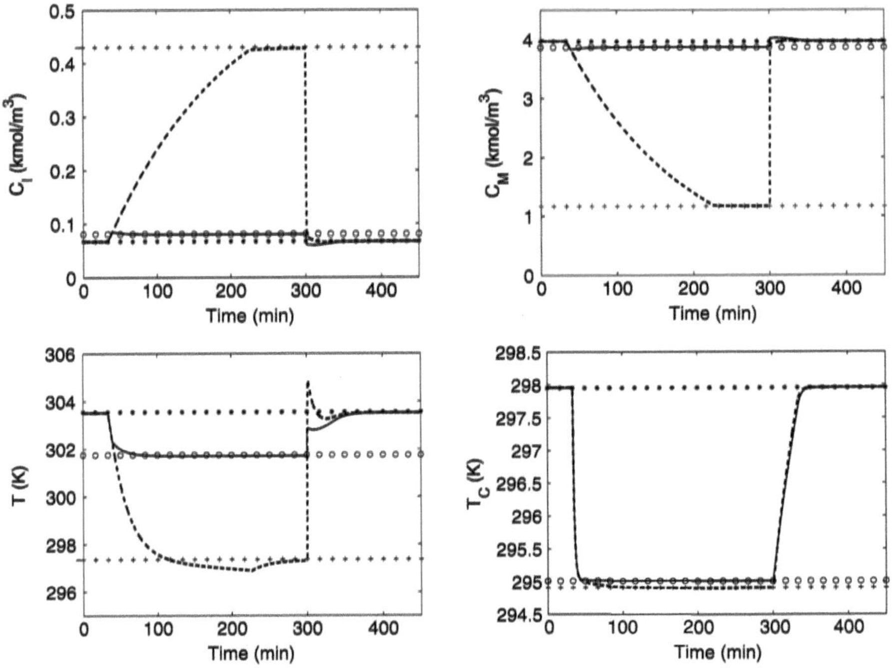

Fig. 5.4 Evolution of the state profiles for the styrene polymerization process for an arbitrarily chosen safe-park point (*dashed lines*) and under the proposed safe-park mechanism (*solid lines*). Fault occurs at 33.3 min and is rectified at 300 min. The nominal equilibrium point N and the safe-park points S_5 and S_1 are denoted by the markers ⋆, ∘, and +, respectively

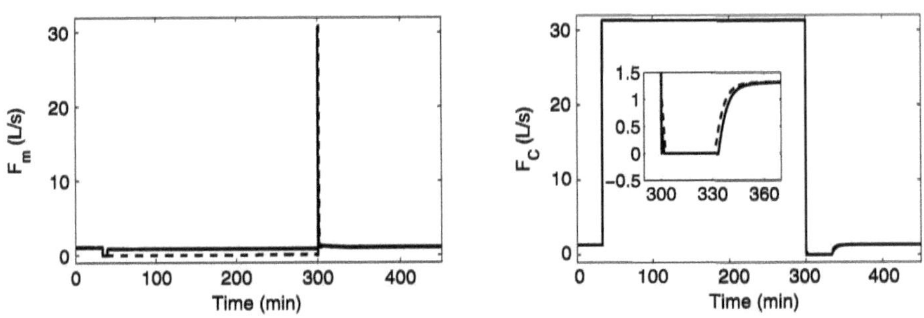

Fig. 5.5 The input profiles for the styrene polymerization process for an arbitrarily chosen safe-park point (*dashed lines*) and under the proposed safe-park mechanism (*solid lines*). Fault occurs at 33.3 min, resulting in the coolant flow rate being stuck at the maximum value during this time, and is rectified at 300 min

points N, S_5, and S_1 are denoted in Figs. 5.4–5.5 by the markers ⋆, ∘, and +, respectively. The closed-loop trajectories and input profiles when the safe-park points S_5 and S_1 are picked are shown by solid and dashed lines, respectively. The closed-loop costs for the two points is also shown in Table 5.4. Once again, even in the presence of uncertainty and disturbances, the closed-loop costs follow the same trend as the estimates, yielding a low cost for the 'optimal' safe park point and demonstrating the robustness of the proposed safe-parking framework.

5.5 Conclusions

In this chapter, we considered the problem of control of nonlinear process systems subject to input constraints and faults in the control actuators. A safe-parking framework was developed for handling faults that preclude the possibility of continued operation at the nominal equilibrium point. First, Lyapunov-based model predictive controllers that allow for an explicit characterization of the stability region subject to constraints on the manipulated input were designed. The stability region was utilized in selecting 'safe-park' points from the safe-park candidates (equilibrium points subject to failed actuators). Specifically, a candidate parking point was termed a safe-park point if (i) the process state at the time of failure resides in the stability region of the safe-park candidate (subject to depleted control action) and (ii) the safe-park candidate resides within the stability region of the nominal control configuration. Performance considerations, such as ease of transition from and to the safe-park point and cost of running the process at the safe-park point, were then quantified and utilized in choosing the optimal safe-park point. The proposed framework was illustrated using a chemical reactor example and its robustness with respect to parametric uncertainty and disturbances was demonstrated via a styrene polymerization process.

Chapter 6
Fault Diagnosis and Robust Safe-Parking

6.1 Introduction

In Chap. 5, we presented a safe-parking framework to handle faults with the assumption that an appropriate fault-detection and isolation mechanism was in place, which allowed the choice of the safe-parking mechanism that utilized the remaining functioning actuators. Moreover, we assumed that the actuator reverts to a known fail-safe position which determines the 'size' of the fault vector. In this chapter, we relax the assumption on the knowledge of the location and magnitude of the fault, necessitating appropriate fault detection and isolation and diagnosis mechanisms. For the existing model-based FDI approaches, FDI is often achieved by generating residuals through the system model and input/output data. Under fault-free conditions, the magnitudes of residuals are small. A fault is reported when a residual breaches the user-specified threshold. Due to the presence of plant–model mismatch, residuals that are sensitive to faults but insensitive to uncertainty and disturbances are desired. Relatively less attention has been paid to the problem of fault diagnosis (not only isolating the fault but also estimating its magnitude), in part due to the nature of the fault-tolerant control techniques described in Chaps. 3–5, that rely on inherent robustness, existence of backup control configurations or actuators reverting to known fail-safe positions.

Motivated by the above considerations, in this chapter we consider the problem of designing an integrated fault diagnosis and safe-parking framework to deal with actuator faults in nonlinear systems. The remainder of the chapter is organized as follows. In Sect. 6.2, the class of systems considered and a control design used to illustrate the safe-parking framework are presented. A model-based fault diagnosis design is first developed under state feedback control in Sect. 6.3.1 and then generalized to handle state estimation errors in Sect. 6.3.2. In the proposed methodology, the fault information is obtained by estimating the outputs of the actuators and comparing them with the corresponding prescribed control inputs. In Sect. 6.4, the safe-parking framework developed previously (to handle the case where the failed actuator reverts to a known fixed value) for fault-tolerant control is extended to handle the case where an actuator seizes at an arbitrary value. The estimate of the failed

P. Mhaskar et al., *Fault-Tolerant Process Control*, DOI 10.1007/978-1-4471-4808-1_6,

actuator position provided by the fault diagnosis design is used to choose a safe-park point at which the system operates temporarily during fault repair, from those generated off-line for a series of design values of the failed actuator position. The discrepancy between the actual value of the failed actuator position and the corresponding design value is handled through the robustness of the control design. The efficacy of the integrated fault diagnosis and safe-parking framework is demonstrated through a chemical reactor example in Sect. 6.5. Finally, Sect. 6.6 presents some concluding remarks.

6.2 Preliminaries

In this section, we present the system description and a robust control design, which will be used to illustrate the safe-parking framework in Sect. 6.5.

6.2.1 System Description

Consider a nonlinear system subject to actuator faults with the following state-space description:

$$\begin{aligned}
&\dot{x} = f\big(x,\theta(t)\big) + G(x)\big[u(t)+\tilde{u}(t)\big],\\
&u(t)\in\mathcal{U}, \qquad \theta(t)\in\Theta, \qquad\qquad (6.1)\\
&u(t)+\tilde{u}(t) = u(t_k)+\tilde{u}(t_k)\in\mathcal{U} \quad \text{for all } t\in[t_k,t_{k+1}), k=0,\dots,\infty,
\end{aligned}$$

where $x=[x_1,\dots,x_n]^{\mathrm{T}}\in\mathbb{R}^n$ is the vector of state variables, $u=[u_1,\dots,u_m]^{\mathrm{T}}\in\mathbb{R}^m$ is the vector of prescribed control inputs given by the control law and $\tilde{u}=[\tilde{u}_1,\dots,\tilde{u}_m]^{\mathrm{T}}\in\mathbb{R}^m$ is the unknown fault vector for the actuators, with the actual control input $u+\tilde{u}$ implemented to the plant taking values in a nonempty compact convex set $\mathcal{U}:=\{u\in\mathbb{R}^m : u_{\min}\le u\le u_{\max}\}$ that contains 0, where $u_{\min}=[u_{1,\min},\dots,u_{m,\min}]^{\mathrm{T}}\in\mathbb{R}^m$ and $u_{\max}=[u_{1,\max},\dots,u_{m,\max}]^{\mathrm{T}}\in\mathbb{R}^m$ denote the lower and upper bounds (constraints) on the vector of manipulated variables, respectively, and $\theta=[\theta_1,\dots,\theta_q]^{\mathrm{T}}\in\mathbb{R}^q$ is the vector of (possibly time-varying) uncertain variables taking values in a nonempty compact convex set $\Theta:=\{\theta\in\mathbb{R}^q:\theta_{\min}\le\theta\le\theta_{\max}\}$ that contains 0, where $\theta_{\min}=[\theta_{1,\min},\dots,\theta_{q,\min}]^{\mathrm{T}}\in\mathbb{R}^q$ and $\theta_{\max}=[\theta_{1,\max},\dots,\theta_{q,\max}]^{\mathrm{T}}\in\mathbb{R}^q$ denote the lower and upper bounds on the vector of uncertain variables, respectively. It is assumed that the functions $f(x,\theta)=[f_i(x,\theta)]_{n\times 1}$ and $G(x)=[g_{ij}(x)]_{n\times m}$ are locally Lipschitz in their arguments, and $f(x,\theta)$ is differentiable with respect to θ ($i=1,\dots,n$; $j=1,\dots,m$). The origin is an equilibrium point of the nominal system (the system of Eq. (6.1) with $\tilde{u}(t)\equiv 0$ and $\theta(t)\equiv 0$) for $u=0$, i.e., $f(0,0)=0$. The control input is prescribed at discrete times $t_k:=k\Delta$, $k=0,\dots,\infty$, where Δ denotes the period during which the control action is kept constant. The faults considered are such that an actuator seizes at an arbitrary position. It is assumed that the corrupted input to the plant is constant during each time interval, that is, $u(t) + \tilde{u}(t) = u(t_k) + \tilde{u}(t_k)$ for all $t \in [t_k, t_{k+1})$. Note that $-u_{\min}$ (or $-\theta_{\min}$) does not have to be equal to $u_{\max}$ (or $\theta_{\max}$), and we have that $\|u\| \leq u_b$ and $\|\theta\| \leq \theta_b$, where $u_b = \|[\max\{-u_{1,\min}, u_{1,\max}\}, \dots, \max\{-u_{m,\min}, u_{m,\max}\}]^{\mathrm{T}}\|$ and $\theta_b = \|[\max\{-\theta_{1,\min}, \theta_{1,\max}\}, \dots, \max\{-\theta_{q,\min}, \theta_{q,\max}\}]^{\mathrm{T}}\|$.

6.2.2 Lyapunov-Based Predictive Control

To illustrate the safe-parking framework for FTC, the Lyapunov-based predictive controller described in Sect. 2.8 is adapted under Assumption 6.1 below and used as an example of a robust control design with a well characterized stability region.

Assumption 6.1 For the system of Eq. (6.1), $f_i(x, \theta)$, $i = 1, \dots, n$, is monotonic with respect to θ_j, $j = 1, \dots, q$, for any $x \in \mathbb{R}^n$ and $\theta_l \in [\theta_{l,\min}, \theta_{l,\max}]$, $l = 1, \dots, q$ and $l \neq j$.

Remark 6.1 In many practical systems, the form of $f(x, \theta)$ is known and the uncertain variables affect $f(x, \theta)$ monotonically, as required in Assumption 6.1. For example, in the Arrhenius law of reaction rates, the parametric uncertainty includes errors in the pre-exponential constant and the activation energy. The reaction rate is monotonically increasing with respect to the pre-exponential constant, while it is monotonically decreasing with respect to the activation energy. Other uncertainty includes the enthalpy of reaction and the heat transfer coefficient. In addition to the parametric uncertainty, θ also models the unknown disturbances entering the system. Typical disturbances include errors in the temperature and concentration of a feed stream, or the temperature of a cooling stream, which also affect the value of $f(x, \theta)$ monotonically. While we work with Assumption 6.1 to simplify the presentation, it should be noted that a more general assumption can be stated as follows: There exist known functions $f_l(x)$ and $f_u(x)$ such that $f_l(x) \leq f(x, \theta) \leq f_u(x)$ for all $\theta \in \Theta$.

Consider the system of Eq. (6.1) under fault-free conditions, for which a control Lyapunov function $V(x)$ exists and Assumption 6.1 holds. Let Π denote a set of states where $\dot{V}(x(t))$ can be made negative by using the allowable values of the constrained input:

$$\Pi = \Big\{x \in \mathbb{R}^n : \sup_{\theta \in \Theta} L_f V(x, \theta) + \inf_{u \in \mathcal{U}} L_G V(x) u \leq -\varepsilon V(x)\Big\}, \tag{6.2}$$

where $L_G V(x) = [L_{g_1} V(x), \dots, L_{g_m} V(x)]$, with g_i being the ith column of G, and ε is a positive real number. It is assumed that $L_f V(x, \theta)$ and $L_G V(x)$ are locally

Lipschitz. To estimate the upper bound on $L_f V(x,\theta)$, let $\theta_{i,l} = [\theta_{i,1,l},\dots,\theta_{i,q,l}]$ and

$$\theta_{i,u} = [\theta_{i,1,u},\dots,\theta_{i,q,u}], \quad i = 1,\dots,n, \quad \text{where } \theta_{i,j,l} = \begin{cases} \theta_{j,\max} & \text{if } \frac{df_i}{d\theta_j} \le 0, \\ \theta_{j,\min} & \text{if } \frac{df_i}{d\theta_j} > 0 \end{cases}$$

and

$$\theta_{i,j,u} = \begin{cases} \theta_{j,\min} & \text{if } \frac{df_i}{d\theta_j} \le 0, \\ \theta_{j,\max} & \text{if } \frac{df_i}{d\theta_j} > 0, \end{cases} \qquad j = 1,\dots,q.$$

Note that $\theta_{i,l}$ and $\theta_{i,u}$ are the instances of θ that make $f_i(x,\theta)$ take its minimum and maximum values for given x, respectively. Let

$$\tilde{\theta}_i = \begin{cases} \theta_{i,l}, & \frac{\partial V}{\partial x_i} \le 0, \\ \theta_{i,u}, & \frac{\partial V}{\partial x_i} > 0, \end{cases} \qquad i = 1,\dots,n.$$

It follows that $\sum_{i=1}^n \frac{\partial V}{\partial x_i} f_i(x,\tilde{\theta}_i)$ is an estimate of the upper bound on $L_f V(x,\theta)$. Note that $\inf_{u\in\mathcal{U}} L_G V(x)u$ can be computed in a similar way. The robust controller of [95] possesses a stability region, an estimate of which is given by

$$\{x \in \Pi' : V(x) \le c\}, \tag{6.3}$$

where Π' is an estimate of Π by replacing $\sup_{\theta\in\Theta} L_f V(x,\theta)$ with $\sum_{i=1}^n \frac{\partial V}{\partial x_i} \times f_i(x,\tilde{\theta}_i)$ and c is a positive (preferably the largest possible) constant.

The Lyapunov-based predictive controller adapted from [95] takes the following form:

$$u^*(\cdot) = \operatorname{argmin}\{J(x,t,u(\cdot)) | u(\cdot) \in S\}, \tag{6.4a}$$

$$\text{s.t.} \quad \dot{x} = f(x,0) + G(x)u, \tag{6.4b}$$

$$L_G V(x(t))u(t) \le -\sum_{i=1}^n \frac{\partial V}{\partial x_i} f_i(x,\tilde{\theta}_i) - \varepsilon V(x(t)), \tag{6.4c}$$

$$x(\tau) \in \Pi' \quad \text{for all } \tau \in [t, t+\Delta), \tag{6.4d}$$

where $S = S(t,T)$ is a family of piecewise continuous functions (functions continuous from the right), with T denoting the control horizon, mapping $[t, t+T)$ into $\mathcal{U}$. A control $u(\cdot)$ in S is characterized by the sequence $\{u(t_k)\}$ and satisfies $u(\tau) = u(t_k)$ for all $\tau \in [t_k, t_k+\Delta)$. The objective function is given by

$$J(x,t,u(\cdot)) = \int_t^{t+T} \left[\|x^u(s;x,t)\|_{Q_w}^2 + \|u(s)\|_{R_w}^2\right] ds, \tag{6.5}$$

where Q_w is a positive semi-definite symmetric matrix, R_w is a strictly positive definite symmetric matrix, and $x^u(s;x,t)$ denotes the solution of Eq. (6.4b), due to control $u(\cdot)$, with the initial state x at time t. In accordance with the receding horizon implementation, the minimizing control $u^*(\cdot)$ is then applied to the system over $[t, t + \Delta)$ and the same procedure is repeated at the next instant.

The stability property of the control law of Eqs. (6.4a)–(6.4d) can be formulated as follows: Given any positive real number d, there exists a positive real number Δ^* such that if $\Delta \in (0, \Delta^*]$ and $x(0) \in \Omega$, then $x(t) \in \Omega$ for all $t \geq 0$ and $\limsup_{t\to\infty} \|x(t)\| \leq d$ (see [95] for further details on the control design). Finally, note that while the control law of Eqs. (6.4a)–(6.4d) is used as an example of a control design for illustration, the proposed results hold under any control law (which we refer to as $RC(x)$) satisfying Assumption 6.2 below.

Assumption 6.2 For the system of Eq. (6.1) under fault-free conditions, there exist a robust control law $RC(x)$ and a set $\Omega \subseteq \mathbb{R}^n$ such that given any positive real number d, there exist positive real numbers Δ^* and T_f such that if $\Delta \in (0, \Delta^*]$ and $x(0) \in \Omega$, then $x(t) \in \Omega$ for all $t \geq 0$ and $\|x(t)\| \leq d$ for all $t \geq T_f$.

6.3 Fault Detection and Diagnosis Structure

In this section, we first propose a fault diagnosis design under state feedback control in Sect. 6.3.1, and then generalize it to handle state estimation errors in Sect. 6.3.2.

6.3.1 Fault Diagnosis Under State Feedback Control

In this section, under the assumption of full state feedback, we design an FDI scheme using constant thresholds and then for a special case, devise an FDD scheme using time-varying thresholds. With the assumption that $m \leq n$, the system of Eq. (6.1) can be decomposed into two coupled subsystems that we denote as a diagnosable subsystem and the remainder of the original system, with states denoted by $x_d \in \mathbb{R}^m$ and $x_{\bar{d}} \in \mathbb{R}^{n-m}$, respectively. Accordingly, we have $f(x,\theta) = [f_d(x,\theta)^{\mathrm{T}}, f_{\bar{d}}(x,\theta)^{\mathrm{T}}]^{\mathrm{T}}$ and $G(x) = [G_d(x)^{\mathrm{T}}, G_{\bar{d}}(x)^{\mathrm{T}}]^{\mathrm{T}}$. The system of Eq. (6.1) can then be written as follows:

$$\dot{x}_d = f_d(x,\theta) + G_d(x)\big[u(t) + \tilde{u}(t)\big], \tag{6.6a}$$

$$\dot{x}_{\bar{d}} = f_{\bar{d}}(x,\theta) + G_{\bar{d}}(x)\big[u(t) + \tilde{u}(t)\big]. \tag{6.6b}$$

The key idea of the proposed methodology is to estimate the outputs of the actuators by using the system model and state measurements, and then compare them with the corresponding prescribed control inputs to construct input-based residuals. To this end, consider the time interval $[t_k, t_{k+1})$, with t_{k+1} being the current time.

Integrating both sides of Eq. (6.6a) over $[t_k, t_{k+1})$ gives the following equation:

$$\begin{aligned} x_d(t_{k+1}) &= x_d(t_k) + \int_{t_k}^{t_{k+1}} \{ f_d(x,\theta) + G_d(x)[u(t) + \tilde{u}(t)] \} \, dt \\ &= x_d(t_k) + F_{d,k} + G_{d,k}[u(t_k) + \tilde{u}(t_k)], \end{aligned} \tag{6.7}$$

where $F_{d,k} = \int_{t_k}^{t_{k+1}} f_d(x,\theta)\,dt$ and $G_{d,k} = \int_{t_k}^{t_{k+1}} G_d(x)\,dt$. Let $x_{d,i}$, $f_{d,i}$, $F_{d,i,k}$, and $G_{d,i,k}$ denote the ith element or row of x_d, f_d, $F_{d,k}$, and $G_{d,k}$, respectively, for $i = 1, \ldots, m$. We say that the subsystem of Eq. (6.6a) is diagnosable if it satisfies Assumption 6.3 below.

Assumption 6.3 For the system of Eq. (6.1), $m \le n$ and $G_{d,k}$ is invertible for $k = 0, \ldots, \infty$.

Remark 6.2 To illustrate the idea behind Assumption 6.3, consider a scalar system described by $\dot{x} = x + u_1 + 2u_2$, where $x, u_1, u_2 \in \mathbb{R}$. For this system, it is impossible to differentiate between faults in u_1 and u_2 because the number of state variables is eclipsed by that of the input variables (i.e., $m > n$). Alternatively, it is possible that inputs affect states in the same manner through different channels. For example, consider the system described by $\dot{x} = x + \left[\begin{smallmatrix} 1 & 1 \\ 2 & 2 \end{smallmatrix}\right] u$, where $x, u \in \mathbb{R}^2$. For this case, the definition of a new variable $v = u_1 + u_2$ leads to an equivalent system of the form $\dot{x} = x + [1, 2]^{\mathrm{T}} v$. Although the number of state variables is equal to that of the input variables in the original system, any fault in u_1 or u_2 can be seen as a fault in v, thereby impeding fault isolation. A simple example of a diagnosable system is given by $\dot{x} = x + \left[\begin{smallmatrix} 1 & 2 \\ 2 & 1 \end{smallmatrix}\right] u$, where $x, u \in \mathbb{R}^2$. In this example, u_2 affects x_1 more than u_1, and u_1 affects x_2 more than u_2, thereby satisfying the condition that the inputs affect the state dynamics uniquely through different channels.

Remark 6.3 In [115], the isolation of faults relies on the assumption that there exists a state variable such that its evolution is directly and uniquely affected by the potential fault. Specifically, it requires that for every input u_j, $j = 1, \ldots, m$, there exist a state x_l, $i \in \{1, \ldots, n\}$ such that with x_l as an output, the relative degree of x_l with respect to u_j and only with respect to u_j is equal to 1. In other words, $g_{i,j}(x) \neq 0$ for all $x \in \mathbb{R}^n$ and $g_{i,l}(x) \equiv 0$ for $l = 1, \ldots, m$ and $l \neq j$. In this case, $G_d(x)$ is a diagonal matrix with non-zero elements on its diagonal. Therefore, $G_{d,k}$ is invertible. Assumption 6.3, however, only requires that $G_{d,k}$ be invertible, and $G_d(x)$ could be a non-diagonal matrix.

Let $[G_{d,k}^{-1}]_i$ denote the ith row of $G_{d,k}^{-1}$ and $[G_{d,k}^{-1}]_{ij}$ denote the jth element of $[G_{d,k}^{-1}]_i$. It follows from Eq. (6.7) that

$$u_i(t_k) + \tilde{u}_i(t_k) = \left[G_{d,k}^{-1}\right]_i \left[x_d(t_{k+1}) - x_d(t_k) - F_{d,k}\right]. \tag{6.8}$$

For $i = 1, \ldots, m$, define the residuals as

$$r_{i,k} = \left|\left[G_{d,k}^{-1}\right]_i \left[x_d(t_{k+1}) - x_d(t_k) - \bar{F}_{d,k}\right] - u_i(t_k)\right|, \tag{6.9}$$

where $\bar{F}_{d,k} = \int_{t_k}^{t_{k+1}} f_d(x, 0)\,dt$. Note that $[G_{d,k}^{-1}]_i[x_d(t_{k+1}) - x_d(t_k) - \bar{F}_{d,k}]$ is the estimate of the actual input to the plant by using the nominal system model. Substituting $u_i(t_k)$ in Eq. (6.8) into Eq. (6.9) gives $r_{i,k} = \|[G_{d,k}^{-1}]_i(F_{d,k} - \bar{F}_{d,k}) + \tilde{u}_i(t_k)|$. The FDI scheme using constant thresholds is formalized in Theorem 6.1 below.

Theorem 6.1 *Consider the system of Eq.* (6.1), *for which Assumption* 6.3 *holds. Assume that* $\|[G_{d,k}^{-1}]_i^{\mathrm{T}}\| \le K_{g,i}$ *for* $k = 0, \ldots, \infty$, *where* $K_{g,i}$ *is a positive real number. Then, there exists* $\delta_i > 0$ *such that if* $r_{i,k} > \delta_i$, *then* $\tilde{u}_i(t_k) \neq 0$.

Proof Since $f_d(x, \theta)$ is locally Lipschitz in θ, there exists $L_f > 0$ such that

$$\left\| f_d(x, \theta) - f_d(x, 0)\right\| \le L_f \theta_b. \tag{6.10}$$

If $\tilde{u}_i(t_k) = 0$, it follows that

$$\begin{aligned} r_{i,k} &= \left|\left[G_{d,k}^{-1}\right]_i (F_{d,k} - \bar{F}_{d,k})\right| = \left|\left[G_{d,k}^{-1}\right]_i \int_{t_k}^{t_{k+1}} \left[f_d(x, \theta) - f_d(x, 0)\right] dt\right| \\ &\le K_{g,i} L_f \theta_b \Delta. \end{aligned} \tag{6.11}$$

It means that for $\delta_i = K_{g,i} L_f \theta_b \Delta$, if $\tilde{u}_i(t_k) = 0$, then $r_{i,k} \le \delta_i$. Therefore, $r_{i,k} > \delta_i$ implies that $\tilde{u}_i(t_k) \neq 0$. This concludes the proof of Theorem 6.1. □

Remark 6.4 Theorem 6.1 shows that there exists a uniform bound on the absolute error between the estimate of the input to the plant and the prescribed control input for each manipulated variable. This result establishes a sufficient condition for FDI: If the bound is breached, then an actuator fault must have taken place. The design allows for 'small' faults, which are indistinguishable from the effect of the system uncertainty, to go undetected; however, such faults, since they essentially have the same effect as the system uncertainty, may be handled by the robustness of the control design.

We then consider a case where Assumption 6.1 is satisfied and derive time-varying bounds (in the discrete-time domain) on the outputs of the actuators for FDD. To this end, we first derive bounds on $F_{d,k}$. Define $\theta_{d,i,l}$ and $\theta_{d,i,u}$ in the same way as $\theta_{i,l}$ and $\theta_{i,u}$ were defined in Sect. 6.2.2, for $i = 1, \ldots, m$. It follows that

$$\int_{t_k}^{t_{k+1}} f_{d,i}(x, \theta_{d,i,l})\,dt \le F_{d,i,k} \le \int_{t_k}^{t_{k+1}} f_{d,i}(x, \theta_{d,i,u})\,dt. \tag{6.12}$$

Let $f_{d,i,k,l} = \int_{t_k}^{t_{k+1}} f_{d,i}(x, \theta_{d,i,l})\,dt$ and $f_{d,i,k,u} = \int_{t_k}^{t_{k+1}} f_{d,i}(x, \theta_{d,i,u})\,dt$ denote the lower and upper bounds on $F_{d,i,k}$, respectively. The FDD scheme using time-varying thresholds is formalized in Theorem 6.2 below.

Theorem 6.2 *Consider the system of Eq.* (6.1), *for which Assumptions* 6.1 *and* 6.3 *hold. Then, there exist* $u_{i,k,l}$ *and* $u_{i,k,u}$ *such that if* $u_i(t_k) \notin [u_{i,k,l}, u_{i,k,u}]$, *then* $\tilde{u}_i(t_k) \neq 0$, *and* $u_i(t_k) + \tilde{u}_i(t_k) \in [u_{i,k,l}, u_{i,k,u}]$.

Proof It follows from Eq. (6.8) that

$$\begin{aligned} u_i(t_k) + \tilde{u}_i(t_k) &= \big[G_{d,k}^{-1}\big]_i \big[x_d(t_{k+1}) - x_d(t_k)\big] - \sum_{j=1}^{m} \big[G_{d,k}^{-1}\big]_{ij} F_{d,j,k} \\ &\geq \big[G_{d,k}^{-1}\big]_i \big[x_d(t_{k+1}) - x_d(t_k)\big] - \sum_{j=1}^{m} \big[G_{d,k}^{-1}\big]_{ij} F_{d,j,k,l}, \end{aligned} \tag{6.13}$$

where

$$F_{d,j,k,l} = \begin{cases} f_{d,j,k,l} & \text{if } [G_{d,k}^{-1}]_{ij} \leq 0, \\ f_{d,j,k,u} & \text{if } [G_{d,k}^{-1}]_{ij} > 0, \end{cases} \quad j = 1, \ldots, m.$$

Let $F_{d,k,l} = [F_{d,1,k,l}, \ldots, F_{d,m,k,l}]^{\mathrm{T}}$. Then, we have that

$$u_i(t_k) + \tilde{u}_i(t_k) \geq \big[G_{d,k}^{-1}\big]_i \big[x_d(t_{k+1}) - x_d(t_k) - F_{d,k,l}\big]. \tag{6.14}$$

Similarly, we have that

$$u_i(t_k) + \tilde{u}_i(t_k) \leq \big[G_{d,k}^{-1}\big]_i \big[x_d(t_{k+1}) - x_d(t_k) - F_{d,k,u}\big], \tag{6.15}$$

where $F_{d,k,u} = [F_{d,1,k,u}, \ldots, F_{d,m,k,u}]^{\mathrm{T}}$, with

$$F_{d,j,k,u} = \begin{cases} f_{d,j,k,u} & \text{if } [G_{d,k}^{-1}]_{ij} \leq 0, \\ f_{d,j,k,l} & \text{if } [G_{d,k}^{-1}]_{ij} > 0, \end{cases} \quad j = 1, \ldots, m.$$

Let $u_{i,k,l} = [G_{d,k}^{-1}]_i [x_d(t_{k+1}) - x_d(t_k) - F_{d,k,l}]$ and $u_{i,k,u} = [G_{d,k}^{-1}]_i [x_d(t_{k+1}) - x_d(t_k) - F_{d,k,u}]$. Thus, $u_{i,k,l} \leq u_i(t_k) + \tilde{u}_i(t_k) \leq u_{i,k,u}$, and $u_{i,k,l} \leq u_i(t_k) \leq u_{i,k,u}$ if $\tilde{u}_i(t_k) = 0$. Therefore, $u_i(t_k) \notin [u_{i,k,l}, u_{i,k,u}]$ implies that $\tilde{u}_i(t_k) \neq 0$. This concludes the proof of Theorem 6.2. □

Remark 6.5 In Theorem 6.2, the monotonic property of the right-hand side of the state equation with respect to the uncertain variables is utilized to generate time-varying bounds on the actual input to the plant. In the absence of faults, the actual input is equal to its prescribed value, which should reside within the set dictated by the estimated bounds on the actual input, for each manipulated variable. If the prescribed value breaches these bounds for some manipulated variable, the only way that it can happen is when the actual input is no longer equal to the prescribed value, resulting in the detection and isolation of a fault. Note that while faults that do not lead to $u_i(t_k) \notin [u_{i,k,l}, u_{i,k,u}]$ cannot be detected, they may be handled through the robustness of the control design. Note also that beyond FDI, the fault diagnosis scheme provides an estimate of the output of the failed actuator.

The FDD procedure for the case where an actuator seizes at an arbitrary position is summarized as follows:

1. At time t_{k+1}, $k = 0, \ldots, \infty$, compute $u_{i,k,l}$ and $u_{i,k,u}$, $i = 1, \ldots, m$.
2. Let

$$r_{b,i,k} := \begin{cases} 1 & \text{if } u_i(t_k) \notin [u_{i,k,l}, u_{i,k,u}], \\ 0 & \text{otherwise}, \end{cases} \tag{6.16}$$

where $r_{b,i,k}$ denotes a binary residual for u_i. If n_d nonzero residuals for u_i are monitored consecutively, where n_d is a design parameter for FDD, report a fault at time $t_d = t_{k+1}$ for the actuator that corresponds to u_i and choose $\bar{u}_{i,l} = \max \bigcup_{j \in \{k+1-n_d, \ldots, k\}} \{u_{i,j,l}\} \cup \{u_{i,\min}\}$ as the lower bound and $\bar{u}_{i,u} = \min \bigcup_{j \in \{k+1-n_d, \ldots, k\}} \{u_{i,j,u}\} \cup \{u_{i,\max}\}$ as the upper bound on the failed actuator position, respectively. Otherwise, repeat step 1.

6.3.2 Handling State Estimation Errors for Fault Diagnosis

In many practical situations, it is not economical to measure all the system states, or in some situations, only part of the system states are inherently measurable, which necessitates output feedback control by using state estimators. In this section, we generalize the fault diagnosis scheme of Sect. 6.3.1 to handle state estimation errors, with the focus on the problem of FDD (and not the state estimator design). To this end, we assume the existence of a state estimator (observer or predictor) which can provide the state estimate, denoted by $\hat{x}(t)$ at time t, that is accurate enough (at least for some time even after an actuator fault takes place) to perform fault diagnosis (see Remark 6.6 for examples of such observers). This is formalized in Assumption 6.4 below [115].

Assumption 6.4 For the system of Eq. (6.1), there exists a state estimator such that, given positive real numbers e and $\tilde{u}_b$, there exists $t_e > 0$ such that if $\|\tilde{u}(t)\| \leq \tilde{u}_b$, then $\|x(t) - \hat{x}(t)\| \leq e$ for all $t \in [t_e, \infty)$. Furthermore, there exists $T_d > 0$ such that if $\|\tilde{u}(t)\| > \tilde{u}_b$ for some $t_f > t_e$, then $\|x(t) - \hat{x}(t)\| \leq e$ for all $t \in [t_e, t_f + T_d]$.

The key idea of the FDD design for the case with state estimation errors is to use the state estimate and the bounds on uncertainty and the estimation errors to determine the bounds on $u(t_k) + \tilde{u}(t_k)$ as in Sect. 6.3.1, which is formalized in Theorem 6.3 below. To this end, let $\hat{F}_{d,k} = \int_{t_k}^{t_{k+1}} f_d(\hat{x}, \theta(t))\, dt$, $\hat{G}_{d,k} = \int_{t_k}^{t_{k+1}} G_d(\hat{x})\, dt$, $\hat{F}_{d,i,k}$ denote the ith element of $\hat{F}_{d,k}$, and $\hat{G}_{d,i,k}$ denote the ith row of $\hat{G}_{d,k}$. The lower and upper bounds on $\hat{F}_{d,i,k}$, denoted by $\hat{f}_{d,i,k,l}$ and $\hat{f}_{d,i,k,u}$, can be computed in the same way as $f_{d,i,k,l}$ and $f_{d,i,k,u}$ in Sect. 6.3.1 by using $\hat{x}$ instead of x.

Theorem 6.3 *Consider the system of Eq.* (6.1) *subject to state estimation errors, for which Assumptions* 6.1 *and* 6.4 *hold. Assume that* $m \leq n$ *and* $\hat{G}_{d,k}$ *is invertible for* $k = 0, \ldots, \infty$. *Then, for* $[t_k, t_{k+1}] \subseteq [t_e, t_f + T_d]$, *there exist* $\gamma = [\gamma_1, \ldots, \gamma_m]^{\mathrm{T}} > 0$, $\hat{u}_{i,k,l}(\gamma)$, *and* $\hat{u}_{i,k,u}(\gamma)$, *such that if* $u_i(t_k) \notin [\hat{u}_{i,k,l}(\gamma), \hat{u}_{i,k,u}(\gamma)]$, *then* $\tilde{u}_i(t_k) \neq 0$, *and* $u_i(t_k) + \tilde{u}_i(t_k) \in [\hat{u}_{i,k,l}(\gamma), \hat{u}_{i,k,u}(\gamma)]$.

Proof It follows from Eq. (6.7) that $F_{d,i,k} = x_{d,i}(t_{k+1}) - x_{d,i}(t_k) - G_{d,i,k}[u(t_k) + \tilde{u}(t_k)]$. Similarly, define $\tilde{F}_{d,i,k} = \hat{x}_{d,i}(t_{k+1}) - \hat{x}_{d,i}(t_k) - \hat{G}_{d,i,k}[u(t_k) + \tilde{u}(t_k)]$, where $\hat{x}_{d,i}$ denotes the estimate of $x_{d,i}$. Since $\|x(t) - \hat{x}(t)\| \leq e$ for all $t \in [t_k, t_{k+1}]$ under Assumption 6.4 and $G(x)$ is locally Lipschitz, there exists $L_{g,i} > 0$ such that $\|\hat{G}^{\mathrm{T}}_{d,i,k} - G^{\mathrm{T}}_{d,i,k}\| \leq L_{g,i}\Delta e$. It follows that

$$\begin{aligned} |\tilde{F}_{d,i,k} - F_{d,i,k}| &\leq \left|\hat{x}_{d,i}(t_{k+1}) - x_{d,i}(t_{k+1})\right| + \left|\hat{x}_{d,i}(t_k) - x_{d,i}(t_k)\right| \\ &\quad + \left|(\hat{G}_{d,i,k} - G_{d,i,k})\left[u(t_k) + \tilde{u}(t_k)\right]\right| \\ &\leq e + L_{g,i}u_b\Delta e, \end{aligned} \tag{6.17}$$

which leads to

$$F_{d,i,k} - (2 + L_{g,i}u_b\Delta)e \leq \tilde{F}_{d,i,k} \leq F_{d,i,k} + (2 + L_{g,i}u_b\Delta)e. \tag{6.18}$$

Since $f_d(x, \theta)$ is locally Lipschitz in x, there exists $L_{f,i} > 0$ such that $|F_{d,i,k} - \hat{F}_{d,i,k}| \leq L_{f,i}\Delta e$, which leads to

$$\hat{F}_{d,i,k} - L_{f,i}\Delta e \leq F_{d,i,k} \leq \hat{F}_{d,i,k} + L_{f,i}\Delta e. \tag{6.19}$$

Note that $\hat{f}_{d,i,k,l} \leq \hat{F}_{d,i,k} \leq \hat{f}_{d,i,k,u}$. Then, Eq. (6.18) and Eq. (6.19) yield

$$\hat{f}_{d,i,k,l} - \gamma_i \leq \tilde{F}_{d,i,k} \leq \hat{f}_{d,i,k,u} + \gamma_i, \tag{6.20}$$

where $\gamma_i = (2 + L_{f,i}\Delta + L_{g,i}u_b\Delta)e$. Since $\hat{G}_{d,k}$ is invertible, we have $u_i(t_k) + \tilde{u}_i(t_k) = [\hat{G}^{-1}_{d,k}]_i[\hat{x}_d(t_{k+1}) - \hat{x}_d(t_k) - \tilde{F}_{d,k}]$, where $[\hat{G}^{-1}_{d,k}]_i$ denotes the ith row of $\hat{G}^{-1}_{d,k}$, $\hat{x}_d = [\hat{x}_{d,1}, \ldots, \hat{x}_{d,m}]^{\mathrm{T}}$, and $\tilde{F}_{d,k} = [\tilde{F}_{d,1,k}, \ldots, \tilde{F}_{d,m,k}]^{\mathrm{T}}$. Now, with the bounds on $\tilde{F}_{d,i,k}$ computed, the rest of the proof proceeds along the same lines as the proof of Theorem 6.2. This concludes the proof of Theorem 6.3. □

Remark 6.6 In the context of output feedback control, the fault diagnosis scheme of Theorem 6.3 requires that the structure of the system allow the design of a state estimator that can provide an accurate enough state estimate. Examples of such estimators include a high-gain state observer (see Chaps. 3 and 4) and a reduced-order nonlinear observer developed in [130].

6.4 Robust Safe-Parking for Fault-Tolerant Control

In this section, we consider the problem of fault-handling for the case where an actuator seizes at an arbitrary position (and does not revert to the pre-designed fail-safe position). The key idea of the proposed approach is to design safe-park point candidates off-line for a series of the output values of the potential failed actuator, and upon FDD, choose a safe-park point online such that the system can be stabilized at the chosen safe-park point by the robust control law, which can handle the error between the actual value of the failed actuator position and its design counterpart.

Specifically, we design safe-park point candidates for M actuator positions of u_i denoted by $\bar{u}_{s,i,j} \in [u_{i,\min}, u_{i,\max}]$, $j = 1, \dots, M$. When designing the control law and characterizing the stability region of a safe-park point candidate, a design uncertain variable of magnitude δ_s (over and above the uncertain variables in the system description), is used to account for the error between the actual value of the failed actuator position, denoted by $\bar{u}_{i,f}$, and the one used to design the safe-park point candidate ($\bar{u}_{s,i,j}$). Let u_{nom} and $u_{s,i,j}$ denote the control laws to stabilize the system at the nominal equilibrium point x_{nom} and a safe-park point $x_{s,i,j}$, respectively, yielding Ω_{nom} and $\Omega_{s,i,j}$ as their stability regions. The schematic in Fig. 6.1 shows the integrated fault diagnosis and safe-parking framework, which is formalized in Theorem 6.4 below (the proof of this theorem follows a similar line of argument as in [57] and is omitted).

Theorem 6.4 *Consider the system of Eq.* (6.1) *under a control law $RC(x)$ satisfying Assumption* 6.2. *Let t_f be the time when a fault takes place, t_d the time when it is detected and diagnosed, and t_r the time when it is repaired. If $x(0) \in \Omega_{\text{nom}}$, $[\bar{u}_{i,l}, \bar{u}_{i,u}] \subseteq [\bar{u}_{s,i,j} - \delta_s, \bar{u}_{s,i,j} + \delta_s]$, $x(t_d) \in \Omega_{s,i,j}$, and $B_{d,s,i,j} \subseteq \Omega_{\text{nom}}$, where $B_{d,s,i,j}$ is a closed ball of radius d around $x_{s,i,j}$, then the switching rule*

$$u(t) = \begin{cases} u_{\text{nom}}(t), & 0 \le t < t_d, \\ u_{s,i,j}(t), & t_d \le t < t_s, \\ u_{\text{nom}}(t), & t_s \le t, \end{cases} \tag{6.21}$$

where $t_s \ge t_r$ is such that $x(t_s) \in \Omega_{\text{nom}}$, guarantees that $x(t) \in \Omega_{\text{nom}}$ $\forall t \in [0, t_f] \cup [t_s, \infty)$ and there exists a positive real number T_f such that $\|x(t)\| \le d$ for all $t \ge T_f$.

Remark 6.7 Upon the confirmation of a fault, the safe-parking mechanism described by Theorem 6.4 is activated to shift the control objective from operating the system at the nominal equilibrium point to maintaining it at a suboptimal but admissible operating point. Note that a safe-park point is chosen from the candidates generated for the design value of the failed actuator position $\bar{u}_{s,i,j}$ such that the range $[\bar{u}_{s,i,j} - \delta_s, \bar{u}_{s,i,j} + \delta_s]$ designed off-line contains the range $[\bar{u}_{i,l}, \bar{u}_{i,u}]$ identified on-line for the failed actuator position, as illustrated in Fig. 6.2. Since $[\bar{u}_{i,l}, \bar{u}_{i,u}]$ contains the actual value of the failed actuator position $\bar{u}_{i,f}$, it is guaranteed that such a safe-park point candidate is a feasible equilibrium point subject to the fault. Note also that an arbitrarily chosen safe-park point candidate is not guaranteed to be a feasible equilibrium point in the presence of the fault. Therefore, the fault information provided by the fault diagnosis design is essential in choosing a safe-park point.

Remark 6.8 The remaining conditions dictating the choice of a safe-park point follow from the safe-parking framework designed for a fail-safe position in Chap. 5.

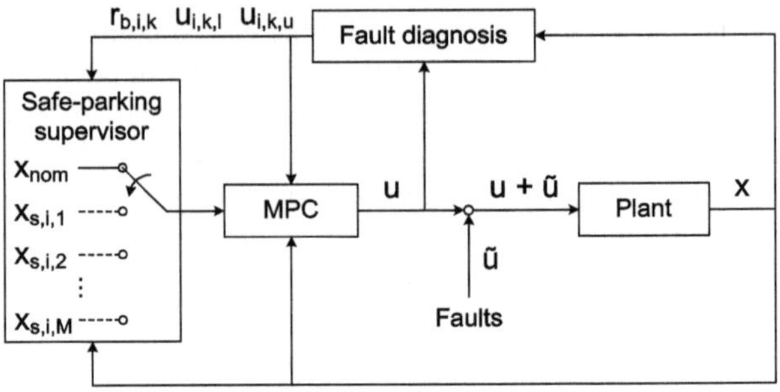

Fig. 6.1 Schematic of the integrated fault diagnosis and safe-parking framework

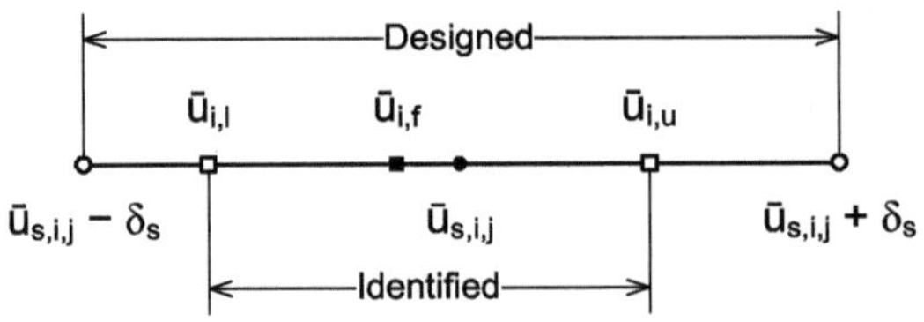

Fig. 6.2 Schematic illustrating the choice of a safe-park point. The range $[\bar{u}_{s,i,j} - \delta_s, \bar{u}_{s,i,j} + \delta_s]$ is designed off-line for the actuator position $\bar{u}_{s,i,j}$ with the robustness margin δ_s. The range $[\bar{u}_{i,l}, \bar{u}_{i,u}]$ is identified online, which contains the actual value of the failed actuator position $\bar{u}_{i,f}$

In particular, to make sure that the system can be driven to the temporary operating point, it requires that the system state should reside within the stability region of the safe-park point at the time of fault confirmation. Note that t_s denotes a time when the system state is within the stability region of the nominal equilibrium point after the fault is repaired. If it is already within the stability region of the nominal equilibrium point at the time of fault repair, then $t_s = t_r$. Otherwise, the control action is implemented to drive the system state to the safe-park point until it reaches the stability region of the nominal equilibrium point. Note in general that the possibility of finding safe-park points and resuming normal operation can be enhanced by the use of control designs (or Lyapunov functions) that yield as large a stability region for the nominal (and safe-park) operation as possible. The size of the stability region remains case-specific; however, the ability to explicitly characterize the stability region (provided by the control design used in this chapter) is useful in ascertaining the ability of the controller to best utilize the available control effort and design the safe-parking framework.

6.5 Simulation Example

In this section, we illustrate the proposed fault diagnosis techniques and the generalized safe-parking framework via a continuous-stirred tank reactor example, as shown in Fig. 6.3, where three parallel irreversible elementary exothermic reactions of the form A $\xrightarrow{k_1}$ B, A $\xrightarrow{k_2}$ U, and A $\xrightarrow{k_3}$ R take place, with A being the reactant

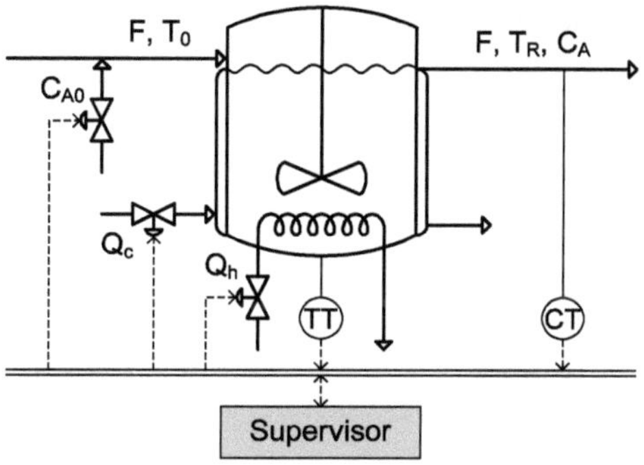

Fig. 6.3 Schematic of the chemical reactor example

species, B the desired product, and U and R the undesired byproducts. The feed to the reactor consists of reactant A at a flow rate F, concentration C_{A0}, and temperature T_0. Under standard assumptions, the mathematical model of the process can be derived from material and energy balances, which takes the following form:

$$\begin{aligned} \dot{C}_A &= \frac{F}{V}(C_{A0} - C_A) - \sum_{i=1}^{3} R_i(C_A, T_R), \\ \dot{T}_R &= \frac{F}{V}(T_0 - T_R) + \sum_{i=1}^{3} \frac{(-\Delta H_i)}{\rho c_p} R_i(C_A, T_R) + \frac{Q}{\rho c_p V}, \end{aligned} \tag{6.22}$$

where $R_i(C_A, T_R) = k_{i0} e^{-E_i/RT_R} C_A$ for $i = 1, 2, 3$, C_A is the concentration of species A in the reactor, T_R is the temperature of the reactor, Q is the rate of heat input to the reactor, V is the volume of the reactor, k_{i0}, E_i, and ΔH_i are the pre-exponential constant, the activation energy, and the enthalpy of reaction i, respectively, and c_p and ρ are the heat capacity and density of the reacting mixture, respectively. The process parameters can be found in Table 6.1.

Under fault-free conditions, the control objective is to stabilize the reactor at the unstable equilibrium point $(C_A, T_R) = (3.50\ \text{kmol/m}^3,\ 405.0\ \text{K})$, denoted by N in Fig. 6.4, by manipulating C_{A0} and Q, where $0 \le C_{A0} \le 6\ \text{kmol/m}^3$ and $-8 \times 10^5\ \text{kJ/hr} \le Q \le 8 \times 10^5\ \text{kJ/hr}$. The manipulated variable $Q = Q_c + Q_h$, where Q_c and Q_h denote cooling and heating, respectively, with $-8 \times 10^5\ \text{kJ/hr} \le Q_c \le 0$ and $0 \le Q_h \le 8 \times 10^5\ \text{kJ/hr}$. The nominal steady-state values of the manipulated variables are $C_{A0} = 4.25\ \text{kmol/m}^3$ and $Q = -6.55 \times 10^4\ \text{kJ/hr}$. The simulations are conducted under a 0.5 % error in the pre-exponential constant (k_{10}) for the main reaction and sinusoidal disturbances in the feed temperature (T_0) with an amplitude of 3 K and a period of 0.2 hr. The bounds on the errors in k_{10} and T_0 used in the monitoring and control design are ± 1.5 % and ± 5 K, respectively. The concentration and temperature measurements are assumed to have a truncated gaussian noise with a standard deviation of 0.01 kmol/m^3 and 0.1 K for the parent normal distribution, respectively. The lower and upper truncation points are -0.02 kmol/m^3 and 0.02 kmol/m^3 for the concentration, and -0.2 K and 0.2 K for the temperature,

Table 6.1 Process parameters for the chemical reactor example

Parameter	Value	Unit
F	4.998	m^3/hr
T_0	300.0	K
V	1.0	m^3
R	8.314	kJ/kmol K
k_{10}	3.0×10^6	hr^{-1}
k_{20}	3.0×10^5	hr^{-1}
k_{30}	3.0×10^5	hr^{-1}
E_1	5.00×10^4	kJ/kmol
E_2	7.53×10^4	kJ/kmol
E_3	7.53×10^4	kJ/kmol
ΔH_1	-5.0×10^4	kJ/kmol
ΔH_2	-5.2×10^4	kJ/kmol
ΔH_3	-5.4×10^4	kJ/kmol
c_p	0.231	kJ/kg K
ρ	1000.0	kg/m^3

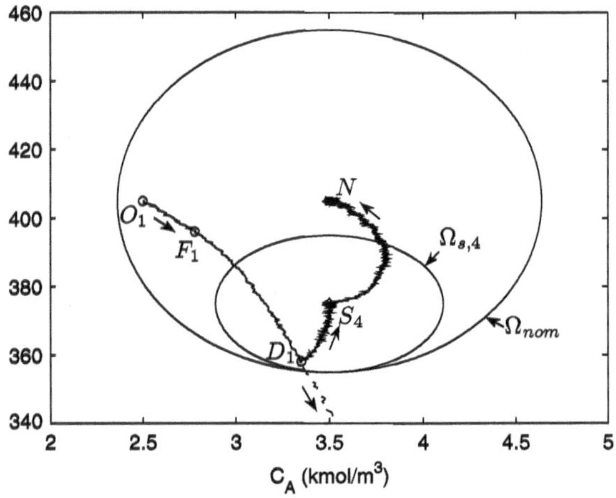

Fig. 6.4 Closed-loop state trajectories for the chemical reactor example where the process starts from O_1 and the cooling valve fails at F_1. The *solid line* shows the case where the fault is confirmed at D_1, the process is stabilized at the safe-park point S_4, and nominal operation is resumed upon fault repair. The *dashed line* shows process instability when no fault-handling mechanism is implemented. The *arrows* show the directions of the trajectories

respectively. The measurements are filtered before performing fault diagnosis and control calculations as $x_f(t_{k+1}) = 0.25x_f(t_k) + 0.75x_m(t_{k+1})$, where x_f and x_m denote the filtered state and noisy measurement, respectively.

To demonstrate the efficacy of the integrated fault diagnosis and safe-parking framework, we consider a failure in the actuator used to control Q_c. The safe-park point candidates are shown in Table 6.2 for 6 actuator positions of Q_c with a robustness margin $\delta_s = 1.25 \times 10^4$ kJ/hr. In the control law of Eq. (6.4a)–(6.4d), an execution time $\Delta = 0.025$ hr $= 1.5$ min and a prediction horizon of 2Δ are used, with $Q_w = \left[\begin{smallmatrix} 1 & 0 \\ 0 & 10 \end{smallmatrix}\right]$ and $R_w = \left[\begin{smallmatrix} 10^5 & 0 \\ 0 & 10^{-6} \end{smallmatrix}\right]$. The Lyapunov function used to character-

Table 6.2 Safe-park point candidates, steady-state values of the manipulated variables, and Lyapunov functions for the chemical reactor example

Safe-park point candidates	Q_c (10^4 kJ/hr)	C_A (kmol/m^3)	T_R (K)	C_{A0} (kmol/m^3)	Q (10^4 kJ/hr)	P $V(x)=x^{T}Px$
S_1	-6.55 ± 1.25	3.50	380	3.78	2.21	$\begin{bmatrix} 2.7 & 0 \\ 0 & 2.5\times10^{-3} \end{bmatrix}$
S_2	-5.73 ± 1.25	3.85	375	4.10	2.40	$\begin{bmatrix} 2.7 & 0 \\ 0 & 2.5\times10^{-3} \end{bmatrix}$
S_3	-4.91 ± 1.25	3.50	380	3.78	2.21	$\begin{bmatrix} 2.7 & 0 \\ 0 & 3.5\times10^{-3} \end{bmatrix}$
S_4	-4.10 ± 1.25	3.50	375	3.73	2.97	$\begin{bmatrix} 2.7 & 0 \\ 0 & 2.5\times10^{-3} \end{bmatrix}$
S_5	-3.28 ± 1.25	3.50	375	3.73	2.97	$\begin{bmatrix} 2.7 & 0 \\ 0 & 3.5\times10^{-3} \end{bmatrix}$
S_6	-2.46 ± 1.25	3.85	375	4.10	2.40	$\begin{bmatrix} 5.0 & 0 \\ 0 & 7.0\times10^{-3} \end{bmatrix}$

ize the stability region and to prescribe the control input for the nominal equilibrium point is chosen as $V(x) = x^{T}Px$, where $P = \begin{bmatrix} 7.72\times10^{-1} & 0 \\ 0 & 4\times10^{-4} \end{bmatrix}$, and those for the safe-park point candidates can be found in Table 6.2. It is assumed that there are 20 samplings during one execution period (i.e., the sampling time is 4.5 s). The trapezoidal rule is used to compute the integrals for the estimation of the bounds on the actual input to the plant. To account for measurement noise, the lower and upper bounds on the estimates of C_{A0} and Q implemented to the plant under state feedback control are relaxed by a magnitude of 0.32 kmol/m^3 and 1848 kJ/hr (inferred from process data under healthy conditions), respectively.

We first consider a case where full state measurements are available and the process starts from an initial condition at O_1 (2.50 kmol/m^3, 405.0 K). The actuator fails at time $t_f = 0.05$ hr, with the process state at F_1 (2.78 kmol/m^3, 396.1 K). The output value of the failed actuator is $\bar{u}_f = -4.19 \times 10^4$ kJ/hr (the same as it was at time t_f^-) during fault repair. The FDD scheme can be explained by Fig. 6.5, where the prescribed inputs are marked by crosses, the actual inputs marked by circles, and the estimated bounds on the actual inputs marked by error bars. Note that a fault is declared when the prescribed value breaches the bounds identified from state measurements. It can be seen that the fault in Q_c is first declared at 0.1 hr (i.e., there is a two-step time delay). Upon the first alarm, the actuator for Q_h is disabled (i.e., the prescribed value of Q_h is 0) to allow FDD for Q_c until the fault is confirmed to be true or false (this step is necessitated by the fact that the FDD scheme cannot differentiate between faults in Q_c and Q_h since they affect the system in an identical fashion). The fault is confirmed at time $t_d = 0.175$ hr after 4 consecutive alarms (i.e., $n_d = 4$), with the process state at D_1 (3.35 kmol/m^3, 358.1 K). The binary residuals for the manipulated variables C_{A0} and Q are shown in Figs. 6.6(a) and 6.6(b), respectively, while the residuals of the manipulated variables obtained by using the nominal process model are shown in Figs. 6.6(c) and 6.6(d), where the thresholds (see the dashed lines) are 0.5 kmol/m^3 and 1.5×10^4 kJ/hr, respectively. It can be seen that similar results are obtained by the FDI designs using constant and time-varying thresholds, with no false alarms generated.

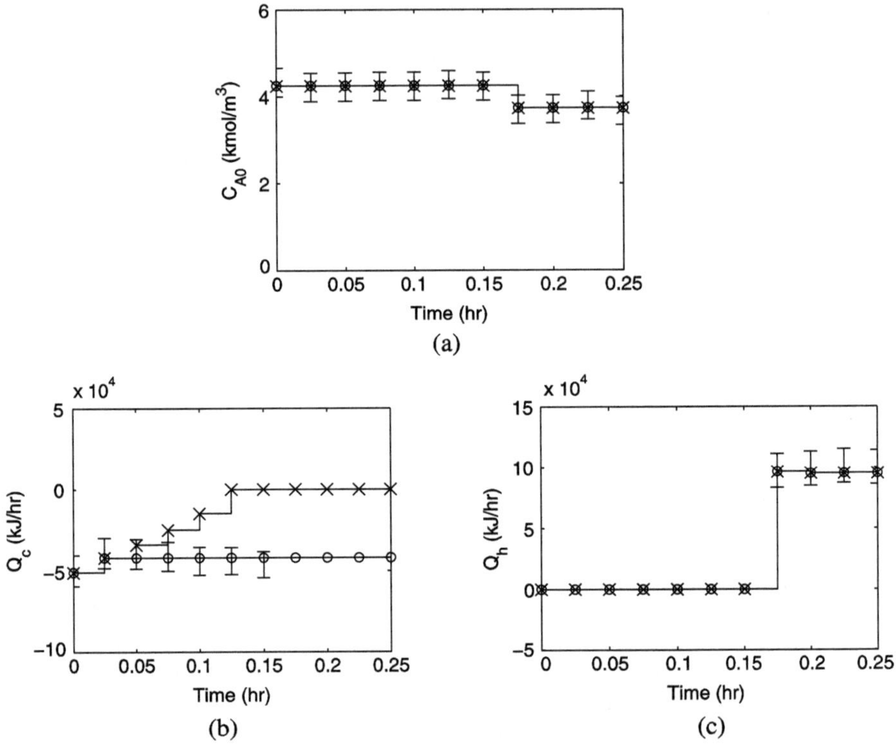

Fig. 6.5 Illustration of the FDD scheme of Theorem 6.2 for the chemical reactor example. The cooling valve fails at time 0.05 hr. The fault is first detected and isolated at 0.1 hr and confirmed at 0.175 hr after 4 consecutive alarms. *Crosses* denote the prescribed inputs, *circles* denote the implemented inputs, and *error bars* denote the estimated bounds on the actual inputs for C_{A0} (**a**), Q_c (**b**), and Q_h (**c**)

Beyond FDI, the fault diagnosis scheme also identifies the lower and upper bounds on the actual value of the failed actuator position, which are -5.00×10^4 kJ/hr and -3.81×10^4 kJ/hr, respectively. This information is then used to choose a safe-park point. By referring to Table 6.2, it is found that the safe-park point candidate $S_4(3.50\ \text{kmol/m}^3,\ 375\ \text{K})$ is designed for the case where the cooling valve seizes at some value in $[-5.35 \times 10^4\ \text{kJ/hr}, -2.85 \times 10^4\ \text{kJ/hr}]$, which contains $[-5.00 \times 10^4\ \text{kJ/hr}, -3.81 \times 10^4\ \text{kJ/hr}]$. Note that the process state at time t_d is also within the stability region of S_4, denoted by $\Omega_{s,4}$. Therefore, S_4 is chosen as a safe-park point. As shown by the solid line in Fig. 6.4, if the safe-parking strategy is implemented, the process is first stabilized at S_4, and nominal operation is resumed upon fault repair. The absence of an appropriately designed fault-handling framework, however, results in process instability, as shown by the dashed line in Fig. 6.4. The corresponding state and input profiles are shown in Fig. 6.7.

We then consider a case where concentration measurements are only available every 10Δ. For this case, we study the problem of estimating the output of the

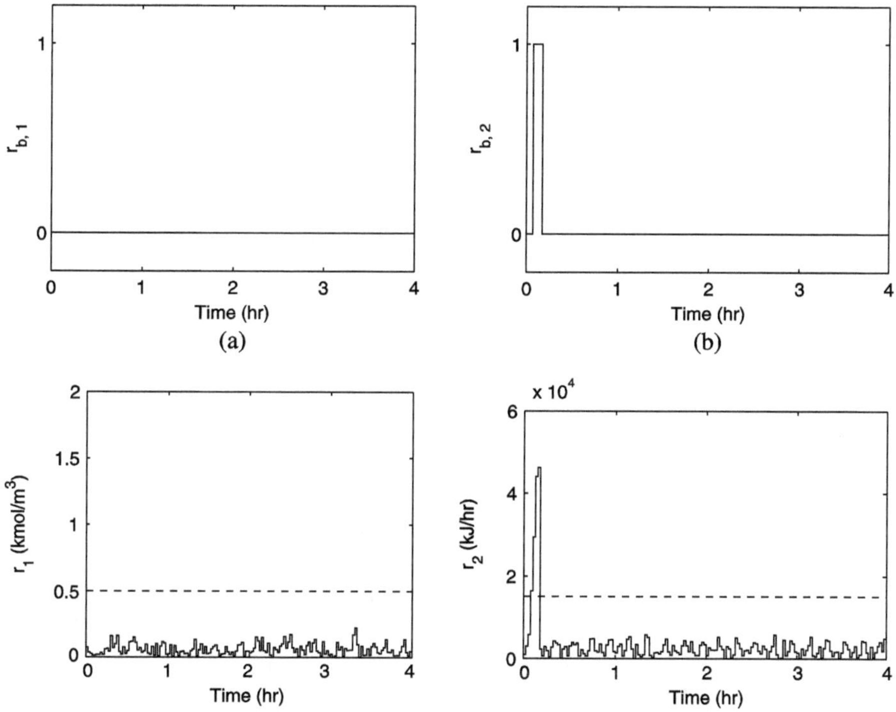

Fig. 6.6 Binary residuals (**a**)–(**b**) defined by Eq. (6.16) and residuals (**c**)–(**d**) defined by Eq. (6.9) for manipulated variables C_{A0} and Q, respectively, in the chemical reactor example

failed actuator and using its estimate to implement the safe-parking operation, with the focus on the diagnosis of the fault magnitude for a fault in Q. The concentration between consecutive measurements is predicted by using the nominal process model and temperature measurements as follows:

$$\begin{aligned} \dot{\hat{C}}_A &= \frac{F}{V}(C_{A0} - \hat{C}_A) - \sum_{i=1}^{3} R_i(\hat{C}_A, T_R), \\ \hat{C}_A(10k\Delta) &= C_A, \end{aligned} \tag{6.23}$$

where $\hat{C}_A$ denotes the estimate of the concentration, which is set to its true value each time an asynchronous measurement is available. In the fault diagnosis design, $\gamma = [0.04, 0.2]^T$ is used to relax the bounds on the estimate of the actual input to the plant. As shown in Fig. 6.8, the process starts from O_2 (4.25 kmol/m^3, 390 K). The fault in Q_c takes place at time $t_f = 0.05$ hr, with the actuator frozen at -2.59×10^4 kJ/hr and the process state at F_2 (4.14 kmol/m^3, 389.1 K). The fault is first detected and isolated at time 0.125 hr and confirmed after 4 consecutive alarms at time $t_d = 0.2$ hr, as shown in Fig. 6.9, with the process state

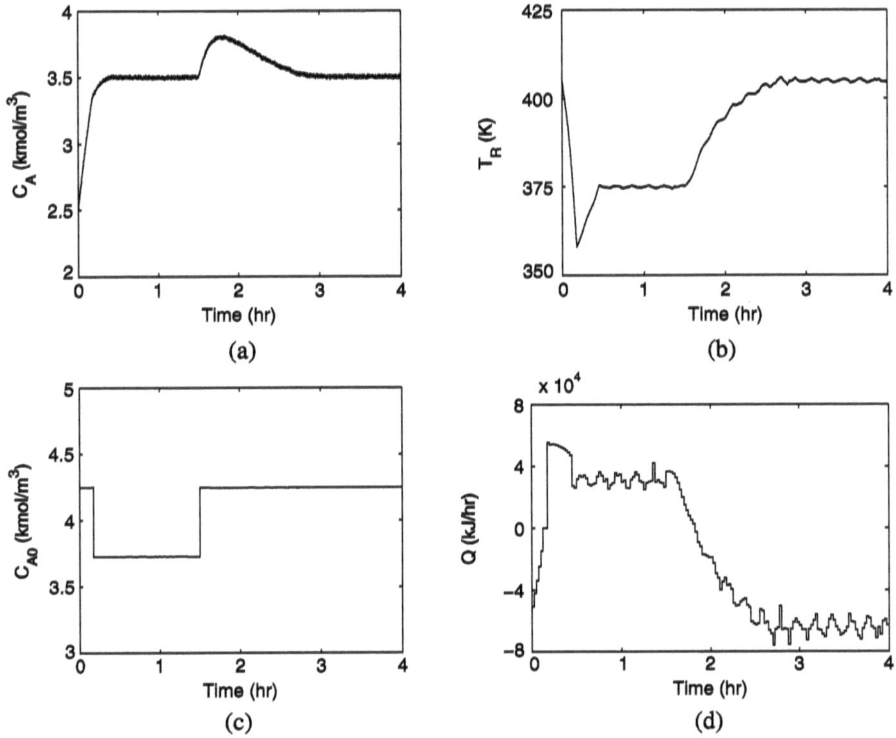

Fig. 6.7 Closed-loop state (**a**)–(**b**) and input (**c**)–(**d**) profiles for the chemical reactor example. The safe-parking operation starts from 0.175 hr, and nominal operation is resumed at 1.5 hr

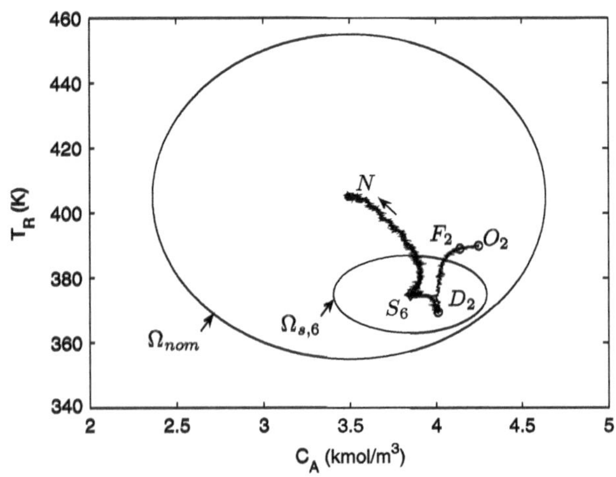

Fig. 6.8 Closed-loop state trajectory for the chemical reactor example with asynchronous concentration measurements where the process starts from O_2 and the cooling valve fails at F_2. The fault is confirmed at D_2, the process is stabilized at the safe-park point S_6, and nominal operation is resumed upon fault repair. The *arrow* shows the direction of the trajectory

at D_2 (4.01 kmol/m^3, 369.4 K). It can be seen from Fig. 6.9 that the estimate of the failed actuator output is $[-3.60 \times 10^4, -2.17 \times 10^4]$, which is a subset of

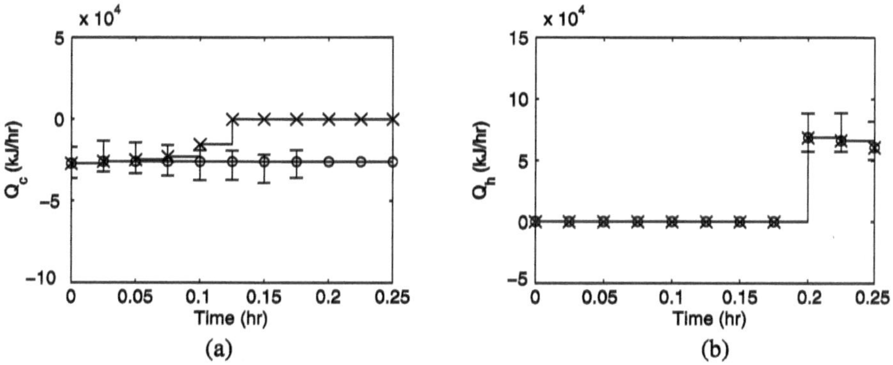

Fig. 6.9 Illustration of the FDD scheme of Theorem 6.3 for the chemical reactor example with asynchronous concentration measurements. The cooling valve fails at time 0.05 hr. The fault is first detected and isolated at 0.125 hr and confirmed at 0.2 hr after 4 consecutive alarms. *Crosses* denote the prescribed inputs, *circles* denote the implemented inputs, and *error bars* denote the estimated bounds on the actual inputs for Q_c (**a**) and Q_h (**b**)

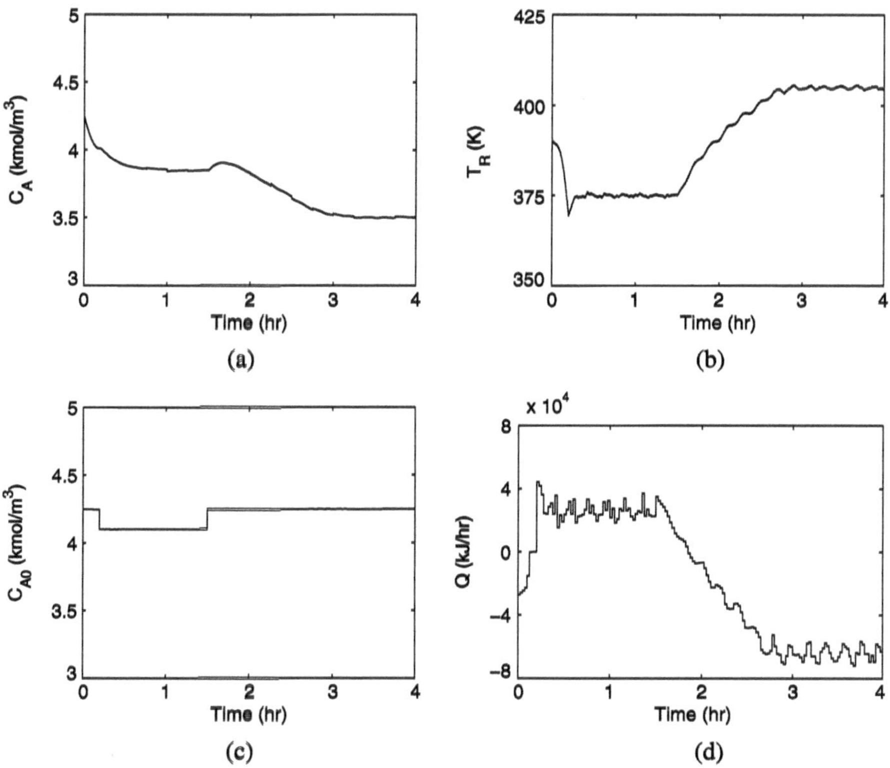

Fig. 6.10 Closed-loop state (**a**)–(**b**) and input (**c**)–(**d**) profiles for the chemical reactor example with asynchronous concentration measurements. The safe-parking operation starts from 0.2 hr, and nominal operation is resumed at 1.5 hr

$[-3.71 \times 10^4, -1.21 \times 10^4]$ designed for S_6 (3.85 kmol/m^3, 375 K) in Table 6.2. Because D_2 also resides within the stability region of S_6, denoted by $\Omega_{s,6}$, S_6 is chosen as a safe-park point. As shown in Fig. 6.8, the process operates at S_6 during fault repair until nominal operation is resumed at $t_r = 1.5$ hr. The corresponding state and input profiles are depicted in Fig. 6.10.

6.6 Conclusions

In this chapter, we considered the problem of designing an integrated fault diagnosis and fault-handling framework to deal with actuator faults in nonlinear systems. A model-based fault diagnosis design was first proposed, which cannot only identify the failed actuator, but also estimate the fault magnitude. The fault information is obtained by estimating the outputs of the actuators and comparing them with the corresponding prescribed control inputs. This methodology was developed under state feedback control and generalized to deal with state estimation errors. Then, the safe-parking framework developed previously (to handle the case where the failed actuator reverts to a known fixed value) for fault-tolerant control was extended to handle the case where an actuator seizes at an arbitrary value. The estimate of the failed actuator position provided by the fault diagnosis design is used to choose a safe-park point, at which the system operates temporarily during fault repair, from those generated off-line for a series of design values of the failed actuator position. The discrepancy between the actual value of the failed actuator position and the corresponding design value is handled through the robustness of the control design. The efficacy of the integrated fault diagnosis and safe-parking framework was demonstrated through a chemical reactor example.

Chapter 7
Utilizing FDI Insights in Controller Design and PID Monitoring

7.1 Introduction

In the previous chapters, the key concepts for FDI were developed and focused on the problem of actuator fault-detection and isolation under a well-performing controller. The FDI schemes were designed independently from the feedback control law and were then applied on top of the closed-loop system operating under a feedback control law that was previously designed without consideration of the possible faults that might occur. The key ideas of fault-detection and isolation can, however, be utilized to design a holistic control structure where the controller is designed to aid the FDI problem and the FDI ideas are utilized to monitor controller performance. In this chapter, we present results along both the above themes.

Figure 7.1(a) shows an independently designed feedback control law and an FDI scheme which are combined only in the final closed-loop system. The paradigm shift proposed in this chapter is illustrated in Fig. 7.1(b) which demonstrates the idea of designing both the feedback control law and the FDI scheme with each other in mind. With the controller design taking into account the FDI scheme, faults may be more easily isolated in the resulting closed-loop system.

The above considerations motivate the development of a data-based method of fault detection and isolation that utilizes the design of the controller to enhance the isolability of the faults in the closed-loop system. Specifically, it is demonstrated in the first part of this chapter that a data-based FDI scheme is able to isolate a given set of faults if the nonlinear closed-loop system satisfies certain isolability conditions in the presence of common-cause process variation. We explicitly characterize this set of isolability conditions and show that it is possible, under certain conditions on the system structure, to design a feedback control law that guarantees that the closed-loop system satisfies the isolability conditions and that the origin of the closed-loop system is asymptotically stable. This is achieved through the use of appropriate nonlinear control laws that effectively decouple the dependency between certain process state variables. The controller enforces a specific structure on the system that makes fault detection and isolation possible without prior knowledge of system behavior under faulty operation. The theoretical results are applied to a CSTR example and

P. Mhaskar et al., *Fault-Tolerant Process Control*, DOI 10.1007/978-1-4471-4808-1_7,

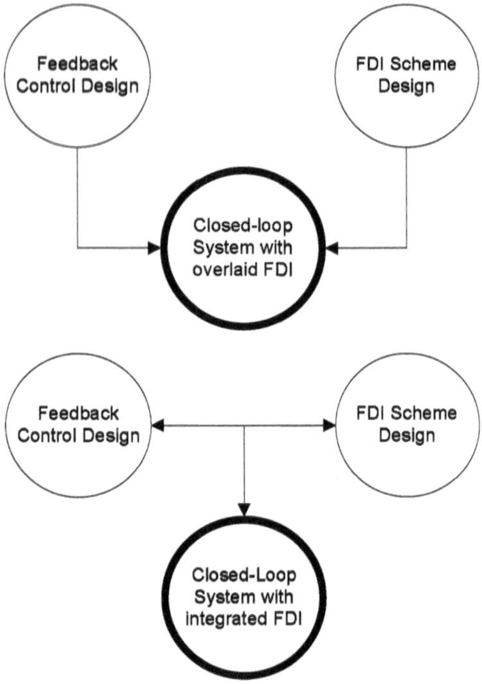

Fig. 7.1 (**a**) (*top*) Common methods of fault diagnosis apply the FDI scheme and feedback control law to the closed-loop system independently from each other. (**b**) (*bottom*) This work proposes integrating the feedback control law design with the FDI scheme in the closed-loop system

to a polyethylene reactor example. It should also be noted that although the examples given in this chapter are presented using a specific method for data-based fault diagnosis, the closed-loop system structure enforced by the presented approach can also be exploited to achieve fault isolation using other data-based fault detection methods.

The importance of controller performance monitoring is well recognized. In the second part of the chapter, we focus on the problem of PID loop monitoring that typically involves lower level controllers with MPC at the higher level providing 'optimal' setpoints. In general, in the calculation of the optimal input trajectories for the manipulated inputs via MPC, the dynamics of the corresponding control actuators that will implement the control actions computed by the MPC are neglected and the MPC-computed control actions are assumed to be directly implemented by the control actuators. However, in practice, these control actuators have their own specific dynamics. As a result of this, there are always discrepancies (i.e., time lags, magnitude differences, etc.) between the actual control actions applied to the process by the control actuators and the control actions requested by the MPC. The mitigation of the influence of these discrepancies in closed-loop performance, relies on the performance of the PID controllers [10]. The representation of this added extra layer of the PID controllers around the control actuators is shown in Fig. 7.2. In this case, the tuning of the PID controllers is critical for the overall control actuator and closed-loop system performance. An actuator with a well-tuned PID controller can effectively implement the actions requested by the MPC; whereas, an actuator

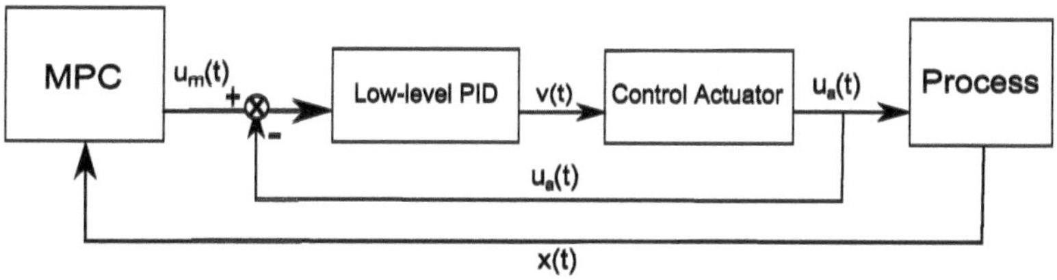

Fig. 7.2 Closed-loop system with MPC as advanced model-based controller and low-level PID controller implemented to regulate the control actuators

with a poorly-tuned PID controller may reduce the performance of the closed-loop system dramatically or may even cause instability of the closed-loop system.

In the second part of the chapter, we show how the concepts of FDI can be utilized towards monitoring the performance of low-level PID loops. With respect to previous works on the subject, there is indeed a plethora of techniques discussed in the literature on monitoring of the performance and tuning of PID controller parameters. With respect to tuning, methods such as Ziegler–Nichols [64, 184], Cohen–Coon [31], internal model control [149, 165], pole placement [168, 175], and others have been widely used to tune PID controller parameters based on either the estimated plant's transfer function or experimentally-obtained step response and/or frequency response curves. Gain scheduling [145, 181] has also been developed to allow PID controllers to be able to self-tune to accommodate changing operating conditions. Multiple works have also been published on automatic retuning of PID parameters based on the current performance of the PID controller and on-line system identification [4, 20, 127, 147, 154, 157, 165, 183]. On the monitoring front, references [52] and [138] provide a survey of available monitoring techniques. Specifically, minimum variance control [66] has been developed as a tool to assess PID performance, while [160, 161], and [159] utilize statistical process control (SPC) to monitor and provide performance criteria to assess the performance of PID controllers. In another work [148], a monitoring scheme was proposed to determine poor tuning/faults using principal component analysis (PCA) and neural networks. One common feature in all of the works in the PID monitoring field mentioned above is the assumption that measurements of the output of the PID-controlled loop are available.

Motivated by the above considerations, in the second part of this chapter, we will address in Sect. 7.3 the problem of real-time monitoring and retuning of low-level PID controllers in the case where the measurement of the actual control action implemented on the process is unavailable. Specifically, we present a method for monitoring the PID performance via a model-based FDI method [112, 115] coupled with real-time process measurements. Using an estimated transfer function model of the control actuators, model-based FDI can be used to detect the discrepancies between the expected actuation level and the actual actuation level performed by the control actuators. Based on the patterns of the residuals, a poorly-tuned actuator can be isolated and retuned accordingly. An example of a nonlinear reactor–separator process under MPC control with low-level PID controllers around the control actuators is used to demonstrate the approach.

7.2 Controller Enhanced FDI

This part of the chapter focuses on a broad class of nonlinear systems subject to actuator faults and disturbances with the following state-space description:

$$\dot{x} = f(x, u, d), \tag{7.1}$$

where $x \in \mathbb{R}^n$ denotes the vector of process state variables, $u \in \mathbb{R}^m$ denotes the vector of manipulated input variables and $d \in \mathbb{R}^p$ denotes the vector of p possible actuator faults or disturbances. Normal operating conditions are defined by $d = 0$. Each component d_k, $k = 1, \dots, p$, of vector d characterizes the occurrence of a given fault. When fault k occurs, variable d_k can take any value. Therefore, the model of Eq. (7.1) can include a broad class of possible faults ranging from actuator faults to complex process disturbances and failures. The system under normal operating conditions and zero input has an equilibrium point at the origin, i.e., $f(0, 0, 0) = 0$.

Before proceeding with the theoretical development, it is important to state that the presented FDI method brings together model-based analysis and controller design techniques for nonlinear, deterministic ordinary differential equation systems and statistical data-based fault-diagnosis techniques that will be applied to the closed-loop system to diagnose faults that affect the process outside of the region determined by the common-cause process variation. To this end, we will first state the isolability conditions for the closed-loop system that need to be enforced by the appropriate control laws on the basis of the nonlinear deterministic system of Eq. (7.1). Subsequently, we will introduce additive autocorrelated noise in the right-hand side of Eq. (7.1) and additive Gaussian noise in the measurements of the vector x to compute the region of operation of the process variable, x, under common-cause variance. Finally, we will demonstrate that the enforcement of an isolable structure in the closed-loop system by an appropriate feedback law allows isolating specific faults whose effect on the closed-loop system leads to sustained process operation outside of the region of common-cause variance.

Under the assumptions of single-fault occurrence and available measurements for all of the process state variables, a data-based fault detection and isolation technique is presented based on the structure of the system in closed-loop with a state feedback controller $u(x)$. The conditions (denoted as isolability conditions) under which this technique can be applied are provided. The main objective is to design a state feedback controller $u(x)$ such that the origin of the system of Eq. (7.1) in closed-loop with this controller is asymptotically stable under normal operating conditions, i.e., $d(t) = 0$, and that the closed-loop system satisfies the isolability conditions needed to apply the presented FDI method. It is shown that for certain systems, the controller can be designed to guarantee that these conditions are satisfied, as well as to stabilize the closed-loop system.

Referring to the assumption that only a single fault occurs at any specific time instance, note that this is a logical assumption from a practical point of view. Namely, it is more likely that a single control actuator (e.g., an automatic valve) will fail at a single time instance during the process operation than it is that two or more control actuators will fail at exactly the same instance of time. Referring to the assumption that measurements of the process state variables are available, note that this assumption is made to simplify the development. In principle, this assumption can be relaxed by using model-based state estimator design techniques for nonlinear systems (e.g., [28]) to construct dynamic systems which yield estimates of the unmeasured states from the output measurements; however, the detailed development of the results for this case is outside the scope of this book. Finally, we focus our attention on general actuator faults and disturbances and do not explicitly consider sensor faults since this issue will be addressed in Chaps. 8 and 9 (see also [43, 105, 106, 143, 169, 170]). Note that with the general way in which the faults d_k are modeled, it is possible to represent virtually any fault because d_k is not restricted in any way and may be any time-varying signal; however, to achieve data-based detection and isolation of the fault d_k in the closed-loop system in the presence of noise in the state equations and measurements (noise which is introduced to model common-cause process variance), $d_k(t)$ should be sufficiently large in a way that is stated precisely in Sect. 7.2.2.

In order to present the FDI method, it is necessary to define the incidence graph of a system and its reduced representation. The following definitions are motivated by standard results in graph theory [65]. This kind of graph-theoretic analysis has been applied before in the context of feedback control of nonlinear systems (see, for example, [35]).

Definition 7.1 The incidence graph of an autonomous system $\dot{x} = f(x)$ with $x \in \mathbb{R}^n$ is a directed graph defined by n nodes, one for each state, x_i, of the system. A directed arc with origin in node x_i and destination in node x_j exists if and only if $\frac{\partial f_j}{\partial x_i} \neq 0$.

The incidence graph of a system shows the dependence of the time derivatives of its states. Figure 7.3 shows the incidence graph of the following system:

$$\begin{aligned} \dot{x}_1 &= -2x_1 + x_2 + d_1, \\ \dot{x}_2 &= -2x_2 + x_1 + d_2, \\ \dot{x}_3 &= -2x_3 + x_1 + d_3 \end{aligned} \tag{7.2}$$

when $d_1 = d_2 = d_3 \equiv 0$. A path from node x_i to node x_j is a sequence of connected arcs that starts at x_i and reaches x_j. A path through more than one arc that starts and ends at the same node is denoted as a loop. States that belong to a loop have mutually dependent dynamics, and any disturbance affecting one of them also affects the trajectories of the others. The mutual dependence of the dynamics of the states that belong to a given loop makes data-based isolation of faults that affect the system a

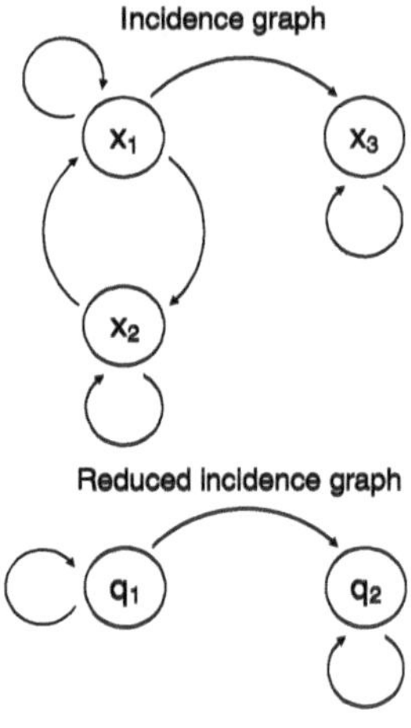

Fig. 7.3 Incidence graph and reduced incidence graph for the system of Eq. (7.2)

difficult task. The following definition introduces the reduced incidence graph of an autonomous system. In this graph, the nodes of the incidence graph belonging to a given loop are united in a single node. This allows identifying which states do not have mutually dependant dynamics.

Definition 7.2 The reduced incidence graph of an autonomous system $\dot{x} = f(x)$ with $x \in \mathbb{R}^n$ is the directed graph of nodes q_i, where $i = 1, \dots, N$, that has the maximum number of nodes, N, and satisfies the following conditions:

- To each node q_i there corresponds a set of states $X_i = \{x_j\}$. These sets of states are a partition of the state vector of the system, i.e.,

$$\bigcup X_i = \{x_1, \dots, x_n\}, \qquad X_i \cap X_j = \emptyset, \quad \forall i \neq j.$$

- A directed arc with origin q_i and destination q_j exists if and only if $\frac{\partial f_l}{\partial x_k} \neq 0$ for some $x_l \in X_i$, $x_k \in X_j$.
- There are no loops in the graph.

In the reduced incidence graph, states that belong to a loop in the incidence graph correspond to a single node. In this way, the states of the system are divided into subsystems that do not have mutually dependent dynamics; that is, there are no loops connecting them. The arcs of the graph indicate if there exists a state corresponding to the origin node that affects a state corresponding to the destination node. Note that the reduced incidence graph can be always obtained, but for strongly coupled systems, it may be defined by a single node, i.e., in the incidence graph there exists a loop that contains all the states of the system. In this case, data-based fault detection and isolation cannot be achieved using the presented method. In the incidence graph of the system of Eq. (7.2) there is a loop that contains states x_1 and x_2. The reduced incidence graph of the system of Eq. (7.2) contains two nodes. Node q_1 corresponds to the states of the loop, that is, $X_1 = \{x_1, x_2\}$. Node q_2 corresponds to $X_2 = x_3$.

Figure 7.3 shows the reduced incidence graph of the system of Eq. (7.2). It can be seen that in the reduced incidence graph there are no loops.

Remark 7.1 In the process model of Eq. (7.1), process and sensor noises are not explicitly taken into account. However, noise is indirectly accounted for in the FDI method below by means of appropriate tolerance thresholds in the decision criteria for fault detection and isolation. The thresholds are generated on the basis of operating data and take into account both sensor and process noise, allowing for an appropriate FDI performance even if the process model and the measurements are corrupted by noise. To demonstrate this point, process and sensor noise are included in the two examples discussed below; see Sect. 7.2.4 for details.

Remark 7.2 Due to the complex nature of faults in nonlinear systems, performing fault isolation with data-based methods alone generally leaves an ambiguous picture. On the other hand, it is possible to perform data-based fault isolation of simple faults using data-based FDI methods (this is discussed and demonstrated in [173] using contribution plots). In some cases, historical data from faulty operation will improve isolation capabilities of data-based methods; however, even with this information, due to overlap in the state-space of the regions corresponding to different faults and incomplete fault libraries, it still may be very difficult to isolate faults in nonlinear process systems.

7.2.1 Data-Based Fault Detection

Data-based methods for fault detection in multivariate systems are well established in statistical process monitoring. This section reviews a standard data-based method of fault detection that will be used in the context of the presented FDI method.

A common approach to monitoring multivariate process performance is based upon the T^2 statistic introduced by Harold Hotelling [70]. This approach allows multivariate processes to be monitored for a shift in the operating mean, $\bar{X}$, using a single test statistic that has a well-defined distribution. The true operating mean can be estimated from past history or chosen based on the known process. Generally, the true process variance is unknown and must be estimated using sampled data. Hotelling's T^2 statistic tests the hypothesis that the current operating mean is the same as $\bar{X}$ with a certain degree of confidence, $\alpha \cdot 100$ %. This is the multivariate generalization of Student's t-distribution. Consider a vector $X \in \mathbb{R}^n$ that is the average of m randomly sampled state measurements. Assuming that X has an n-variate normal distribution with an unknown variance–covariance matrix, Σ, the T^2 statistic can be computed using the operating mean, $\bar{X}$, estimated from historical data, and the estimated covariance matrix, S, estimated from the m measurements contributing to X, as follows:

$$T^2 = m(X - \bar{X})^T S^{-1}(X - \bar{X}). \tag{7.3}$$

Based on the assumption that the measurements in X are normally distributed, the T^2 statistic has the following distribution:

$$T^2 \sim \frac{mn}{(m-n+1)} F(n, m-n+1), \tag{7.4}$$

where $F(n, m-n+1)$ is the F distribution with n and $m-n+1$ degrees of freedom. An upper control limit (UCL) for the T^2 statistic can be calculated by finding the value, T^2_{UCL} on the T^2 distribution for which there is probability α of a greater or equal value occurring, that is, $P(T^2 \geq T^2_{\mathrm{UCL}}) = \alpha$ with

$$T^2_{\mathrm{UCL}} = \frac{mn}{(m-n+1)} F_\alpha(n, m-n+1). \tag{7.5}$$

Note that T^2 is a positive quantity and has no lower control limit. With this definition of the UCL, α is the probability of a Type I error, or false alarm. This implies that at least once every $1/\alpha$ samples there is expected to be a false alarm or, in other words, the average run length (ARL) is equal to $1/\alpha$. Decreasing the value of α will increase the ARL and thus decrease the likelihood of a Type I error. However, this decreases the power of the statistical test. Power is measured as $1-\beta$ where β is the probability of a Type II error, which is that a failure has occurred, but is not detected by the test. Because the focus of this work is on failures that cause significant change in the operating point and assumes a persistent state of failure before declaring a fault, finding the balance between the statistical power of the test and the likelihood of a false alarm is not considered (see Remark 7.6 for further discussion on this issue).

In addition to the method presented above, other methods using Hotelling's T^2 statistic have been established which deviate from the strict definition of the test. In particular, due to the nature of continuous chemical processes, it is sometimes convenient to estimate S from historical data. This assumes that data from future observations will have similar covariance. Methods that use historical data generally have two phases of operation. Phase 1 is for testing during fault-free operation to verify that the process is in control. The following UCL is used for the T^2 statistic in Phase 1 [104]:

$$T^2_{\mathrm{UCL}} = \frac{n(h-1)(m-1)}{hm-h-n+1} F_\alpha(n, hm-h-n+1), \tag{7.6}$$

where h is the number of m-sized samples used to evaluate the covariance matrix S from historical data. Phase 2 is for the normal monitoring of a process for faults with the following control limit:

$$T^2_{\mathrm{UCL}} = \frac{n(h+1)(m-1)}{hm-h-n+1} F_\alpha(n, hm-h-n+1). \tag{7.7}$$

Note that when h is large, these limits are nearly identical. In addition, it is often convenient to use a sample size of $m = 1$ where individual observations are monitored (i.e., [104, 158]). This is commonly used in data-based fault detection and isolation methods (see, for example, [80, 104, 144, 158, 171]). In this scenario, the UCL becomes

$$T^2_{\mathrm{UCL}} = \frac{(h^2-1)n}{h(h-n)} F_\alpha(n, h-n), \tag{7.8}$$

where h is now the total number of historical measurements used to evaluate the covariance matrix S. In the simulation section of this chapter, we use both the traditional method of Hotelling's T^2 statistic by monitoring sampled data sets of size m with the corresponding UCL in Eq. (7.5) where the estimated covariance matrix, S, is evaluated at each step from the m observations, as well as the single observation approach using the control limit from Eq. (7.8) and the appropriate S based on h historical observations.

The T^2 statistic is widely used for fault detection purposes in multivariate processes and can be used for both the full state vector and the transformed state vector in the reduced PCA space. The T^2 statistic for the full state vector does not provide additional information that can be used for isolating the underlying cause of a fault. In some cases, the T^2 statistics of certain subgroups of the state vector (or functions of it) can be monitored in addition to the full vector to assist in fault isolation. In this situation, the process is decomposed into subsystems, generally based on function, structure and/or behavior allowing fault detection and isolation techniques to be applied to subgroups of sensor measurements. The context of the decomposition itself narrows the detection and isolation focus allowing the application of the T^2 statistic for localized detection. As the focus of the process decomposition context narrows, detection approaches isolation. If the focus is narrowed to a particular process component then detection and isolation become one and the same. Examples of work in which decompositions are used for localized FDI are in [135] and [91]. This idea for data-based isolation using the T^2 statistic for each subsystem is also utilized in the context of the presented method in the next section.

Remark 7.3 Note that the methods of fault detection presented in this section will naturally account for process and sensor noise. Thus, the T^2 statistic, which scales the process data by the inverse of the covariance matrix, will be tolerant to the normal amount of process and measurement variation without signalling a fault. However, if the variance of the system were to change during the course of operation, this could signal a fault in the system when using a covariance matrix, S, estimated from historical data. This type of fault will generally not be declared as this work requires a fault large enough to cause persistent failure as discussed in Remark 7.6.

7.2.2 Data-Based Isolation Based on a Fault Signature

Data-based isolation of the underlying cause of a faulty process behavior is, in general, a difficult problem which strongly depends on the structure of the closed-loop system. In systems with multiple possible faults, one-dimensional statistics such as

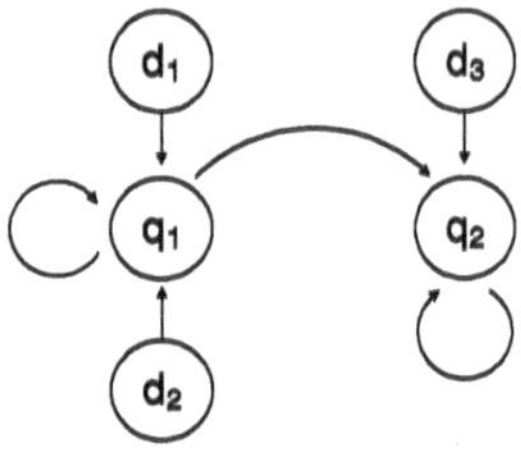

Fig. 7.4 Isolability graph for the system of Eq. (7.2)

the T^2 statistic presented in the previous section cannot be used to perform fault isolation when applied globally. To understand this point in the context of a specific example, consider the system of Eq. (7.2). It can be seen, based upon the structure of the system, that a fault in d_1 or a fault in d_2 will affect the state trajectories of all three states of the system. In this case, the fault will be readily detected, but the T^2 statistic and the state trajectories will not provide further information with which one can reliably determine whether a fault in d_1 or d_2 had occurred. However, if a failure in d_3 were to occur, it can be seen from the system equations that only the state trajectory of state 3 would be affected. With this particular structure, which is that there is no path from the affected state, x_3, to x_1 or x_2, it is possible to isolate the fault d_3 by observing the affected state trajectories at the time of the failure. Thus, it can be seen that under certain conditions, isolation is possible.

The example given above motivates introducing a set of isolability conditions which guarantee that fault isolation is possible based on the state trajectories affected by a given fault. This will also provide guidelines for the design of control laws that guarantee that these conditions are satisfied. In order to precisely state these conditions, the isolability graph of an autonomous system is defined below.

Definition 7.3 The isolability graph of an autonomous system $\dot{x} = f(x, d)$ with $x \in \mathbb{R}^n$, $d \in \mathbb{R}^p$ is a directed graph made of the N nodes of the reduced incidence graph of the system $\dot{x} = f(x, 0)$ and p additional nodes, one for each possible fault d_k. The graph contains all the arcs of the reduced incidence graph of the system $\dot{x} = f(x, 0)$. In addition, a directed arc with origin in fault node d_k and destination to a state node q_j exists if and only if $\frac{\partial f_l}{\partial d_k} \neq 0$ for some $x_l \in X_j$.

Figure 7.4 shows the isolability graph of the system of Eq. (7.2). The isolability graph of an autonomous system subject to p faults shows, in addition to the incidence arcs of the reduced incidence graph, which loops of the system are affected by each possible fault. Based on this graph, it is possible to define the signature of a fault.

Definition 7.4 The signature of a fault d_k of an autonomous system subject to p faults $\dot{x} = f(x, d)$ with $x \in \mathbb{R}^n$, $d \in \mathbb{R}^p$ is a binary vector W^k of dimension N, where N is the number of nodes of the reduced incidence graph of the system. The ith component of W^k, denoted W_i^k, is one if there exists a path in the isolability graph from the node corresponding to fault k to the node q_i corresponding to the set of states X_i, or zero otherwise.

The signature of a fault indicates the set of states that are affected by the fault. If each of the corresponding signatures of the faults is different, then it is possible to isolate the faults using a data-based fault-detection method. Faults d_1 and d_2 in the system of Eq. (7.2) have the same signature, $W^1 = [1\ 1]^T$, because d_1 and d_2 both directly affect q_1 and there is a path from q_1 to q_2. This implies that both faults affect the same states and upon detection of a fault with the signature $W^1 = [1\ 1]^T$, it is not possible to distinguish between them based upon the signature. On the other hand, the signature of fault d_3 in the same system is $W^1 = [0\ 1]^T$ because there is no path to q_1 from q_2, which is the node directly affected by d_3. This implies that the states corresponding to node q_1 are effectively decoupled from fault d_3. This allows distinguishing between a fault in d_3 and a fault in either d_1 or d_2 in the system of Eq. (7.2) based on the profiles of the state trajectories.

In this chapter, we design and implement appropriate feedback laws in the closed-loop system that induce distinct signatures for specific faults to allow their isolation. In the next section, we present methods for the design of controllers that enforce an isolable structure in the closed-loop system. In the remainder of this section, we discuss the issue of determination of the fault signatures for the closed-loop system in the absence and presence of noise in the differential equations and measurements. This determination of the fault signature from process measurements will also lead to a characterization of the type of fault signals, $d_k(t)$, for which isolation can be achieved when common-cause variation is considered for the closed-loop system (caused by the introduction of noise in the differential equations and measurements). Specifically, referring to the deterministic closed-loop system (i.e., no noise is present in the states or in the measurements), the signature of the fault, W^k, for any time-varying signal, $d_k(t)$, can be computed directly from the isolability graph and is independent of the type of time-dependence of $d_k(t)$. In other words, the signal $d_k(t)$ need not satisfy any conditions for its signature to be computed. Once the fault signature is computed, then fault isolation is immediate in the deterministic case by checking whether or not the signature of the system corresponds to a defined fault. However, in the presence of noise in the states and measurements, $d_k(t)$ has to be sufficiently large to have an effect that leads to operation of the process states outside of the range expected due to common-cause variance for a sufficiently large period of time to allow isolation of the fault, based on its signature, from other causes that can lead to violations of the upper control limit for a small period of time. Specifically, in the presented method, the following statistics based on the state trajectories of the system of Eq. (7.1) in closed-loop with a given feedback controller $u(x)$ in the presence of noise in the states and measurements are monitored:

- T^2 statistic based on the full state x with upper control limit T^2_{UCL}.
- T_i^2 statistic with $i = 1, \dots, N$ based on the states $x_j \in X_i$, where X_i are the sets of states corresponding to each one of the nodes of the reduced incidence graph. To each T_i^2 statistic a corresponding upper control limit T^2_{UCLi} is assigned.

The fault detection and isolation procedure then follows the steps given below:

1. A fault is detected if $T^2(t) > T^2_{\text{UCL}}$, $\forall t,\ t_f \le t \le T_P$, where T_P is chosen so that the window $T_P - t_f$ is large enough to allow fault isolation with a desired degree of confidence and depends on the process time constants and potentially on available historical information of the process behavior.
2. A fault that is detected can be isolated if the signature vector of the fault $W(t_f, T_P)$ can be built as follows:

$$T_i^2(t) > T^2_{\text{UCLi}}, \quad \forall t,\ t_f \le t \le T_P \rightarrow W_i(t_f, T_P) = 1;$$

$$T_i^2(t) \not> T^2_{\text{UCLi}}, \quad \forall t,\ t_f \le t \le T_P \rightarrow W_i(t_f, T_P) = 0.$$

   In such a case, fault d_k is detected at time T_P if $W(t_f, T_P) = W^k$. If two or more faults are defined by the same signature, isolation between them is not possible on the basis of the fault signature obtained from the isolability graph.

The conditions in steps 1 and 2 above state that the fault $d_k(t)$ has to be sufficiently large in order to be detected and isolated.

Remark 7.4 States for which there is no path from a given fault node to the corresponding subsystem node in the isolability graph are not affected by changes in the value of d_k; thus, they are effectively decoupled from the fault d_k. The FDI method can be applied if the signatures of the closed-loop system faults are different. This is the isolability condition. Note that the signature of a fault depends on the structure of the closed-loop system, in particular, on the isolability graph. For example, if the reduced incidence graph has only one node, isolation is not possible. In the following section, we propose to design the feedback controller $u(x)$ to guarantee that the reduced incidence graph of the closed-loop system has more than one node, that there exist faults with different signatures, and that the origin of the closed-loop system is asymptotically stable.

Remark 7.5 The concept of the "signature of a fault" employed in this section can be generalized in the context of monitoring the evolution of a set of variables defined as functions of the state. In particular, given any variable change, the isolability graph can be obtained in the new state space and the signature defined on the basis of the new state variables. In the next section, an example of this idea is provided for input/output linearizable, nonlinear systems where the signature of a fault is given in a partially linearized state space.

Remark 7.6 The upper control limit is chosen taking into consideration common-cause variance, including process and sensor noise, in order to avoid false alarms. Thus, small disturbances or failures may go undetected if the magnitude and effect of the disturbance is similar to that of the inherent process variance. For this reason, it was stated in the fault detection and isolation procedure that a fault d_k must be "sufficiently large" in order for $T_i^2(t)$ to exceed the threshold T^2_{UCLi}, $\forall t,\ t_f \le t \le T_P$. It is assumed that if a fault d_k is not large enough to cause $T_i^2(t)$ to exceed the threshold T^2_{UCLi}, $\forall t,\ t_f \le$

$t \le T_P$ (where t_f is the time in which $T_i^2(t_f) \ge T_{\text{UCL}}^2$ for the first time) then the fault is not "sufficiently large" and its effect on the closed-loop system, from the point of view of faulty behavior, is not of major consequence. Therefore, such a d_k is not considered to be a fault. However, it should be noted that a fault d_k that is large enough to cause the T^2 derived from the full state vector, x, to cross the upper control limit signaling a fault may not be large enough to signal a fault in all of the affected subgroups. In this case, it is possible to have a false isolation. This is investigated in the simulation case studies section. Finally, the condition $T_i^2(t) \ngtr T_{\text{UCLi}}^2$, $\forall t,\ t_f \le t \le T_P$, allows violation of the UCL in the full state vector and individual subsystems due to other causes for a short period of time. However, such violations do not modify the fault signature $W(t_f, T_P)$ if T_P is chosen to be sufficiently large.

Remark 7.7 We would like to point out that the isolability conditions are not restrictive from a practical point of view. These conditions are not restrictive in the sense that it is generally possible to induce at least some degree of decoupling within any given system. For example, any system with a relative degree $r \le n$ can be decoupled using the method presented in the next section based on feedback linearization. Systems such as this are very common in practice. However, while the isolability conditions can generally be met for one or a few faults in almost any system, it can be difficult to isolate all faults within any given system using this method alone.

7.2.3 Controller Enhanced Isolation

7.2.3.1 Enforcing an Isolable Closed-Loop System Structure Through Controller Design

In general, control laws are designed without taking into account the FDI scheme that will be applied to the closed-loop system. We propose to design an appropriate nonlinear control law to allow isolation of given faults using the method presented in the previous section by effectively decoupling the dependency between certain process state variables to enforce the fault isolability conditions in the closed-loop system. As explained in the previous section, this requires that the structure of the isolability graph of the closed-loop system be such that at least one or more faults be partially decoupled from one or more nodes on the isolability graph. The main idea is to obtain an isolability graph of the closed-loop system which provides a different signature for each fault. The achievement of this key requirement can be accomplished by a variety of nonlinear control laws. In general, providing a systematic procedure to design a controller that guarantees both closed-loop stability and satisfaction of the isolability conditions for any nonlinear process is not possible. The specific form of the controller depends on the structure of the open-loop system and such a controller may not exist. One general procedure that can be followed, however, is to decouple a set of states from the rest. Recursively applying this decoupling technique, appropriate closed-loop isolability graphs can be obtained in certain cases. As an example of this design approach, we first provide a controller that can be applied to nonlinear systems with the following state space description:

$$\begin{aligned} \dot{x}_1 &= f_{11}(x_1) + f_{12}(x_1, x_2) + g_1(x_1, x_2)u + d_1, \\ \dot{x}_2 &= f_2(x_1, x_2) + d_2, \end{aligned} \tag{7.9}$$

where $x_1 \in \mathbb{R}$, $x_2 \in \mathbb{R}^n$, $u \in \mathbb{R}$ and $g_1(x_1, x_2) \neq 0$ for all $x_1 \in \mathbb{R}$, $x_2 \in \mathbb{R}^n$. With a state feedback controller of the form

$$u(x_1, x_2) = -\frac{f_{12}(x_1, x_2) - v(x_1)}{g_1(x_1, x_2)}, \tag{7.10}$$

the closed-loop system takes the form

$$\begin{aligned} \dot{x}_1 &= f_{11}(x_1) + v(x_1) + d_1, \\ \dot{x}_2 &= f_2(x_1, x_2) + d_2, \end{aligned} \tag{7.11}$$

where $v(x_1)$ has to be designed in order to achieve asymptotic stability of the origin of the x_1 subsystem when $d_1 = 0$. Note that explicit stabilizing control laws that provide explicitly-defined regions of attraction for the closed-loop system have been developed using Lyapunov techniques for specific classes of nonlinear systems, particularly input-affine nonlinear systems; please, see Chap. 2 (see also [28, 45, 46, 78]) for results in this area. The origin of the closed-loop system is asymptotically stable if $\dot{x}_2 = f_2(x_1, x_2)$ is input-to-state stable with respect to x_1; please, see Chap. 2 for discussion on ISS. In this case, the presented controller guarantees asymptotic stability of the closed-loop system, as well as different signatures for faults d_1 and d_2. Note that the reduced incidence graph is defined by two nodes corresponding to both states and the signatures are given by $W^1 = [1\ 1]^T$ and $W^2 = [0\ 1]^T$.

The controller design method discussed above provides a basic tool for obtaining control laws that provide closed-loop stability and satisfy the isolability constraints. The main idea is to force decoupling in a first controller design step (in this case $u(x)$) and then ensure closed-loop stability in a second (in this case $v(x)$). Additionally, the next section provides a systematic controller design for a particular class of nonlinear systems. This procedure along with the class of systems under consideration are introduced in the following subsection.

7.2.3.2 Input/Output Linearizable Nonlinear Systems

In this subsection, we focus on a class of process systems modeled by single-input single-output nonlinear systems with multiple possible faults which have the following state-space description

$$\begin{aligned} \dot{x} &= f(x) + g(x)u + \sum_{k=1}^{p} w_k(x)d_k, \\ y &= h(x), \end{aligned} \tag{7.12}$$

where $x \in \mathbb{R}^n$ is the state, $u \in \mathbb{R}$ is the input, $y \in \mathbb{R}$ is the controlled output and $d_k \in \mathbb{R}$ represents a possible fault. It is assumed that f, g, h, and w_k are sufficiently smooth functions, that is, all necessary derivatives exist and are continuous functions of x, and that a set of p possible faults has been identified. Each of these faults is characterized by an unknown input to the system d_k that can model actuator failures and disturbances. As before, this definition of d_k is not restricted by value and may be time-varying, and thus, it can model a very broad class of faults. The system has an equilibrium point at $x = 0$ when $u(t) = 0$, $d_k(t) \equiv 0$ and $h(0) = 0$. Note that in general this equilibrium point may correspond to a given set-point of the output.

The main control objective is to design a feedback control law $u(x)$ such that the origin is an asymptotically stable equilibrium point of the closed-loop system, and moreover, the closed-loop system satisfies the isolability conditions. Feedback linearization is used to accomplish this task. First, it is necessary to review the definition of the relative degree of the output, y, with respect to the input, u, in the system of Eq. (7.12) (see also Sect. 2.5).

Definition 7.5 (Cf. [72]) Referring to the system of Eq. (7.12), the relative degree of the output, y, with respect to the input, u, is the smallest integer, $r \in [1, n]$, for which

$$\begin{aligned} L_g L_f^i h(x) &= 0, \quad i = 0, \ldots, r-2, \\ L_g L_f^{r-1} h(x) &\neq 0. \end{aligned}$$

A system with an input relative degree $r \leq n$ is input-output linearizable. If $r = n$ the entire input-state dynamics can be linearized. If $r < n$, the feedback controller can be chosen so that a linear input–output map is obtained from an external input, v, to the output, y, even though the state equations are only partially linearized (see also, [72]). To be specific, if the system of Eq. (7.12) has input relative degree $r < n$, then there exists a coordinate transformation (see [72]) $(\zeta, \eta) = T(x)$ such that the representation of the system of Eq. (7.12) with $d_k = 0$ for all $k = 1, \ldots, p$ (that is, the system without faults), in the (ζ, η) coordinates, takes the form

$$\begin{aligned} \dot{\zeta}_1 &= \zeta_2, \\ &\vdots \\ \dot{\zeta}_{r-1} &= \zeta_r, \\ \dot{\zeta}_r &= L_f^r h(x) + L_g L_f^{r-1} g(x) u, \\ \dot{\eta}_1 &= \Psi_1(\zeta, \eta), \\ &\vdots \\ \dot{\eta}_{n-r} &= \Psi_{n-r}(\zeta, \eta), \end{aligned} \tag{7.13}$$

where $y = \zeta_1$, $x = T^{-1}(\zeta, \eta)$, $\zeta = [\zeta_1, \dots, \zeta_r]^T$ and $\eta = [\eta_1, \dots, \eta_{n-r}]^T$. Choosing $u(x)$ in an appropriate way, the dynamics of ζ can be linearized and controlled properly using linear control theory. The stability of the closed-loop system, however, can only be assured if the inverse dynamics ($\dot{\eta} = \Psi(\zeta, \eta)$) satisfy additional stability assumptions. In particular, the inverse dynamics must be input-to-state stable with respect to ζ. If this is the case, then an appropriate control law can be designed for the input–output subsystem that guarantees stability of the entire closed-loop system. In the following theorem, we review one example of an input–output feedback-linearizing controller. The controller presented, under the assumption of no faults, guarantees asymptotic stability of the closed-loop system.

Theorem 7.1 (Cf. [72]) *Consider the system of Eq. (7.12) with $d_k = 0$ for all $k = 1, \dots, p$ under the feedback law*

$$u(x) = \frac{1}{L_g L_f^{r-1} h(x)} \big[K T_\zeta(x) - L_f^r h(x) \big], \tag{7.14}$$

where $\zeta = T_\zeta(x)$. Assume K is chosen such that the matrix $A + BK$ has all of its eigenvalues in the left-hand side of the complex plane where

$$A = \begin{bmatrix} 0_{r-1} & I_{r-1} \\ 0 & 0_{r-1}^T \end{bmatrix}, \qquad B = \begin{bmatrix} 0_{r-1} \\ 1 \end{bmatrix}.$$

I_{r-1} is the $(r-1) \times (r-1)$ identity matrix and 0_{r-1} is the $(r-1) \times 1$ zero vector. Then, if the dynamic system $\dot{\eta} = \Psi(\zeta, \eta)$ is locally input-to-state stable (ISS) with respect to ζ, the origin of the closed-loop system is locally asymptotically stable.

We prove that under certain assumptions, if the state-feedback law given in Eq. (7.14) is used, then the faults of system of Eq. (7.12) can be isolated into two different groups: those that affect the output and those that do not affect the output. The main idea is that the isolability graph of the closed-loop system in the coordinates (ζ, η) provides different signatures for the faults depending on their relative degree, which is defined below (this definition was introduced in [34] in the context of feedforward/feedback control of nonlinear systems with disturbances, but it is employed here to address a completely different issue).

Definition 7.6 (Cf. [34]) Referring to the system of Eq. (7.12), the relative degree, $\rho_k \in [1, n]$, of the output, y, with respect to the fault d_k is the smallest integer for which

$$\begin{aligned} &L_{w_k} L_f^i h(x) = 0, \quad i = 0, \dots, \rho_k - 2, \\ &L_{w_k} L_f^{\rho_k - 1} h(x) \neq 0. \end{aligned} \tag{7.15}$$

The definition of the relative degree of a fault is analogous to that of the relative degree of the input, but instead of relating the output to the input, this definition of relative degree relates the output to a particular fault. If a feedback-linearizing controller is used, then the faults can be divided into two different groups: those with a relative degree ρ_k that is greater than the relative degree r and those with a relative degree ρ_k that is less than or equal to r. When a fault occurs, the faults of the first group will not affect the output, y, while those of the latter will.

To show this point, taking into account Definitions 7.5 and 7.6, there exists (see [72]) a coordinate transformation $(\zeta, \eta) = T(x)$ such that the representation of the system of Eq. (7.12) with $d_j = 0$ for all $d_j \neq d_k$ (that is, the system subject only to fault d_k), in the (ζ, η) coordinates, takes the form

$$\begin{aligned}
\dot{\zeta}_1 &= \zeta_2, \\
&\vdots \\
\dot{\zeta}_{r-1} &= \zeta_r, \\
\dot{\zeta}_r &= L_f^r h(x) + L_g L_f^{r-1} h(x) u, \\
\dot{\eta}_1 &= \Psi_1(\zeta, \eta, d_k), \\
&\vdots \\
\dot{\eta}_{n-r} &= \Psi_{n-r}(\zeta, \eta, d_k),
\end{aligned}$$

where $y = \zeta_1$, $x = T^{-1}(\zeta, \eta)$, $\zeta = [\zeta_1, \dots, \zeta_r]^T$ and $\eta = [\eta_1, \dots, \eta_{n-r}]^T$. Following the definition of the state-feedback law of Eq. (7.14), the following state-space representation is obtained for ζ:

$$\dot{\zeta} = (A + BK)\zeta.$$

This dynamical system is independent of d_k. Therefore, the trajectory of the output y is independent of the fault d_k. This result, however, does not hold if the relative degree ρ_k of the fault d_k is equal to or smaller than r. In this case, the coordinate change does not eliminate the dependence of the output on the fault d_k. Applying the same coordinate change $(\zeta, \eta) = T(x)$, the dynamics of the system of Eq. (7.12) with $d_j = 0$ for all $d_j \neq d_k$ (that is, the system subject to fault d_k), in the (ζ, η) coordinates, takes the form

$$\begin{aligned}
\dot{\zeta}_1 &= \zeta_2 + \Phi_1(d_k), \\
&\vdots \\
\dot{\zeta}_{r-1} &= \zeta_r + \Phi_{r-1}(d_k), \\
\dot{\zeta}_r &= L_f^r h(x) + L_g L_f^{r-1} h(x) u + \Phi_r(d_k), \\
\dot{\eta}_1 &= \Psi_1(\zeta, \eta, d_k), \\
&\vdots
\end{aligned}$$

$$\dot{\eta}_{n-r} = \Psi_{n-r}(\zeta, \eta, d_k),$$

where $y = \zeta_1$, $x = T^{-1}(\zeta, \eta)$, $\zeta = [\zeta_1, \dots, \zeta_r]^T$ and $\eta = [\eta_1, \dots, \eta_{n-r}]^T$. In this case, when the fault occurs, the output is affected. In summary, if controller of Eq. (7.14) is used, the possible faults of the system of Eq. (7.12) are divided into two groups, each with a different signature. When a fault occurs, taking into account whether the trajectory of the output is affected or not, one can determine which group the fault belongs to. Note that if only two faults are defined and $\rho_1 \le r$ and $\rho_2 > r$, then the fault is automatically isolated.

Remark 7.8 The feedback linearizing control laws presented in this subsection are designed to enforce a linear input/output structure in the closed-loop system. Although the external input, $v = K\zeta$, may be designed to stabilize the resulting linear closed-loop system optimally, the total control action u is not optimal with respect to a closed-loop performance index (cost) that includes a penalty on the control action.

7.2.4 Simulation Case Studies

In this section, the presented approach for integrated FDI and controller design is applied to two chemical process examples. First, we consider a CSTR example and utilize feedback linearization to design a nonlinear controller that yields a closed-loop system for which the isolability conditions hold. Second, we consider a polyethylene reactor example and design a nonlinear control law, based on the general method of Sect. 7.2.3.1, that yields a closed-loop system for which the isolability conditions hold. In both cases, we demonstrate that data-based fault detection and isolation is achieved under feedback control laws that enforce isolability in the closed-loop system, an outcome that is not possible, in general, when other feedback control designs that do not enforce the required structure are used.

7.2.4.1 Application to a CSTR Example

The first example considered is a well-mixed CSTR in which a feed component A is converted to an intermediate species B and finally to the desired product C, according to the reaction scheme

$$\mathrm{A} \overset{1}{\rightleftharpoons} \mathrm{B} \overset{2}{\rightleftharpoons} \mathrm{C}.$$

Both steps are elementary, reversible reactions and are governed by the following Arrhenius relationships

$$r_1 = k_{10} e^{\frac{-E_1}{RT}} C_A, \qquad r_{-1} = k_{-10} e^{\frac{-E_{-1}}{RT}} C_B,$$

$$r_2 = k_{20} e^{\frac{-E_2}{RT}} C_B, \qquad r_{-2} = k_{-20} e^{\frac{-E_{-2}}{RT}} C_C,$$

where k_{i0} is the pre-exponential factor and E_i is the activation energy of the ith reaction where the subscripts $1, -1, 2, -2$ refer to the forward and reverse reactions of steps 1 and 2. R is the gas constant while C_A, C_B, and C_C are the molar concentrations of species A, B, and C, respectively. The feed to the reactor consists of pure A at flow rate F, concentration C_{A0} and temperature T_0. The state variables of the system include the concentrations of the three main components C_A, C_B, and C_C as well as the temperature of the reactor, T. Using first principles and standard modeling assumptions, the following mathematical model of the process is obtained

$$\begin{aligned}
\dot{C}_A &= \frac{F}{V}(C_{A0} - C_A) - r_1 + r_{-1} + d_1, \\
\dot{C}_B &= -\frac{F}{V}C_B + r_1 - r_{-1} - r_2 + r_{-2}, \\
\dot{C}_C &= -\frac{F}{V}C_C + r_2 - r_{-2}, \\
\dot{T} &= \frac{F}{V}(T_0 - T) + \frac{(-\Delta H_1)}{\rho c_p}(r_1 - r_{-1}) + \frac{(-\Delta H_2)}{\rho c_p}(r_2 - r_{-2}) + u + d_2,
\end{aligned} \tag{7.16}$$

where V is the reactor volume, ΔH_1 and ΔH_2 are the enthalpies of the first and second reactions, respectively, ρ is the fluid density, c_p is the fluid heat capacity, d_1 and d_2 denote faults/disturbances and $u = Q/\rho c_p$ is the manipulated input, where Q is the heat input to the system. The values of the parameters of the process model of Eq. (7.16) are given in Table 7.1.

The system of Eq. (7.16) is modeled with sensor measurement noise and autoregressive process noise. The sensor measurement noise was generated using a zero-mean normal distribution with standard deviation σ_M applied to the measurements of all the process states. The autoregressive process noise was generated discretely as $w_k = \phi w_{k-1} + \xi_k$, where $k = 0, 1, \dots,$ is the discrete time step, ϕ is the autoregressive coefficient, and ξ_k is obtained at each sampling step using a zero-mean normal distribution with standard deviation σ_p. Table 7.2 provides the values of the noise parameters for each state of the system of Eq. (7.16). Because of the dynamic nature of the process and the autocorrelated process noise, it is expected that the state trajectories will be serially correlated. Although the distribution of the state measurements in open-loop operation may not be normal (Gaussian), the influence of feedback control is such that the measurements under closed-loop operation are approximately normal (see also [104]). Figure 7.5 shows the distribution of the state measurements of the closed-loop system of Eq. (7.16) under the feedback-linearizing control law in fault-free operation over a long period of time compared with a Gaussian distribution. Note that although the long-term distribution is approximated well by a normal distribution, this will not hold true for short-term operation, a point that will affect the choice of test statistic to be applied. The controlled output, y, of the system is defined as the concentration of the desired product C_C. This particular definition of the output, while meaningful from the point of view of regulating the desired product concentration, will be also useful in the context

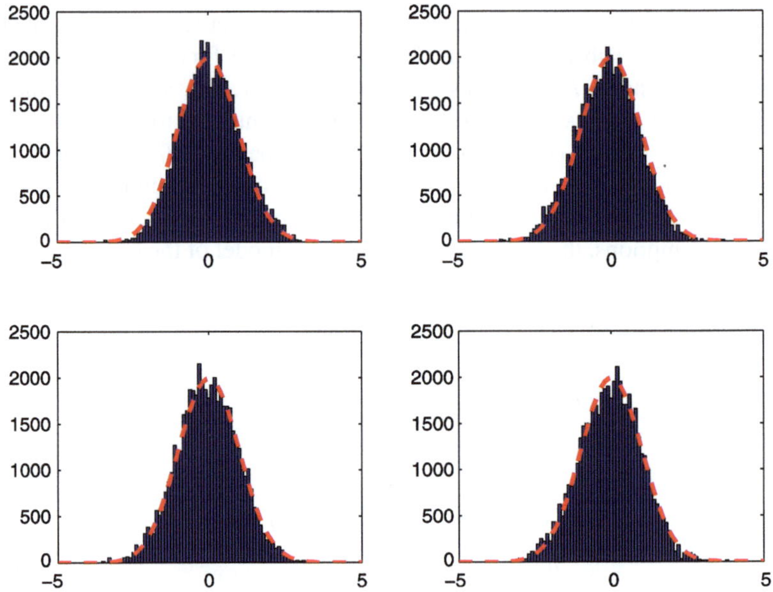

Fig. 7.5 CSTR example. Distribution of normalized, fault-free operating data compared with a normal distribution of the same mean and variance

Table 7.1 CSTR example process parameters

F	$1\ [\mathrm{m^3/h}]$	V	$1\ [\mathrm{m^3}]$
k_{10}	$1.0 \cdot 10^{10}\ [\mathrm{min^{-1}}]$	E_1	$6.0 \cdot 10^4$ [kJ/kmol]
k_{-10}	$1.0 \cdot 10^{10}\ [\mathrm{min^{-1}}]$	E_{-1}	$7.0 \cdot 10^4$ [kJ/kmol]
k_{20}	$1.0 \cdot 10^{10}\ [\mathrm{min^{-1}}]$	E_2	$6.0 \cdot 10^4$ [kJ/kmol]
k_{-20}	$1.0 \cdot 10^{10}\ [\mathrm{min^{-1}}]$	E_{-2}	$6.5 \cdot 10^4$ [kJ/kmol]
ΔH_1	$-1.0 \cdot 10^4$ [kJ/kmol]	R	8.314 [kJ/kmol K]
ΔH_2	$-0.5 \cdot 10^4$ [kJ/kmol]	T_0	300 [K]
C_{A0}	4 [kmol/m^3]	ρ	1000 [kg/m^3]
c_p	0.231 [kJ/kg K]		

of fault isolation. We consider only faults d_1 and d_2, which represent undesired changes in C_{A0} (disturbance) and T_0/Q (disturbance/actuator fault), respectively. For example, if C_{A0} changes by ΔC_{A0} then $d_1 = \frac{F}{V}\Delta C_{\mathrm{A0}}$. These changes may be the consequence of an error in external control loops. In this system, the input u appears in the temperature dynamics and is of relative degree 2 with respect to the output, $y = C_{\mathrm{C}}$. The fault d_1 appears only in the dynamics of C_{A} and is of relative degree 3 with respect to the output, $y = C_{\mathrm{C}}$. Finally, fault d_2 is of relative degree 2.

The control objective is to regulate the system at the equilibrium point

$$C_{\mathrm{C}s} = 0.9471\ \mathrm{kmol/m^3}, \qquad T_s = 312.6\ \mathrm{K}, \qquad u_s = 0\ \mathrm{K/s}, \tag{7.17}$$

Table 7.2 CSTR example noise parameters

	σ_m	σ_p	ϕ
C_A	1E–2	1E–2	0.9
C_B	1E–2	1E–2	0.9
C_C	1E–2	1E–2	0.9
T	1E–1	1E–1	0.9

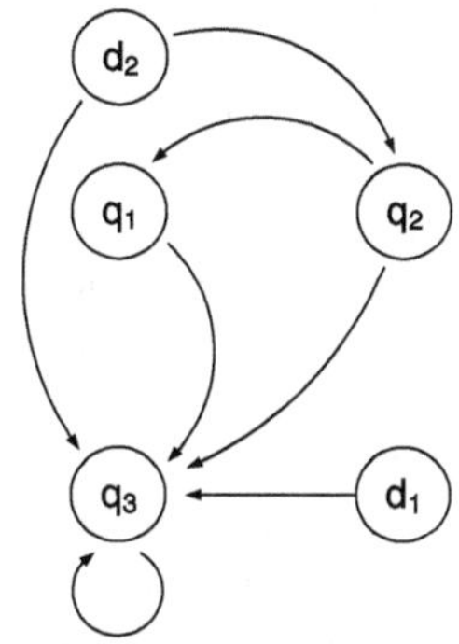

Fig. 7.6 Isolability graph for the system of Eq. (7.16). $v_1 = \{\zeta_1\}$, $v_2 = \{\zeta_2\}$, and $v_3 = \{\eta\}$

where the subscript s refers to the steady state value at equilibrium. To this end, we consider two different feedback controllers: a controller based on feedback linearization and a proportional controller (it is important to point out that the conclusions of this simulation study would continue to hold if the proportional controller is replaced by proportional-integral-derivative control, model-predictive control, or any other controller that does not achieve decoupling of the controlled output, $y = C_C$, from the fault, d_1, in the closed-loop system). The feedback-linearizing controller takes the form of Eq. (7.14) with

$$K = [-1 \quad -1].$$

Note that the state variables are shifted so that the origin represents the desired set point. The proportional controller takes the form

$$u = (T_s - T).$$

In the closed-loop system operating under the feedback-linearizing control law, according to the results of previous section, faults with a relative degree higher than that of the input (i.e., $\rho_k > 2$) will not affect the output in the event of a failure. Therefore, because d_1 has a relative degree of 3, it will not affect the behavior of the output. Conversely, because fault d_2 is of relative degree 2, its effect cannot be decoupled from the output. This result is illustrated in Fig. 7.6. The nodes in this figure are $q_1 = \zeta_1$, $q_2 = \zeta_2$, and $q_3 = \{\eta_1, \eta_2\}$, where $\zeta_1 = C_C$, $\zeta_2 = \dot{\zeta}_1$, and $\{\eta_1, \eta_2\}$ are combinations of C_A, C_B, and T such that $[\zeta; \eta] = T(C_A, C_B, C_C, T)$ is an invertible transformation. The isolability graph of this system in the transformed coordinates shows that each of the states in the ζ subsystem is a separate node and that the states in the η subsystem form a single additional node. Although there are multiple nodes in the ζ subsystem, because each is directly affected by d_1, the effect is the same as if they were a single node. Moreover, since there is no path from the η subsystem node to any of the ζ subsystem nodes and d_2 only affects the η subsystem node directly, the signatures for faults d_1 and d_2 will be unique and thus isolable. Additionally, it should be noted that the trajectory of ζ_1 follows that of the output, C_C, and the ζ subsystem is not affected by the other states. Thus, monitoring the output, C_C, as one subsystem and the remaining states as a second subsystem is equivalent to monitoring the subsystems formed in the transformed space.

The isolability property stated above, however, does not hold for the closed-loop system under proportional control. In that case, when a fault occurs (whether it be d_1 or d_2), the output is affected by the presence of the fault. These theoretical predictions were tested by simulating the system of Eq. (7.16) in closed-loop under both proportional control and feedback-linearizing control. In both cases, the system was initially operating at the steady-state of Eq. (7.17) with a failure appearing at time $t = 0.5$ hr.

Based upon the structure of the closed-loop system under feedback-linearizing control, the state vector was divided into two subvectors, $X_1 = \{C_C\}$ and $X_2 = \{C_A, C_B, T\}$ as discussed above. Hotelling's statistic (Eq. (7.3)) for the full state vector (T^2) and each of the subvectors (T_1^2 and T_2^2) were monitored to detect and evaluate the presence of a fault. Detection was performed based on the T^2 statistic violating the upper control limit T^2_{UCL} defined in Eq. (7.5) using $m = 10$ randomly sampled measurements at intervals of $\Delta t = -\ln(\xi)/W_s$ where ξ is a uniformly distributed random variable from 0 to 1 and W_s is the sample rate of 1 sample per minute. Similarly, isolation was done based on the detection of a violation of the UCL in T_1^2 and T_2^2 and the known fault signatures computed from the isolability graph, $W_1 = [0\ 1]$ and $W_2 = [1\ 1]$. Additionally, the same data was tested with a sample size $m = 1$ and the upper control limits as defined in Eq. (7.8). In this case a much higher sampling rate was used (20 samples per minute) because there was no need to capture a larger time scale (see Remark 7.9). As described in the section on data-based fault detection, the method of single observations relies on the covariance matrix S calculated from historical data under common-cause variation only and the method of $m = 10$ observations uses a covariance matrix S obtained from the new observations being analyzed in each sample.

The closed-loop system was simulated under proportional and feedback-linearizing control. Noise in the states and measurements was included as discussed above. A fault in d_1 was introduced as a step change of magnitude 1 $\mathrm{kmol/m^3\ s}$. Figure 7.7 shows the state trajectories for the closed-loop system under the proportional and the feedback-linearizing controller. Figure 7.8 shows the T^2 statistics for the system under feedback-linearizing control, calculated from $m = 10$ randomly sampled state measurements using the T^2_{UCL} from Eq. (7.5) with confidence level $\alpha = 0.001$ and degrees of freedom $(3, 8)$ for T_1^2, $(1, 10)$ for T_2^2, and $(4, 7)$ for T^2. Also, the data is prone to greater false alarms, because over the short window of 10 observations the trajectories are much more serially correlated and can be susceptible to

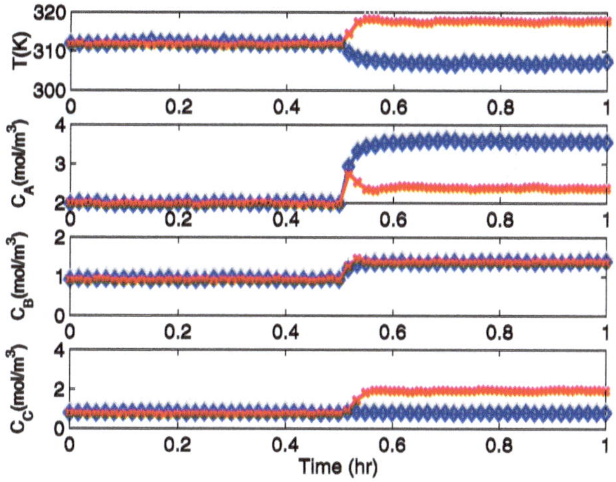

Fig. 7.7 CSTR example. State trajectories of the closed-loop system under feedback-linearizing (⋄) and P (×) control with a fault d_1 at $t = 0.5$ hr

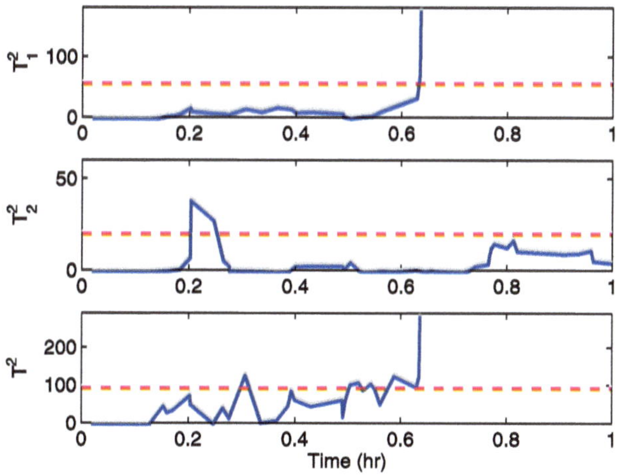

Fig. 7.8 CSTR example. Closed-loop system under feedback-linearizing control with sample size $m = 10$. Statistics T^2, T_1^2, and T_2^2 (*solid*) with T_{UCL} (*dashed*) with a failure in d_1 at $t = 0.5$ hr

almost singular covariance matrices, leading to large T^2 values for small deviations from the mean. Figure 7.9 shows the T^2 statistic for the same results, calculated instead from individual observations ($m = 1$) using the UCL from Eq. (7.8) with confidence level $\alpha = 0.01$ and degrees of freedom (3,2997), (1,2999), and (4,2996) for T_1^2, T_2^2, and T^2, respectively. Observe that the moving average of $m = 10$ observations causes a delay in the fault detection time compared to the case where $m = 1$.

In both methods, the T^2 statistic exceeds the upper control limit T^2_{UCL}, signaling a failure, around $t = 0.5$ hr. The T_1^2 value remained below its threshold while the T_2^2 value exceeded $T^2_{\mathrm{UCL}2}$. This shows that the output (subvector 1) was not affected by the failure. In the case of proportional control with a failure in d_1, the T^2 statistic accurately shows that the failure occurred around time $t = 0.5$ hr. Figures 7.10 and 7.11 show the results $m = 10$ and $m = 1$, respectively. However, in

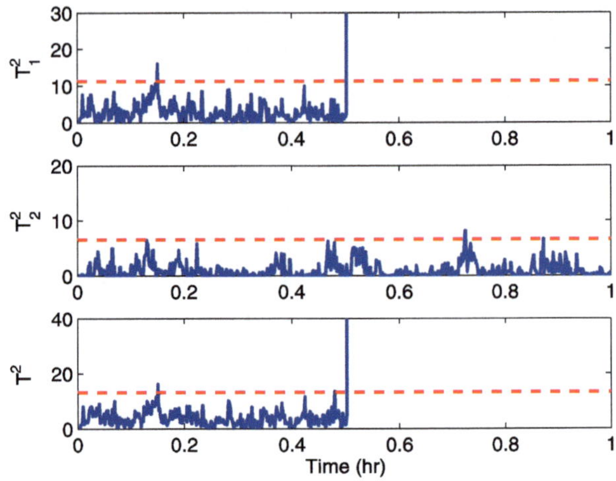

Fig. 7.9 CSTR example. Closed-loop system under feedback-linearizing control with sample size $m = 1$. Statistics T^2, T_1^2, and T_2^2 (*solid*) with T_{UCL} (*dashed*) with a failure in d_1 at $t = 0.5$ hr

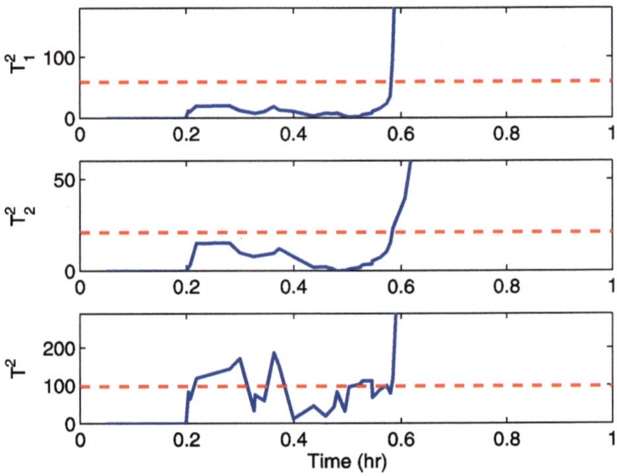

Fig. 7.10 CSTR example. Closed-loop system under proportional control with sample size $m = 10$. Statistics T^2, T_1^2, and T_2^2 (*solid*) with T_{UCL} (*dashed*) with a failure in d_1 at $t = 0.5$ hr

this simulation, all of the state trajectories were affected by the failure resulting in values of T_1^2 and T_2^2 that exceeded the upper control limits. In the case of a failure in d_2, introduced as a step change of magnitude 1 K/s both proportional control and feedback-linearizing control show failures in T^2 at $t = 0.5$ hr as well as in both subsystems T_1^2 and T_2^2 see Fig. 7.12 and Fig. 7.13. Looking at T_1^2 and T_2^2 in Fig. 7.9 and Fig. 7.12, it is clear that fault d_1 did not affect the output whereas d_2 did. In this situation, where only one fault in each group is considered, it is possible to successfully identify the failure in Fig. 7.9 as d_1. However, for proportional control, all of the states were affected by each failure (see Fig. 7.11 and Fig. 7.13) leaving an unclear picture as to the cause of the fault.

A Monte Carlo simulation study was performed by randomly varying the fault sizes and the amount of variance in the process and measurement noise in order to verify that the method performs as expected in a broad range circumstances. In

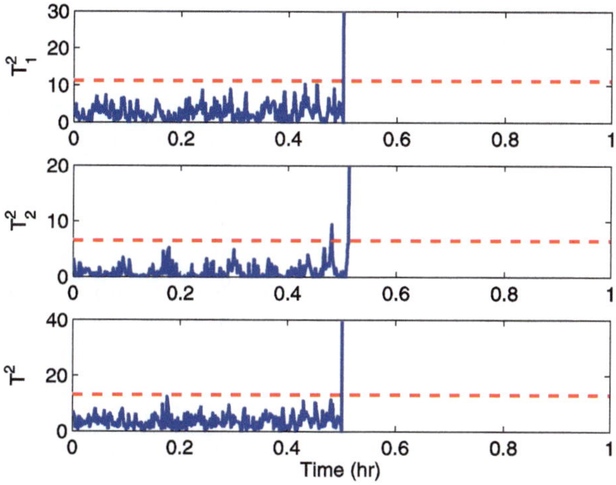

Fig. 7.11 CSTR example. Closed-loop system under proportional control with sample size $m = 1$. Statistics T^2, T_1^2, and T_2^2 (*solid*) with T_{UCL} (*dashed*) with a failure in d_1 at $t = 0.5$ hr

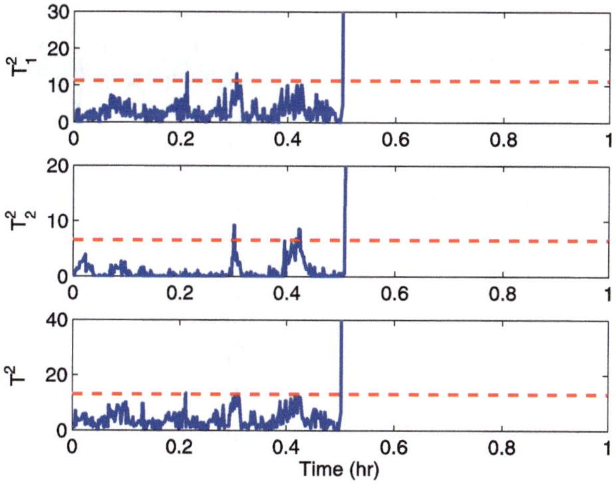

Fig. 7.12 CSTR example. Closed-loop system under feedback-linearizing control with sample size $m = 1$. Statistics T^2, T_1^2, and T_2^2 (*solid*) with T_{UCL} (*dashed*) with a failure in d_2 at $t = 0.5$ hr

total, 500 simulations were run, each with uniformly distributed random values of fault size, process noise variance, and sensor noise variance. Only a fault in d_1 was considered with values ranging from 0 to 3 kmol/m^3 s. The standard deviation of the process noise σ_p and the sensor noise σ_m ranged from 0 to twice the values reported in Table 7.2. A single observation T^2 statistic was used with the associated UCL. The results of these simulations were that from 500 runs, faults were detected when $d_1 > 0.21$ with an average initial detection time of 30.7 min. Out of the 500 runs, a single run was detected by the T^2 statistic, but showed no failure in either T_1^2 or T_2^2.

Finally, to follow-up on the point of Remark 7.8, while the feedback-linearizing controller is not an optimal controller, Fig. 7.14 shows that the control action requested by the feedback-linearizing controller is not excessive and is comparable to that of the control action requested by the proportional controller.

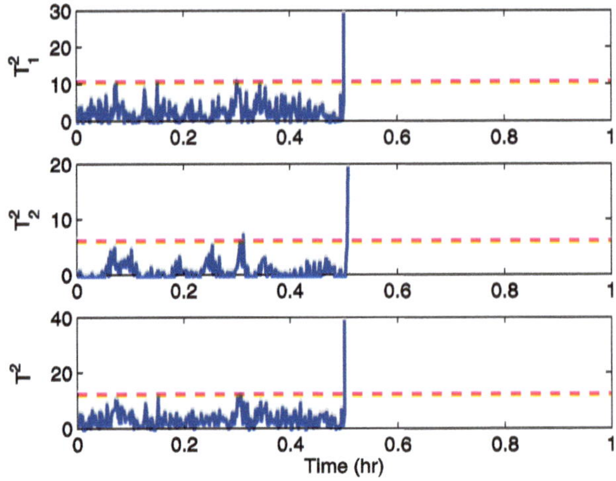

Fig. 7.13 CSTR example. Closed-loop system under proportional control with sample size $m = 1$. Statistics T^2, T_1^2, and T_2^2 (*solid*) with T_{UCL} (*dashed*) with a failure in d_2 at $t = 0.5$ hr

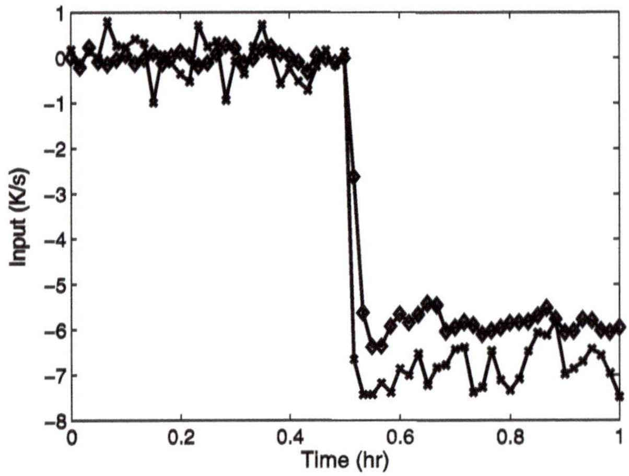

Fig. 7.14 CSTR example. Manipulated input profiles for both the proportional controller (◇) and the feedback-linearizing controller (×) with a failure in d_1 at time $t = 0.5$ hr

Remark 7.9 The simulation results showed that the traditional setting for Hotelling's T^2 statistic which calls for using m randomly sampled observations and a covariance matrix based upon the sampled data was less accurate than the method of individual observations. This is due to the fact that the data is not normally distributed on a short timescale. A small number of observations in a sample can lead to an almost singular S, while on the other hand, the predicted distribution for a large number of observations per sample becomes increasingly narrow which reveals the fact that the data over a short period are in fact serially correlated. While this could be remedied by using a larger sample timescale, this may become inappropriate due to the need to quickly identify faults. However, the single observation method is a reasonable approach because the individual observations hold to the normal distribution over a long period of time.

Table 7.3 Polyethylene reactor example process variables

a_c	active site concentration of catalyst
b_t	overhead gas bleed
B_w	mass of polymer in the fluidized bed
C_{pm1}	specific heat capacity of ethylene
C_v	vent flow coefficient
C_{pw}, C_{pIn}, C_{ppol}	specific heat capacity of water, inert gas, and polymer
E_a	activation energy
F_c, F_g	flow rate of catalyst and recycle gas
F_{In}, F_{M_1}, F_w	flow rate of inert, ethylene, and cooling water
H_f, H_{g0}	enthalpy of fresh feed stream, total gas outflow stream from reactor
H_{g1}	enthalpy of cooled recycle gas stream to reactor
H_{pol}	enthalpy of polymer
H_r	heat liberated by polymerization reaction
H_{reac}	heat of reaction
$[In]$	molar concentration of inerts in the gas phase
k_{d_1}, k_{d_2}	deactivation rate constant for catalyst site 1, 2
k_{p0}	pre-exponential factor for polymer propagation rate
$[M_1]$	molar concentration of ethylene in the gas phase
M_g	mass holdup of gas stream in heat exchanger
$M_r C_{pr}$	product of mass and heat capacity of reactor walls
M_w	mass holdup of cooling water in heat exchanger
M_{W_1}	molecular weight of monomer
P_v	pressure downstream of bleed vent
R, RR	ideal gas constant, unit of J/mol K, m^3 atm/mol K
T, T_f, T_{feed}	reactor, reference, feed temperature
T_{g_1}, T_{w_1}	temperature of recycle gas, cooling water stream from exchanger
T_{w_i}	inlet cooling water temperature to heat exchanger
UA	product of heat exchanger coefficient with area
V_g	volume of gas phase in the reactor
V_p	bleed stream valve position
Y_1, Y_2	moles of active site type 1, 2

7.2.4.2 Application to a Polyethylene Reactor

In this subsection, the presented method will be demonstrated using a model of an industrial gas phase polyethylene reactor. The feed to the reactor consists of ethylene, comonomer, hydrogen, inerts, and a catalyst. A recycle stream of unreacted gases flows from the top of the reactor and is cooled by passing through a water-cooled heat exchanger. Cooling rates in the heat exchanger are adjusted by mixing cold and warm water streams while maintaining a constant total cooling water flow rate through the heat exchanger. Mass balances on hydrogen and comonomer have not been considered in this study because hydrogen and comonomer have only mild effects on the reactor dynamics [101]. A mathematical model for this reactor has the following form [33]:

$$
\begin{aligned}
\frac{d[In]}{dt} &= \frac{1}{V_g}\left(F_{\text{In}} - \frac{[In]}{[M_1]+[In]}b_t\right),\\
\frac{d[M_1]}{dt} &= \frac{1}{V_g}\left(F_{M_1} - \frac{[M_1]}{[M_1]+[In]}b_t - R_{M_1}\right),\\
\frac{dY_1}{dt} &= F_c a_c - k_{d_1}Y_1 - \frac{R_{M_1}M_{W_1}Y_1}{B_w} + d_2,\\
\frac{dY_2}{dt} &= F_c a_c - k_{d_2}Y_2 - \frac{R_{M_1}M_{W_1}Y_2}{B_w} + d_2,\\
\frac{dT}{dt} &= \frac{H_f + H_{g1} - H_{g0} - H_r - H_{\text{pol}}}{M_r C_{pr} + B_w C_{\text{ppol}}} + d_1,\\
\frac{dT_{w_1}}{dt} &= \frac{F_w}{M_w}(T_{w_i} - T_{w_1}) - \frac{UA}{M_w C_{pw}}(T_{w_1} - T_{g_1}),\\
\frac{dT_{g_1}}{dt} &= \frac{F_g}{M_g}(T - T_{g_1}) + \frac{UA}{M_g C_{pg}}(T_{w_1} - T_{g_1}) + d_3,
\end{aligned}
\tag{7.18}
$$

where

$$
\begin{aligned}
b_t &= V_p C_v\sqrt{([M_1]+[In])RRT - P_v},\\
R_{M_1} &= [M_1]k_{p0}e^{\frac{-E_a}{R}\left(\frac{1}{T}-\frac{1}{T_f}\right)}(Y_1+Y_2),\\
C_{pg} &= \frac{[M_1]}{[M_1]+[In]}C_{pm1} + \frac{[In]}{[M_1]+[In]}C_{\text{pIn}},\\
H_f &= (F_{M_1}C_{pm1} + F_{\text{In}}C_{\text{pIn}})(T_{\text{feed}} - T_f),\\
H_{g1} &= F_g(T_{g_1} - T_f)C_{pg},\\
H_{g0} &= (F_g + b_t)(T - T_f)C_{pg},\\
H_r &= H_{\text{reac}}M_{W_1}R_{M_1},\\
H_{\text{pol}} &= C_{\text{ppol}}(T - T_f)R_{M_1}M_{W_1}.
\end{aligned}
\tag{7.19}
$$

The definitions for all the variables used in Eqs. (7.18)–(7.19) are given in Table 7.3 and their values can be found in Table 7.4 (see [33, 58]). Under normal operating conditions, the open-loop system behaves in an oscillatory fashion (i.e., the system possesses an open-loop unstable steady-state surrounded by a stable limit cycle). The open-loop unstable steady-state around which the system will be con-

Table 7.4 Polyethylene reactor example parameters and units

V_g	=	500	m^3
V_p	=	0.5	
P_v	=	17	atm
B_w	=	$7 \cdot 10^4$	kg
k_{p0}	=	$85 \cdot 10^{-3}$	m^3/mol s
E_a	=	(9000)(4.1868)	J/mol
C_{pw}	=	$(10^3)(4.1868)$	J/kg K
C_v	=	7.5	mol/atm$^{0.5}$ s
C_{pm1}, C_{pIn}	=	(11)(4.1868), (6.9)(4.1868)	J/mol K
C_{ppol}	=	$(0.85 \cdot 10^3)(4.1868)$	J/kg K
k_{d_1}	=	0.0001	s^{-1}
k_{d_2}	=	0.0001	s^{-1}
M_{W_1}	=	$28.05 \cdot 10^{-3}$	kg/mol
M_w	=	$3.314 \cdot 10^4$	kg
M_g	=	6060.5	mol
$M_r C_{pr}$	=	$(1.4 \cdot 10^7)(4.1868)$	J/K
H_{reac}	=	$(-894 \cdot 10^3)(4.1868)$	J/kg
UA	=	$(1.14 \cdot 10^6)(4.1868)$	J/K s
$F_{\text{In}}, F_{M_1}, F_g$	=	5, 190, 8500	mol/s
F_w	=	$(3.11 \cdot 10^5)(18 \cdot 10^{-3})$	kg/s
F_c^s	=	$\frac{5.8}{3600}$	kg/s
$T_f, T_{\text{feed}}^s, T_{w_i}$	=	360, 293, 289.56	K
RR	=	$8.20575 \cdot 10^{-5}$	m^3 atm/mol K
R	=	8.314	J/mol K
a_c	=	0.548	mol/kg
$u_1^{\max}, u_2^{\max}$	=	$5.78 \cdot 10^{-4}, 3.04 \cdot 10^{-4}$	K/s, mol/s
$[In]_s$	=	439.68	mol/m^3
$[M_1]_s$	=	326.72	mol/m^3
Y_{1_s}, Y_{2_s}	=	3.835, 3.835	mol
$T_s, T_{w_{1_s}}, T_{g_{1_s}}$	=	356.21, 290.37, 294.36	K

trolled is

$$\begin{aligned} &[In]_{ss} = 439.7\ \text{mol/m}^3, && [M_1]_{ss} = 326.7\ \text{mol/m}^3, \\ &Y_{1ss}, Y_{2ss} = 3.835\ \text{mol}, && T_{ss} = 356.2\ \text{K}, \\ &T_{g1ss} = 290.4\ \text{K}, && T_{w1ss} = 294.4\ \text{K}. \end{aligned}$$

Note that with the given parameters, the dynamics of Y_1, Y_2 are identical and will be reported in the results as a single combined state. In this example, we consider three possible faults, d_1, d_2, and d_3 which represent a change in the feed temperature, catalyst deactivation, and a change in the recycle gas flow rate, respectively. The manipulated inputs are the feed temperature, T_{feed}, and the inlet flow rate of ethylene, F_{M_1}. The control objective is to stabilize the system at the open-loop unstable steady state. In addition, in order to apply the presented FDI scheme, the controller must guarantee that the closed-loop system satisfies the isolability conditions. The open-loop system is highly coupled. If the controller does not impose a specific structure, all the states have mutually dependent dynamics (i.e., they consist of one node in the isolability graph as stated in Definition 7.4). In the present work, we propose to design a nonlinear controller to decouple $[In]$, $[M_1]$, and T from (Y_1, Y_2) and from T_{w_1} and T_{g_1}. In this way, the resulting closed-loop system consists of three subsystems (i.e., three nodes in the isolability graph) that do not have mutually dependent dynamics. In addition, the signature of each of the three faults is different, and thus, the fault isolability conditions are satisfied. In order to accomplish this objective, we define the following control laws:

$$\begin{aligned} F_{M_1} &= u_2 V_g + F_{M_{1ss}}, \\ T_{\text{feed}} &= \frac{u_1(M_r C_{pr} + B_W C_{\text{ppol}}) + H_{fss}}{F_{M_1} C_{pm1} + F_{\text{In}} C_{\text{pIn}}} + T_f \end{aligned} \tag{7.20}$$

with

$$\begin{aligned} u_1 &= \frac{H_r - H_{rss} + H_{\text{pol}} - H_{\text{polss}} - H_{g1} + H_{g1ss}}{M_r C_{pr} + B_w C_{\text{ppol}}} + v_1, \\ u_2 &= \frac{R_{M_1} - R_{M_{1ss}}}{V_g} + v_2, \end{aligned} \tag{7.21}$$

where terms with the subscript ss are constants evaluated at the steady state and v_1, v_2 are the external inputs that will allow stabilizing the resulting closed-loop system (see Eq. (7.22)) below. Under the control law of Eq. (7.21), the dynamics of the states, T and $[M_1]$, take the following form in the closed-loop system:

$$\begin{aligned} \frac{d[M_1]}{dt} &= \left(F_{M_1} - \frac{[M_1]}{[M_1]+[In]} b_t - R_{M_{1ss}}\right)\frac{1}{V_g} + v_2, \\ \frac{dT}{dt} &= \frac{H_f + H_{g1ss} - H_{g0} - H_{rss} - H_{\text{polss}}}{M_r C_{pr} + B_w C_{\text{ppol}}} + v_1 + d_1. \end{aligned} \tag{7.22}$$

It can be seen that these states only depend on $[In]$, $[M_1]$, and T. The closed-loop system under the controller of Eq. (7.20) has a reduced incidence graph with three nodes q_1, q_2, and q_3 corresponding to the three partially decoupled subsystems $X_1 = \{[In], [M_1], T\}$, $X_2 = \{Y_1, Y_2\}$, and $X_3 = \{T_{g_1}, T_{w_1}\}$, respectively. The resulting isolability graph for the closed-loop system is shown in Fig. 7.15. This structure leads to each of the three faults d_1, d_2, and d_3 having unique signatures $W^1 = [1\ 1\ 1]^T$, $W^2 = [0\ 1\ 0]^T$, and $W^3 = [0\ 0\ 1]^T$ and allows fault detection and isolation in the closed-loop system using the presented data-based FDI scheme. In open-loop operation, the system has an unstable steady-state with a limit-cycle as shown by [58]. In order to understand the stability properties of the entire closed-loop system, the stability of each subsystem around its equilibrium point was investigated assuming that the remaining states were at their equilibrium points. It can be

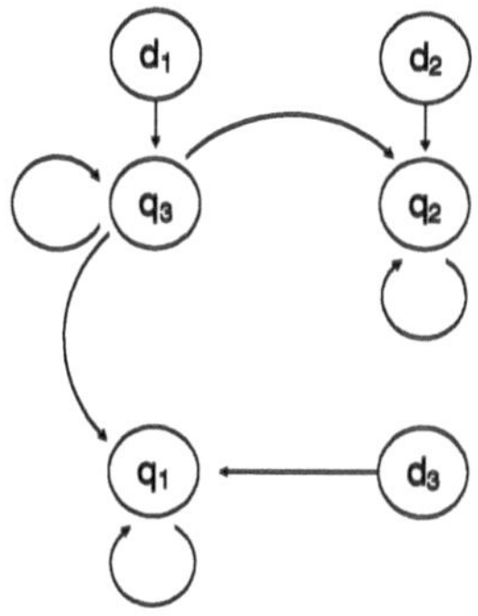

Fig. 7.15 Isolability graph for the system of Eq. (7.18)

seen that both of the uncontrolled subsystems $X_2 = \{Y_1, Y_2\}$, and $X_3 = \{T_{g_1}, T_{w_1}\}$ are stable. This implies that to obtain a stable closed-loop system, the control inputs v_1, v_2 have to be designed to stabilize the subsystem $X_1 = \{[In], [M_1], T\}$. In the present example, two PI controllers are implemented that determine v_1 and v_2 to regulate each state independently. By simulation, the PI controllers have been tuned to stabilize the equilibrium point of the closed-loop system and achieve a reasonable closed-loop response with regard to requested control action and response time. Note that any controller that stabilizes subsystem X_1 can be used. The main objective is to demonstrate the presented data-based FDI method. The PI controllers are defined as follows:

$$\begin{aligned} v_1(t) &= K_1\left(T_{ss} - T + \frac{1}{\tau_1}\int_0^t (T_{ss} - T)\,dt\right), \\ v_2(t) &= K_2\left([M_1]_{ss} - [M_1] + \frac{1}{\tau_2}\int_0^t ([M_1]_{ss} - [M_1])\,dt\right) \end{aligned} \tag{7.23}$$

with $K_1 = 0.005$, $K_1 = 0.0075$, $\tau_2 = 1000$, $\tau_1 = 500$. We will refer to the controller defined by Eqs. (7.20), (7.21), and (7.23) as the "decoupling" controller. Additionally, for comparison purposes, a controller is used that stabilizes the closed-loop system, but does not take into account the isolability conditions of the presented FDI method. Specifically, two PI controllers will be used to regulate T and M_1. This will be denoted as the "PI-only" control law. The inputs F_{M_1} and T_{feed} are defined by Eq. (7.20), but in this case, u_1 and u_2 are evaluated by applying the PI controllers of Eq. (7.23) with the same tuning parameters to the states T and M_1.

The PI-only controller stabilizes the equilibrium point under normal operating conditions, however, all the states are mutually dependent, or in other words the reduced incidence graph consists of only one node. This implies that every fault affects all the state trajectories, making isolation of the fault a difficult task. The presented FDI scheme cannot be applied because the closed-loop system does not satisfy the isolability conditions, i.e., all the system faults have the same signature.

Simulations have been carried out for several scenarios to demonstrate the effectiveness of the presented FDI scheme in detecting and isolating the three faults d_1, d_2, and d_3. In all the simulations, sensor measurement and process noise were included. The sensor measurement noise trajectory was generated using a sample time

Table 7.5 Polyethylene reactor noise parameters

	σ_p	σ_m	ϕ
$[In]$	1E–3	5E–2	0
$[M_1]$	1E–3	5E–2	0.7
Y	1E–3	1E–2	0.7
T	5E–3	5E–2	0.7
T_{g_1}	5E–3	5E–2	0.7
T_{w_1}	5E–3	5E–2	0.7

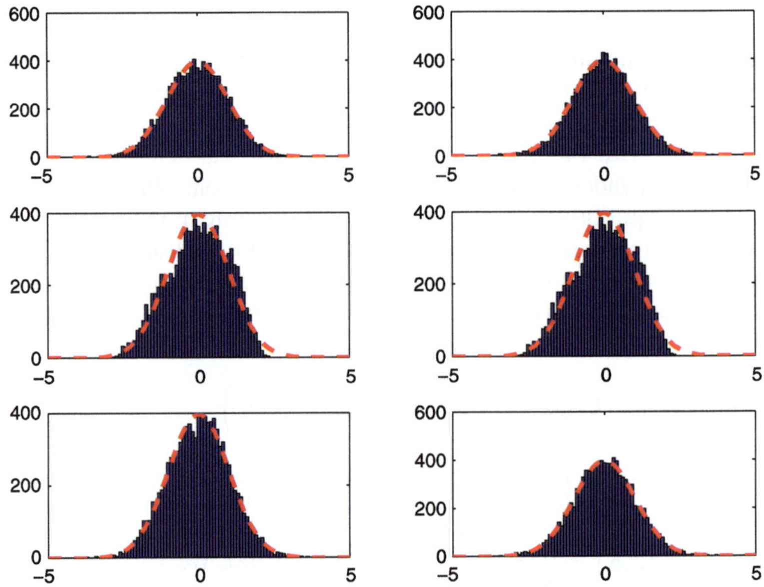

Fig. 7.16 Polyethylene reactor example. Distribution of normalized, fault-free operating data compared with a normal distribution of the same mean and covariance

of ten seconds and a zero-mean normal distribution with standard deviation σ_M. The autoregressive process noise was generated discretely as $w_k = \phi w_{k-1} + \xi_k$, where $k = 0, 1, \ldots,$ is the discrete time step, with a sample time of ten seconds, ϕ is the autoregressive coefficient, and ξ_k is obtained at each sampling step using a zero-mean normal distribution with standard deviation σ_p. The autoregressive process noise is added to the right-hand side of the differential equations for each state and the sensor measurement noise is added to the measurements of each state. Sensor measurement noise and process noise are evaluated independently for each state variable. The process and sensor measurement noise for Y_1 and Y_2 are taken to be equal. Table 7.5 provides the values of the noise parameters for each state of the system of Eq. (7.18). The same assumptions regarding the multivariate normal distribution of the measured process data under closed-loop operation for the CSTR

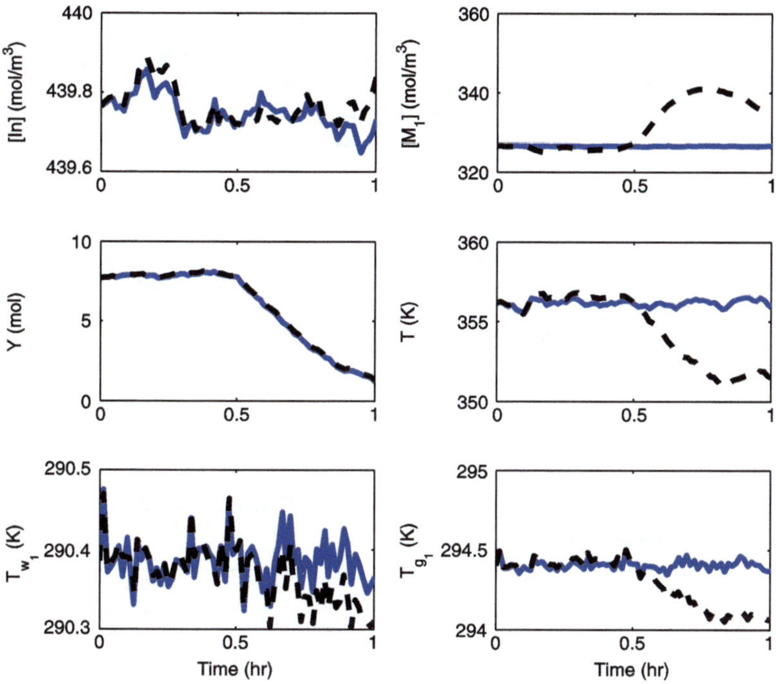

Fig. 7.17 Polyethylene reactor example. State trajectories of the closed-loop system under decoupling (*solid*) and PI-only (*dashed*) controllers with a fault d_2 at $t = 0.5$ hr

example of Sect. 7.2.4.1 apply to this example. Figure 7.16 shows the distribution of the state measurements over a long period of fault-free operation is approximately Gaussian.

For each failure d_k, two simulations have been carried out. One using the decoupling controller and another using the PI-only controller. Both simulations have been carried out using the same sensor measurement and process noise trajectories. Starting from steady-state, the three different failures with values $d_1 = 10$ K/s, $d_2 = -0.002$ mol/s, and $d_3 = 300$ K/s were introduced at time $t = 0.5$ hr. These failures are disturbances in the dynamics of T, Y, and T_{g_1} and represent changes in the feed temperature, catalyst deactivation, and changes in the recycle gas flow rate, respectively. Figures 7.17, 7.18, and 7.19 show the state trajectories of the closed-loop system under the decoupling controller (solid line) and the PI-only controller (dashed line) for each of the three possible faults. It can be seen that for the PI-only controller, each time a fault occurs, all states deviate from the normal operating region around the equilibrium point. This makes isolation a difficult task. However, the closed-loop state trajectories under the decoupling controller demonstrate that when a given fault occurs, not all state trajectories are affected. The decoupling of some states from given faults allows for the isolation of the faults based on the T_i^2 statistics. Specifically, the state trajectories of the closed-loop system under the decoupling controller were monitored using the T^2 statistic based on all the states

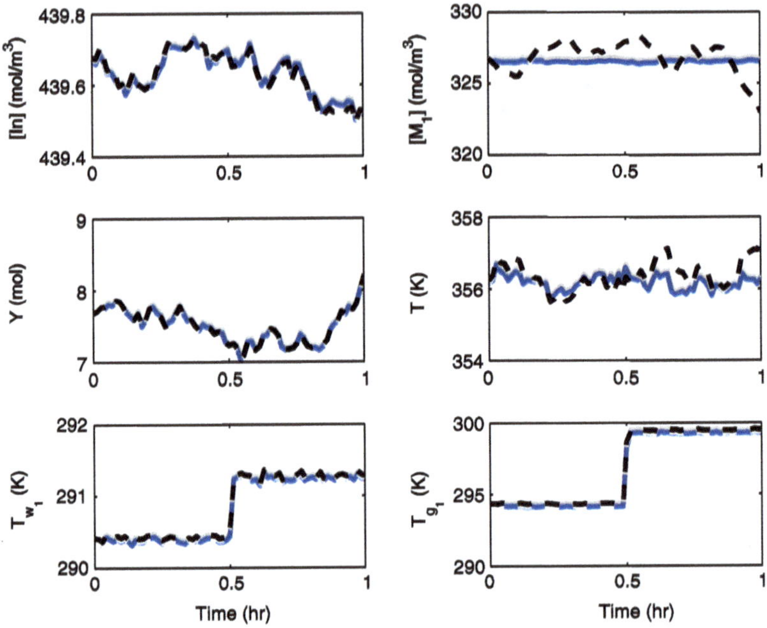

Fig. 7.18 Polyethylene reactor example. State trajectories of the closed-loop system under the decoupling (*solid*) and PI-only (*dashed*) controllers with a fault d_3 at $t = 0.5$ hr

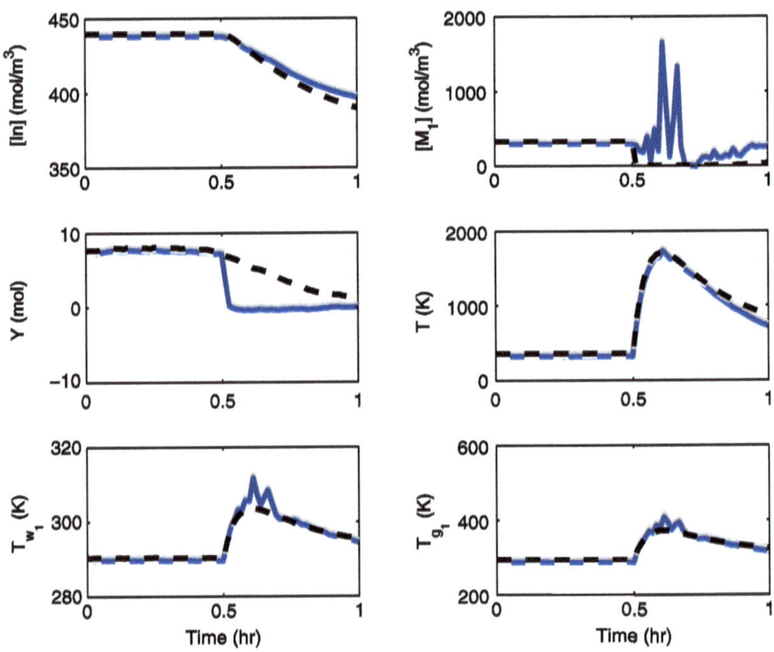

Fig. 7.19 Polyethylene reactor example. State trajectories of the closed-loop system under the decoupling (*solid*) and PI-only (*dashed*) controllers with a fault d_1 at $t = 0.5$ hr

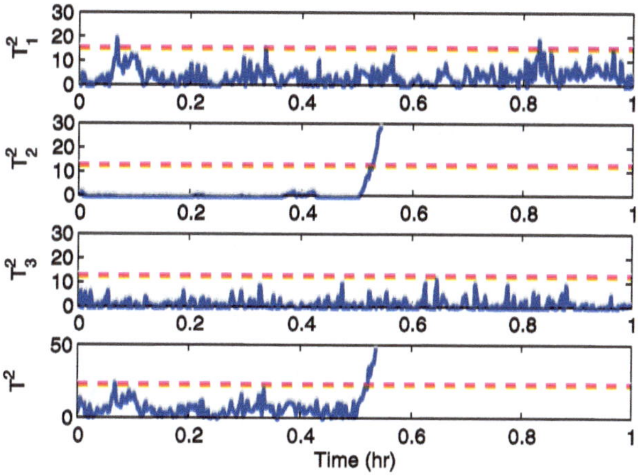

Fig. 7.20 Polyethylene reactor example. Statistics T^2, T_1^2, T_2^2, and T_3^2 (*solid*) with T_{UCL} (*dashed*) of the closed-loop system under the decoupling controller with a failure in d_2 at $t = 0.5$ hr

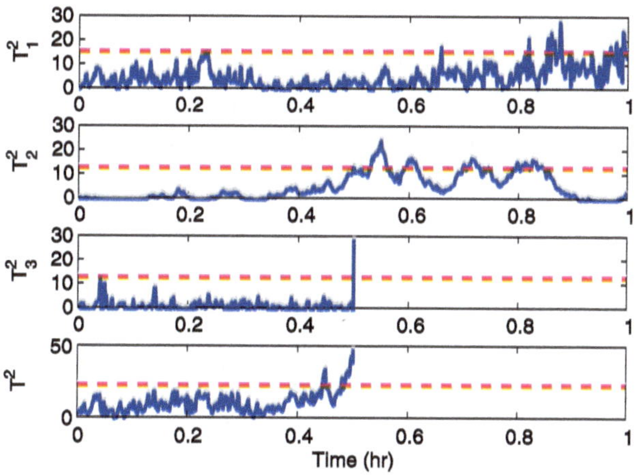

Fig. 7.21 Polyethylene reactor example. Statistics T^2, T_1^2, T_2^2, and T_3^2 (*solid*) with T_{UCL} (*dashed*) of the closed-loop system under the decoupling controller with a failure in d_3 at $t = 0.5$ hr

of the system of Eq. (7.18) and the T_i^2 statistic corresponding to each one of the three subsystems X_1, X_2, and X_3. All statistics were monitored using the single-observation method ($m = 1$) with the upper control limit defined in Eq. (7.8) and the covariance matrix, S, obtained from historical observations. As in the CSTR example, simulations were also run using a multiple observation test statistic ($m = 10$). This method showed similar results in terms of fault detection and isolation to the ones of the single observation statistic and are not presented here for brevity.

Figures 7.20, 7.21, and 7.22 show the trajectories of T^2, T_1^2, T_2^2, and T_3^2 for each different scenario along with the corresponding upper control limits. Each failure is defined by a unique signature that can be isolated based on the monitored statistics. Figure 7.20 shows the statistics corresponding to the simulation with a failure in d_2. The signature of d_2 is $W^2 = [0\ 1\ 0]^T$ because the dynamics of the states corresponding to X_1 and X_3 are not affected by fault d_2, that is, there is no path from the node corresponding to d_2 to the nodes corresponding to X_1 and X_2 in the isolability graph of the closed-loop system. Figure 7.20 clearly shows the fault occurring at

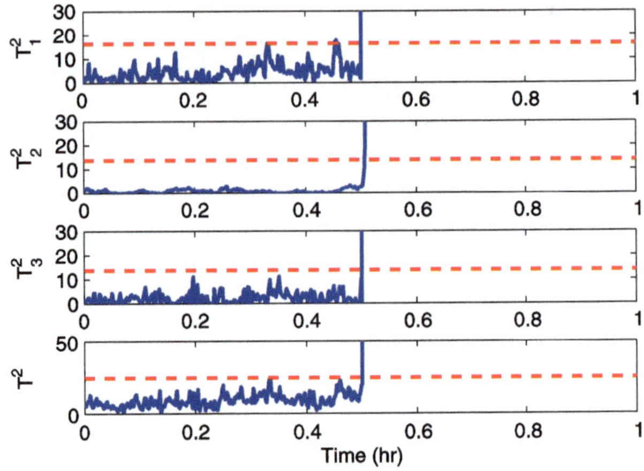

Fig. 7.22 Polyethylene reactor example. Statistics T^2, T_1^2, T_2^2, and T_3^2 (*solid*) with T_{UCL} (*dashed*) of the closed-loop system under the decoupling controller with a failure in d_1 at $t = 0.5$ hr

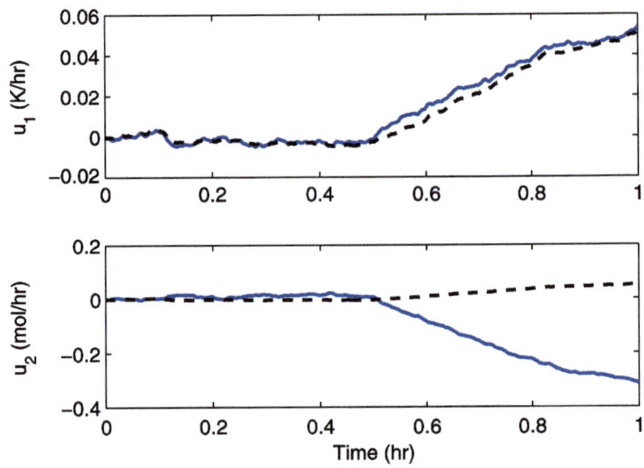

Fig. 7.23 Polyethylene reactor example. Manipulated input profiles for both decoupling (*solid*) and PI-only (*dashed*) control with a fault in d_2 at $t = 0.5$ hr

time $t = 0.5$ hr and the signature that we would expect, that is, only T_2^2 violates the upper control limit. The state trajectories of this faulty scenario of Fig. 7.17 demonstrates that there is a failure affecting Y starting at $t = 0.5$ hr. The failure affects all the state trajectories under PI-only control but affects only Y for the closed-loop system under nonlinear decoupling control. Similarly, a failure in T_{g_1} affects only subsystem X_3. The state trajectories of Fig. 7.18 shows that under PI-only control, all of the states are affected, whereas under decoupling control, only the subsystem $X_3 = \{T_{g_1}, T_{w_1}\}$ is affected. The statistics in Fig. 7.21 show that the signature of the fault is $[0\ 0\ 1]^T = W^3$. The signature of fault d_1 is $W^1 = [1\ 1\ 1]^T$, meaning that this fault affects all the states in the closed-loop system. The state trajectories and the corresponding statistics are shown in Fig. 7.19 and Fig. 7.22. The control action required under the decoupling control law is on the same order of magnitude as that of the PI-only controller. Figure 7.23 shows the manipulated input trajectories for both controllers in the scenario with fault d_2 occurring.

Remark 7.10 Although the method of determining faults by monitoring T_i^2 values was used in this example, other FDI methods could benefit from the fact that the enforced structure separates regions of faulty operation. In the case where the desired structure is only partially achieved due to plant-model mismatch or other uncertainties, it may be necessary to utilize more sophisticated methods of fault detection and isolation (e.g., contribution plots or clustering). It should be noted that even an incomplete decoupling will benefit many of these methods as the regions of faulty operation are still at least partially separated.

7.3 Using FDI for Controller Performance Monitoring

In this part of the chapter, we consider nonlinear process systems with constraints on the inputs described by the following state-space model:

$$\dot{x}(t) = f\big(x(t)\big) + G\big(x(t)\big)u_a(t) + w(t), \tag{7.24}$$

where $x(t) \in \mathbb{R}^{n_x}$ is an n_x-element column vector representing n_x states of the system, $u_a(t) \in U \subseteq \mathbb{R}^{m_u}$ is an m_u-element column vector representing m_u inputs to the system, and $w(t) \in W \subseteq \mathbb{R}^{n_x}$ is an n_x-element column vector representing the process noise to the system. U is a convex set, $f(\cdot)$ is a non-linear sufficiently smooth vector function, and $G(\cdot)$ is a $n_x \times m_u$ matrix whose elements are sufficiently smooth functions that relate the jth input to the ith state with $1 \leq j \leq m_u$ and $1 \leq i \leq n_x$. Without loss of generality, $x = 0$ is assumed to be the equilibrium of the unforced system, i.e., $\dot{x}(t) = 0$ when $x = 0$, $u_a = 0$, and $w = 0$.

Since the central focus of this part of the chapter is on the difference between the requested actuation computed by the model-based controller and the actual actuation level applied to the process by the control actuators, we shall distinguish the two elements by calling the requested actuation $u_m(t)$ and the actual actuation $u_a(t)$. The results in this part of the chapter are illustrated using the LMPC design presented in Sect. 2.8. One assumption about the design of the model-based control system used is that it does not explicitly account for the dynamics of the control actuators and the presence of the process noise. Therefore, the model used for the design of the model-based control system assumes the following dynamics for the process:

$$\dot{\tilde{x}}(t) = f\big(\tilde{x}(t)\big) + G\big(\tilde{x}(t)\big)u_m(t), \tag{7.25}$$

where u_m is the commanded actuation by the high-level MPC.

We make the following assumptions regarding the stability of the closed-loop system. We assume that there exists a Lyapunov-based controller $h(\tilde{x})$ as well as a corresponding Lyapunov function $V(x)$ such that the origin of the nominal closed-loop system under this controller, i.e., system of Eq. (7.25) with $u_m(t) = h(\tilde{x})\ \forall t$, is asymptotically stable. The existence of the controller $h(\tilde{x})$ allows us to formulate an LMPC that inherits the stability properties of $h(\tilde{x})$, and it is described by the following optimization problem:

$$\min_{u_c \in S(\Delta)} \int_0^{N_c \Delta} \left[\hat{x}^T(\tau) Q \hat{x}(\tau) + u_c^T(\tau) R u_c(\tau)\right] d\tau, \tag{7.26a}$$

$$\dot{\hat{x}}(\tau) = f\big(\hat{x}(\tau)\big) + G\big(\hat{x}(\tau)\big) u_c(\tau), \tag{7.26b}$$

$$\hat{x}(0) = x(t_k), \tag{7.26c}$$

$$u_c(\tau) \in U, \tag{7.26d}$$

$$\frac{\partial V(x(t_k))}{\partial x} G\big(x(t_k)\big) u_c(0) \le \frac{\partial V(x(t_k))}{\partial x} G\big(x(t_k)\big) h\big(x(t_k)\big), \tag{7.26e}$$

where $S(\Delta)$ is the family of piece-wise constant functions with sampling period Δ, Q, and R are strictly positive definite symmetric weighting matrices, $x(t_k)$ is the process state measurement obtained at t_k, $\hat{x}$ is the predicted trajectory of the system under the MPC, N_c is the number of steps in the prediction horizon, and V is the Lyapunov function corresponding to the controller $h(\tilde{x})$.

The optimal solution to this optimization problem is denoted by $u_c^*(\tau|t_k)$. The LMPC is implemented following a receding horizon strategy; at each sampling time t_k, a new state measurement $x(t_k)$ is received from the sensors and the optimization problem of Eq. (7.26a)–(7.26e) is solved, and $u_c^*(0|t_k)$ is sent to the actuators and it is implemented for $t \in [t_k, t_{k+1}]$.

As depicted in Fig. 7.2, $u_m(t)$ is sent from the model-based controller as the set-point to the control actuators. PID controllers are installed around these control actuators to help accelerate the actuator's response so that $u_a(t)$ can approach the value of $u_m(t)$ faster. Equation (7.27) below shows the relationship between u_m and u_a in the Laplace domain:

$$u_a(s) = \frac{G_p G_c}{1 + G_p G_c} u_m(s), \tag{7.27}$$

where G_p is the actuator's transfer function and G_c is the PID controller's transfer function. G_c contains 3 parameters: K_c (proportional gain), τ_I (integral time constant), and τ_D (derivative time constant) and takes the following form:

$$G_c = K_c\left(1 + \frac{1}{\tau_I s} + \tau_D s\right). \tag{7.28}$$

The transfer function of the actuator's dynamics, G_p, on the other hand, can be approximated as a first-order transfer function with dead time G_p' as follows:

$$G_p' = K_p \frac{e^{-\tau_d s}}{\tau_p s + 1}, \tag{7.29}$$

where K_p is the actuator's gain, τ_d is the actuator dead time, and τ_p is the actuator's time constant.

The estimation of the actuator's transfer function (G'_p) will be needed by the FDI algorithm below when the actuator's expected behavior is calculated and also at the retuning step when a new set of PID parameters is calculated. The expected actuation level (denoted by $u'_a(t)$) will be used as the benchmark upper limit of how well the control actuators can perform. We note that the parameters of the PID controller should be tuned in such a way that the low-level closed-loop response (i.e., actuator under the PID controller) is fast relative to the sampling time of the MPC controller such that the actual actuator output (control action implemented on the process) is as close as possible to the control action requested by the MPC at each sampling time. A rigorous analysis of this problem can be done using singular perturbation techniques for two-time-scale processes.

Remark 7.11 Note that in the design of the LMPC of Eqs. (7.26a)–(7.26e) and its closed-loop stability analysis, one assumption is that the requested actuation $u_m(t)$ is applied directly to the process by the control actuators. In a practical setting, however, $u_m(t)$ has to go through the dynamics of the PID-controlled actuators before the system is actuated with $u_a(t)$. The central focus of this work is on how to bring $u_a(t)$ to be as close as possible to $u_m(t)$. The relationship between $u_a(t)$ and $u_m(t)$ will be discussed in detail in the next section.

Remark 7.12 Though a Lyapunov-based MPC is used in this chapter as the model-based control system to demonstrate how the problem of low-level PID monitoring and retuning based on process state measurements can be approached, the monitoring and retuning methods presented here can be applied to any type of model-based control system (i.e., geometric control or Lyapunov-based control discussed in Chap. 2, distributed MPC [87, 88], etc.). Specifically, as long as the requested actuation level $u_m(t)$ and the process state measurements are available to the monitoring and retuning system at all times, the same method presented in this work can be applied to detect the deviation of the actual actuation level $u_a(t)$ from the requested actuation level $u_m(t)$.

7.3.1 Monitoring and Retuning of Low-Level PID Loops

We consider the case where there is no access by the monitoring system to the measurements of the actual actuation levels $u_a(t)$ implemented by the control actuators on the process. Therefore, the detection of poor PID tunings must be performed based on the measurements of the states of the process. To this end, an FDI method is used as the main tool to extract actuator behavior from the process state measurements. Specifically, we use exponentially-weighted-moving-average (EWMA) residuals to detect and isolate poorly-tuned PID loops. Once a poorly-tuned actuator is isolated, a model-based tuning rule such as Cohen–Coon or internal model control is applied to the PID controller that regulates the poorly-tuned actuator.

The residuals are constructed from the difference between the expected behavior and the actual behavior of the plant. This is done by comparing the evolution of the actual system obtained from the state measurements against the evolution of the ideal filtered states based on the plant model. The actual closed-loop system state ($x(t)$) evolves in the following manner:

$$\begin{aligned} \dot{x}(t) &= f\big(x(t)\big) + G\big(x(t)\big)u_a(t) + w(t), \\ u_a(s) &= \frac{G_p G_c}{1 + G_p G_c} u_m(s), \end{aligned} \tag{7.30}$$

where $u_m(t)$ is the control action computed by the MPC and $u_a(t)$ is the actual actuation performed by the actuators. The filter state ($\breve{x}(t)$), on the other hand, evolves as follows:

$$\begin{aligned} \dot{\breve{x}}_i(t) &= f_i\big(\hat{x}_i(t)\big) + G_i\big(\hat{x}_i(t)\big)u'_a(t), \\ \hat{x}_i &= [x_1, \dots, x_{i-1}, \breve{x}_i, x_{i+1}, \dots, x_{n_x}]^T, \\ u'_a(s) &= \frac{G'_p G'_c}{1 + G'_p G'_c} u_m(s), \\ \breve{x}(N\Delta_m) &= x(N\Delta_m), \quad \forall N = 0, 1, 2, \dots, \end{aligned} \tag{7.31}$$

where Δ_m is the MPC sampling time, G'_p is the estimated transfer function matrix of the control actuators, G'_c is a well-tuned PID controller transfer function matrix based on the estimated model of the actuator G'_p. This makes $u'_a(t)$ the expected actuation level of $u_a(t)$.

Using Eq. (7.30) and Eq. (7.31), the real-time measurements of $x(t)$ can be compared against the evolution of $\breve{x}(t)$. The residual, or the difference between $x_i(t)$ and $\breve{x}_i(t)$ denoted by $r_i(t)$, is expressed in the following manner:

$$r_i(t) = \big|\breve{x}_i(t) - x_i(t)\big|. \tag{7.32}$$

In the absence of noise and if $G'_p = G_p$, whenever the jth element of u_a deviates from its expected behavior u'_{aj} and the ith-row-jth-column element of the $G(x)$ matrix is nonzero, the ith residual (r_i) would instantaneously become nonzero. In other words, r_i is nonzero only when there is a problem with the actuators that directly affect the ith state of the system (relative degree of 1) [112, 115].

In practice, however, model mismatch, process noise, and measurement noise are always present to some degree. Therefore, in a practical setting, the residuals will be nonzero regardless of the accuracy of the process model used in Eq. (7.31). Thus, before the model-based FDI method can be used in practice, the effects of process and measurement noise levels must first be recorded from fault-free closed-loop process operation data (with both the PID controllers and the MPC being well-tuned). On the basis of these noisy closed-loop system states, the mean and the standard deviation of the residuals are calculated and the thresholds are determined.

Occasional noise spikes can make the residuals exceed the thresholds for a brief period of time even when the actuators are functioning well; this can lead to the common problem of false alarms. To reduce the incidence of false alarms, we define a

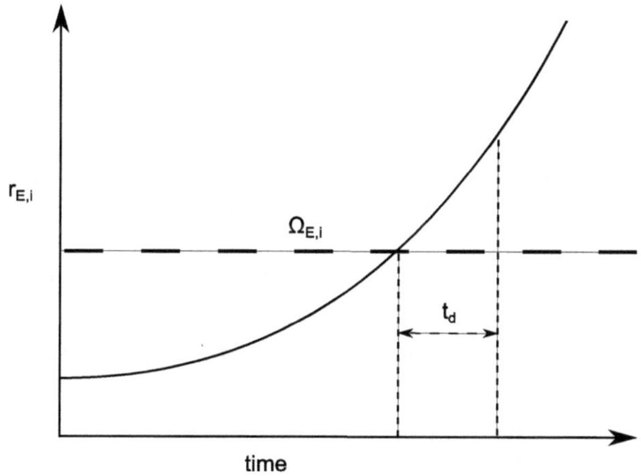

Fig. 7.24 Monitoring scheme of PID response behavior based on the EWMA residuals of the process state. Poor tuning is declared after $r_{E,i}$ exceeds its threshold $\Omega_{E,i}$ continuously for $t = t_d$

modified residual $r_{E,i}, i = 1, \ldots, n_x$, for each residual r_i, calculated at discrete time instants t_k with $t_k = t_0 + k\Delta_r,\ k = 0, 1, 2, \ldots$ and Δ_r being the interval between two consecutive state measurements. The weighted residual is calculated using an exponentially weighted moving average (EWMA) method as follows [23, 24]:

$$r_{E,i}(t_k) = \lambda r_i(t_k) + (1 - \lambda) r_{E,i}(t_{k-1}) \tag{7.33}$$

with $r_{E,i}(t_0) = r_i(t_0)$ and the weighting factor $\lambda \in (0, 1]$. The parameter λ determines the rate at which past data enters into the calculations of the weighted residual. When $\lambda = 1$, $r_{E,i}$ is equivalent to r_i. The typical range of λ is between 0.2 and 0.5 depending on the desired level of sensitivity [24, 89]. Lower values of λ make the $r_E(t)$ curve smoother as potential noise spikes will have a smaller effect on the overall shape of the curve, i.e., instances of false alarm will be reduced. However, in the event where an actual poor tuning occurs, it may be detected and isolated more slowly.

The threshold, denoted by $\Omega_{E,i}$, for fault detection is defined as follows:

$$\Omega_{E,i} = \mu_i + \alpha\sigma_i\sqrt{\frac{\lambda}{2 - \lambda}}, \tag{7.34}$$

where α is a threshold parameter determining how sensitive the FDI is; a typical value of α is an integer value between 1 and 5. The parameters μ_i and σ_i are the mean and the standard deviation of the ith residual during normal operation. Once $r_{E,i}$ exceeds the threshold ($\Omega_{E,i}$) for a fixed amount of time t_d (determined by the user), then poor tuning is declared in the actuator(s) directly affecting the ith state and the retuning algorithm is activated. Figure 7.24 shows the schematic of how the EWMA residuals are used to activate the PID retuning algorithm at the end of waiting time t_d.

Once a poorly-tuned actuator is isolated, a PID tuning method can be applied to the PID controller based on the estimated transfer function of the actuator G'_p.

To help ensuring the stability of the retuning algorithm, we employ a stability constraint. Specifically, whenever retuning is performed, the retuning algorithm makes sure that $\frac{G'_p G_c}{1+G'_p G_c}$ contains only strictly negative poles. In this work, we use Cohen–Coon and internal model control method to retune the PID parameters to demonstrate the approach. If desired, other model-based tuning rules may be used as well. Please, see [149, 154, 165, 183] for other PID tuning methods.

Remark 7.13 One feature that should be noted is that the PID retuning will be initiated if the magnitude of the residuals is above a certain threshold. This means that even if the difference between $u_{aj}(t)$ and $u'_{aj}(t)$ is appreciable but the difference between $\breve{x}_i(t)$ and $x_i(t)$ is smaller than the threshold, the retuner will do nothing. This is a direct result of the fact that the real value of $u_a(t)$ is unknown and has to be estimated from the trajectories of the process states. A scenario like this can also happen when $G_{ij}(\cdot)$ is small.

Remark 7.14 The isolability structure of the system is also critical to the use of the monitoring algorithm proposed here. If, from the patterns of the residuals, a poorly-performing actuator cannot be isolated with high confidence (i.e., two actuators have the same signature because they directly affect the same system state), then all control actuators that may be poorly tuned should be retuned. In principle, it is also possible to use empirical models from input-output data in the MPC design as well as in the monitoring of the PID control loops. One potential problem of using this approach is the difficulty of isolating which specific PID control loop is poorly performing because input/output empirical models can not account for the coupling between different process variables the way state-space first principles models do.

Remark 7.15 In the design of the filter of Eq. (7.31), a well-tuned PID controller, G'_c, is assumed to be known and is used to calculate the benchmark performance of the overall control system. In the case that G'_c is not known, the control action computed by the MPC, u_m, can be used directly in the filter design (i.e., replace u'_a by u_m in Eq. (7.31)) to obtain an estimate of the expected process state evolution. Furthermore, once a poorly-tuned actuator is isolated, retuning of the parameters of PID controller used in this actuator should be carried out to account for changes in operation conditions as well as control actuator wear and tear over time.

7.3.2 Application to a Nonlinear Chemical Process Network

7.3.2.1 Process Description and Modeling

We demonstrate the PID monitoring and retuning methodology presented in the previous section using a three-vessel reactor–separator chemical process network. A schematic of the process is shown in Fig. 7.25. The first two vessels are assumed

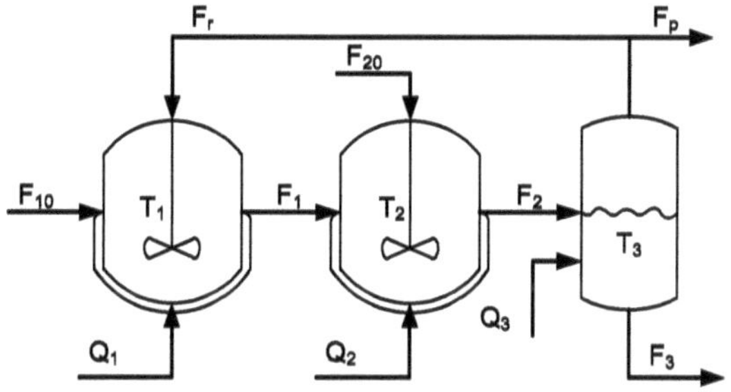

Fig. 7.25 Schematic of the process. Two CSTRs and a flash tank with recycle stream

Table 7.6 Process parameter values

$T_{10} = 300$, $T_{20} = 300$	K
$F_{10s} = 5$, $F_{20s} = 5$, $F_r = 1.9$	m^3/hr
$Q_{1s} = 0$, $Q_{2s} = 0$, $Q_{3s} = 0$	kJ/hr
$V_1 = 1.0$, $V_2 = 0.5$, $V_3 = 1.0$	m^3
$E_1 = 5E4$, $E_2 = 5.5E4$	kJ/kmol
$k_1 = 3E6$, $k_2 = 3E6$	1/hr
$\Delta H_1 = -5E4$, $\Delta H_2 = -5.3E4$	kJ/kmol
$H_{\text{vap}} = 5$	kJ/kmol
$C_p = 0.231$	kJ/kg K
$R = 8.314$	kJ/kmol K
$\rho = 1000$	kg/m^3
$\alpha_A = 2$, $\alpha_B = 1$, $\alpha_C = 1.5$, $\alpha_D = 3$	unitless
$MW_A = 50$, $MW_B = 50$, $MW_C = 50$	kg/kmol s

to be ideal CSTRs, followed by a flash tank separator. There are two fresh feed streams of pure reactant A of concentration C_{A10} to both reactors (with flow rates F_{10} and F_{20} respectively) and a recycle stream (F_r) from the flash tank to the first reactor. Specifically, the overhead vapor from the flash tank is condensed and recycled to the first CSTR, and the bottom product stream is removed. The effluent of vessel 1 is fed to vessel 2 and the effluent from vessel 2 is fed to the flash tank. Each vessel has an external heat input or heat removal system (Q_1, Q_2, and Q_3). The steady-state flow rate and heat input are denoted by F_{10s}, F_{20s}, Q_{1s}, Q_{2s}, and Q_{3s} and their values are given in Table 7.6. There are two parallel chemical reactions considered in this process; first, reactant A is converted to desired product B, and second, A is converted to undesired product C (referred to as reaction 1 and 2, respectively). Under standard modeling assumptions, the dynamic energy and material balance equations that can describe this process take the following form:

$$\frac{dT_1}{dt} = \frac{F_{10}}{V_1}(T_{10} - T_1) + \frac{F_r}{V_1}(T_3 - T_1) + \frac{Q_1}{\rho C_p V_1} + \frac{-(\Delta H_1)}{\rho C_p} k_1 e^{\frac{-E_1}{RT_1}} C_{A1} + \frac{(-\Delta H_2)}{\rho C_p} k_2 e^{\frac{-E_2}{RT_1}} C_{A1}, \tag{7.35a}$$

$$\frac{dC_{A1}}{dt} = \frac{F_{10}}{V_1}(C_{A10} - C_{A1}) + \frac{F_r}{V_1}(C_{Ar} - C_{A1}) - k_1 e^{\frac{-E_1}{RT_1}} C_{A1} - k_2 e^{\frac{-E_2}{RT_1}} C_{A1}, \tag{7.35b}$$

$$\frac{dC_{B1}}{dt} = \frac{-F_{10}}{V_1} C_{B1} + \frac{F_r}{V_1}(C_{Br} - C_{B1}) + k_1 e^{\frac{-E_1}{RT_1}} C_{A1}, \tag{7.35c}$$

$$\frac{dC_{C1}}{dt} = \frac{-F_{10}}{V_1} C_{C1} + \frac{F_r}{V_1}(C_{Cr} - C_{C1}) + k_2 e^{\frac{-E_2}{RT_1}} C_{A1}, \tag{7.35d}$$

$$\begin{aligned}\frac{dT_2}{dt} &= \frac{F_1}{V_2}(T_1 - T_2) + \frac{F_{20}}{V_2}(T_{20} - T_2) + \frac{Q_2}{\rho C_p V_2} + \frac{(-\Delta H_1)}{\rho C_p} k_1 e^{\frac{-E_1}{RT_2}} C_{A2} \\ &\quad + \frac{(-\Delta H_2)}{\rho C_p} k_2 e^{\frac{-E_2}{RT_2}} C_{A2},\end{aligned} \tag{7.35e}$$

$$\frac{dC_{A2}}{dt} = \frac{F_1}{V_2}(C_{A1} - C_{A2}) + \frac{F_{20}}{V_2}(C_{A20} - C_{A2}) - k_1 e^{\frac{-E_1}{RT_2}} C_{A2} - k_2 e^{\frac{-E_2}{RT_2}} C_{A2}, \tag{7.35f}$$

$$\frac{dC_{B2}}{dt} = \frac{F_1}{V_2}(C_{B1} - C_{B2}) - \frac{F_{20}}{V_2} C_{B2} + k_1 e^{\frac{-E_1}{RT_2}} C_{A2}, \tag{7.35g}$$

$$\frac{dC_{C2}}{dt} = \frac{F_1}{V_2}(C_{C1} - C_{C2}) - \frac{F_{20}}{V_2} C_{C2} + k_2 e^{\frac{-E_2}{RT_2}} C_{A2}, \tag{7.35h}$$

$$\frac{dT_3}{dt} = \frac{F_2}{V_3}(T_2 - T_3) - \frac{H_{\text{vap}} F_r}{\rho C_p V_3} + \frac{Q_3}{\rho C_p V_3}, \tag{7.35i}$$

$$\frac{dC_{A3}}{dt} = \frac{F_2}{V_3}(C_{A2} - C_{A3}) - \frac{F_r}{V_3}(C_{Ar} - C_{A3}), \tag{7.35j}$$

$$\frac{dC_{B3}}{dt} = \frac{F_2}{V_3}(C_{B2} - C_{B3}) - \frac{F_r}{V_3}(C_{Br} - C_{B3}), \tag{7.35k}$$

$$\frac{dC_{C3}}{dt} = \frac{F_2}{V_3}(C_{C2} - C_{C3}) - \frac{F_r}{V_3}(C_{Cr} - C_{C3}), \tag{7.35l}$$

where T_1, T_2, and T_3 are the temperatures of vessels 1, 2, and 3, respectively, T_{10} and T_{20} are the temperatures of the feed streams to vessels 1 and 2, respectively, F_{10} and F_{20} are the volumetric feed flow rates into vessels 1 and 2, respectively, and F_1 and F_2 are the volumetric flow rates out of vessels 1 and 2, respectively. F_r is the recycle stream volumetric flow rate from vessel 3 to vessel 1. V_1, V_2, and V_3 are the volumes of the three vessels, Q_1, Q_2, and Q_3 are the heat inputs into the vessels, C_{A1}, C_{B1}, C_{C1}, C_{A2}, C_{B2}, C_{C2}, C_{A3}, C_{B3}, and C_{C3} are the concentrations of A, B, and C in the vessels 1, 2, and 3, respectively, C_{Ar}, C_{Br}, and C_{Cr} are the concentrations of A, B, and C in the recycle stream. ρ is the mass density of the reacting fluid, C_p is the heat capacity of the reacting fluid, k_1 and k_2 are the reaction rate constants of reactions 1 and 2, respectively, E_1 and E_2 are the activation energy of reactions 1 and 2, respectively, ΔH_1 and ΔH_2 are the enthalpies of reactions 1 and 2, respectively, and H_{vap} is the heat of vaporization for the fluid in vessel 3. Finally, R is the universal gas constant.

The composition of the flash tank recycle stream is described by Eqs. (7.36a)–(7.36d) below, which assumes constant relative volatility for each species within the temperature operating range. This assumption allows calculation of the composition in the recycle stream relative to the composition of the liquid holdup in the flash tank. Each tank is assumed to have static holdup and the reactions in the flash tank are considered negligible. Specifically, we have:

$$C_{\mathrm{A}r} = \frac{\alpha_{\mathrm{A}} C_{\mathrm{A3}}}{K}, \tag{7.36a}$$

$$C_{\mathrm{B}r} = \frac{\alpha_{\mathrm{B}} C_{\mathrm{B3}}}{K}, \tag{7.36b}$$

$$C_{\mathrm{C}r} = \frac{\alpha_{\mathrm{C}} C_{\mathrm{C3}}}{K}, \tag{7.36c}$$

$$K = \alpha_{\mathrm{A}} C_{\mathrm{A3}} \frac{MW_{\mathrm{A}}}{\rho} + \alpha_{\mathrm{B}} C_{\mathrm{B3}} \frac{MW_{\mathrm{B}}}{\rho} + \alpha_{\mathrm{C}} C_{\mathrm{C3}} \frac{MW_{\mathrm{C}}}{\rho} + \alpha_{\mathrm{D}} x_{\mathrm{D}}, \tag{7.36d}$$

where α_{A}, α_{B}, α_{C}, and α_{D} are the relative volatility constants of the three reacting species along with the inert species D. MW_{A}, MW_{B}, and MW_{C}, are the molecular weights of the three reacting species. Finally, x_{D} is the mass fraction of the inert species D in the liquid phase of vessel 3. The values of the process parameters are given in Table 7.6.

The system of Eqs. (7.35a)–(7.35l) is solved numerically using explicit Euler method with a time step of $\Delta_p = 0.001$ hr. Process and sensor measurement noise are also used in the process simulation. The sensor measurement noise is generated using a zero-mean normal distribution with a standard deviation of 2.5 K for the three temperature state measurements and 1 kmol/m^3 for the nine concentration state measurements. The process noise is generated similarly and it is included as an additive term in the right-hand-side of the ordinary differential equations of Eqs. (7.35a)–(7.35l) with a zero-mean normal distribution and the same standard deviation values used for the measurement noise. In all three vessels, the heat inputs are used as the manipulated variables for controlling the process network at the operating steady-state. Therefore, the corresponding relative degrees of these variables with respect to the temperatures of the three vessels (reactor 1, reactor 2, and separator) are all one, thereby allowing isolation of poor-tuning in each one of these actuators from process measurements. In addition the second tank's inlet flow rate is chosen as another manipulated variable. The system has one unstable and two stable steady states. The operating steady-state is the *unstable steady-state* shown in Table 7.7.

We focus on the problem of monitoring and retuning of the PID controllers used to regulate the three heat input control actuators to each of the vessels: Q_1, Q_2, Q_3, at the values computed by the MPC in each sampling time. In order to calculate the benchmark performance for each actuator ($u_a'(s)$) and a new set of PID parameters when PID retuning is needed, a first-order approximation of the transfer function

Table 7.7 Operating steady-state (x_s)

T_1	370	K
C_{A1}	3.32	kmol/m^3
C_{B1}	0.17	kmol/m^3
C_{C1}	0.04	kmol/m^3
T_2	435	K
C_{A2}	2.75	kmol/m^3
C_{B2}	0.45	kmol/m^3
C_{C2}	0.11	kmol/m^3
T_3	435	K
C_{A3}	2.88	kmol/m^3
C_{B3}	0.50	kmol/m^3
C_{C3}	0.12	kmol/m^3

of the actuator (G'_p) must be computed. In this example, all actuator dynamics are modeled with first-order transfer functions with time delay. All actuators have the same time constant (τ_p) of 2.82 seconds and time delay (τ_d) of 3.60 seconds, resulting in the following transfer function:

$$G_{\text{actuator}} = \frac{e^{-3.60s}}{2.82s + 1}. \tag{7.37}$$

The control action computed by the MPC is sent to the control actuators every $\Delta_m = 0.01$ hr. Thus, at every sampling time $t = N\Delta_m$, $N = 0, 1, 2, \ldots$, the low-level PID controllers take the MPC command ($u_m(t)$) as the set-point and drive the actual actuation level ($u_a(t)$) to the set-point under the following closed-loop dynamics:

$$u_a(s) = \frac{G_p G_c}{1 + G_p G_c} u_m(s).$$

We choose the following parameters for PID monitoring and retuning. We pick the EWMA parameter λ to be 0.2. The EWMA residual threshold parameter α is chosen to be 5. The waiting time for fault isolation based on the EWMA residual is set to be $t_d = 0.01$ hr.

For the actuators with the transfer function presented in Eq. (7.37), the PID parameters that give the best closed-loop response were found to be the following:

$$\begin{aligned} K_c^* &= 0.648, \\ \tau_I^* &= 5.94 \text{ s}, \\ \tau_D^* &= 0.54 \text{ s}. \end{aligned} \tag{7.38}$$

These parameters were used to calculate G'_c. The poles of $\frac{G'_p G'_c}{1+G'_p G'_c}$ calculated with the parameters above are found to be all negative. This, in conjunction with the

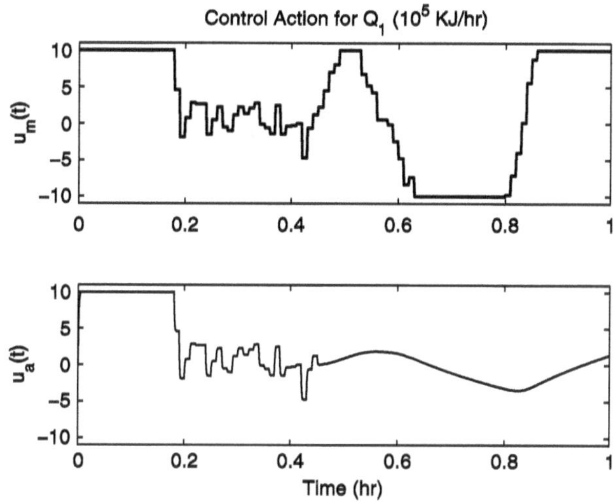

Fig. 7.26 Example 1: Requested actuation level by the MPC ($u_m(t)$) and actual actuation level ($u_a(t)$) when PID retuning is not implemented

approximate transfer function (G'_p) of the actuators of Eq. (7.37), was then used to approximate the ideal actuation performance ($u'_a(s)$) of each control actuator.

7.3.2.2 Simulation Results

In the following two examples, we will illustrate how PID monitoring and retuning are applied to the system.

Example 1 In this example, we start the process from the following initial condition: $x(0) = 0.8x_s$ where x_s is the operating steady-state. All the control actuators are properly tuned with the PID parameters shown in Eq. (7.38). At time $t = 0.45$ hr, we apply poor tuning to the PID controller for the actuator Q_1 with the following parameters:

$$\begin{aligned} K_c &= 0.00909, \\ \tau_I &= 11.9 \text{ s}, \\ \tau_D &= 0.655 \text{ s}. \end{aligned} \tag{7.39}$$

Figure 7.26 shows the comparison between the requested actuation level $u_m(t)$ and the actual actuation level $u_a(t)$ for Q_1 if the monitoring and retuning system is inactive. The EWMA residuals of the temperature in 3 vessels are shown in Fig. 7.27.

With the monitoring system active, Fig. 7.28 shows the evolution of PID response $u_a(t)$ as it is retuned at $t = 0.475$ hr. As shown in Fig. 7.29, at $t = 0.465$ hr, r_{E,T_1} starts exceeding its threshold Ω_{E,T_1}. At this point, the value of r_{E,T_1} starts being monitored closely for $t_d = 0.01$ hr. By the time the system reaches $t = 0.475$ hr, the value of r_{E,T_1} is found to have been above its threshold Ω_{E,T_1} for the entire duration

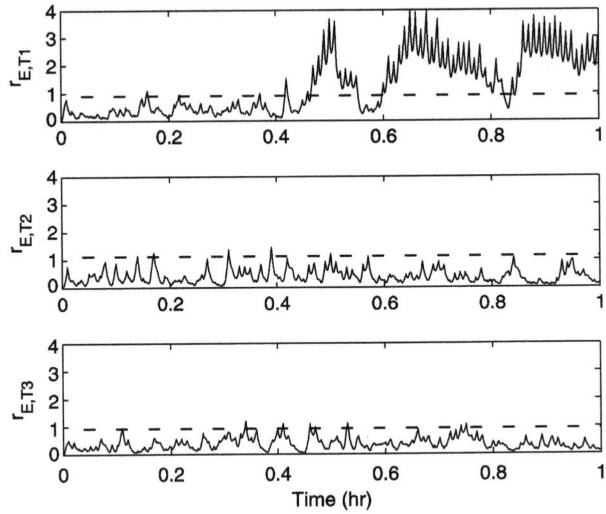

Fig. 7.27 Example 1: Temperature residuals for the 3 vessels computed via EWMA when PID retuning is not implemented. The *dashed lines* represent the EWMA residual thresholds $\Omega_{E,i}$

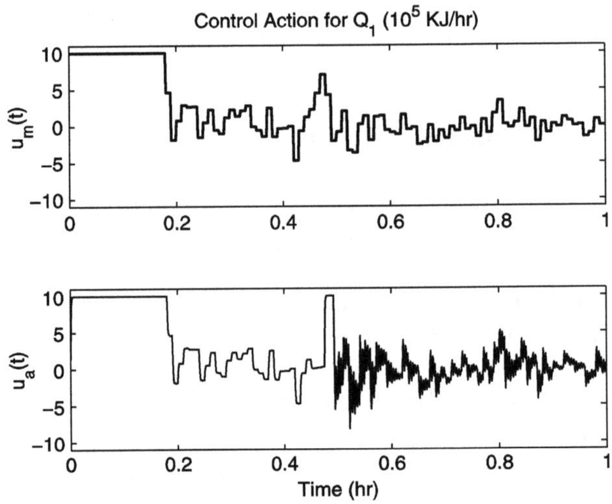

Fig. 7.28 Example 1: Requested actuation level by the MPC ($u_m(t)$) and actual actuation level ($u_a(t)$) when PID retuning is implemented

from $t = 0.465$ hr to $t = 0.475$ hr. Because the process state T_1 is the only state that is directly affected by the control actuator Q_1, given the model-based FDI filter design, any anomaly detected in r_{E,T_1} is the result of a problem with the Q_1 control actuator. Therefore, the actuator Q_1 can be isolated with high confidence as the actuator with poor PID tuning. While other residuals (r_{E,T_2} and r_{E,T_3}) occasionally exceed their thresholds at various time instances during the operation, they do not exceed the thresholds for longer than $t_d = 0.01$ hr. Thus, the monitoring system concludes that their values exceed their thresholds simply because of process and measurement noise.

Once the Q_1 control actuator is isolated as the poorly-tuned actuator, Cohen–Coon tuning method is applied to the controller around Q_1 based on the estimated

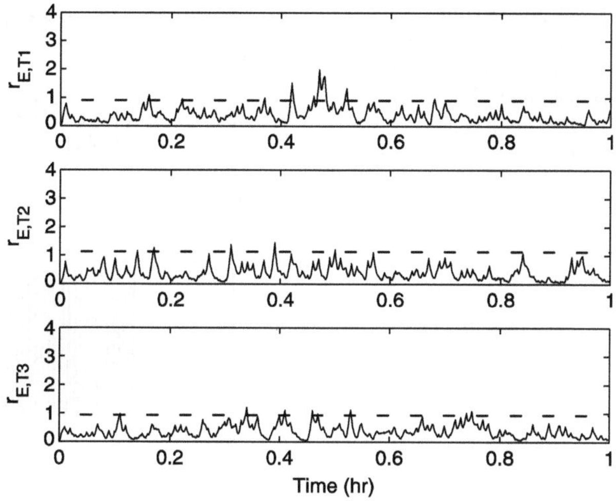

Fig. 7.29 Example 1: Temperature residuals for the 3 vessels computed via EWMA when PID retuning is implemented. The *dashed lines* represent the EWMA residual thresholds $\Omega_{E,i}$

transfer function of the control actuator G'_p. The Cohen–Coon tuning rule is based on the first-order-plus-dead-time estimation of the transfer function of the controlled process. Specifically, the Cohen–Coon tuning rule is as follows [31]:

$$K_c = \frac{\tau_p}{K_p \tau_d}\left(\frac{4}{3} + \frac{\tau_d}{4\tau_p}\right), \tag{7.40a}$$

$$\tau_I = \tau_d \frac{32 + 6\frac{\tau_d}{\tau_p}}{13 + 8\frac{\tau_d}{\tau_p}}, \tag{7.40b}$$

$$\tau_D = \tau_d \frac{4}{11 + 2\frac{\tau_d}{\tau_p}}, \tag{7.40c}$$

where K_p is the actuator's gain, τ_d is the actuator dead time, and τ_p is the actuator's time constant. With this tuning rule and the estimated transfer function of the actuator G'_p presented in Eq. (7.37), the resulting parameters for the PID of Q_1 are as follows:

$$\begin{aligned} K_c &= 1.29, \\ \tau_I &= 6.15 \text{ s}, \\ \tau_D &= 1.06 \text{ s}. \end{aligned} \tag{7.41}$$

After Q_1 is retuned, no more problem can be detected from the EWMA residuals of T_1. In terms of the actual control actuator performance, after being retuned with Cohen–Coon method, $u_a(t)$ tracks $u_m(t)$ quite well; please, see Fig. 7.28.

Example 2 In this example, we will use internal model control tuning rule [149] to tune the PID parameters. We initialize the process model from the following

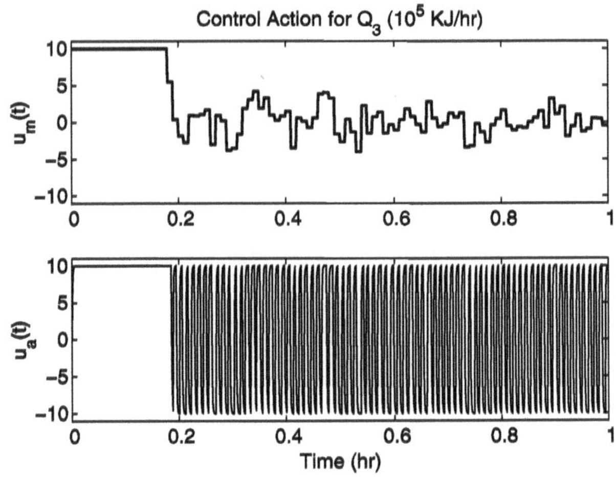

Fig. 7.30 Example 2: Requested actuation level by the MPC ($u_m(t)$) and actual actuation level ($u_a(t)$) when PID retuning is not implemented

initial condition: $x(0) = 0.8x_s$ where x_s is the operating steady-state. All PID controllers start out being properly tuned with the parameters presented in Eq. (7.38). At time $t = 0.1$ hr, a poor PID tuning with the following parameters:

$$\begin{aligned} K_c &= 6.48, \\ \tau_I &= 0.594 \text{ s}, \\ \tau_D &= 5.40 \text{ s} \end{aligned} \tag{7.42}$$

is applied to the PID controller for the control actuator Q_3. Figure 7.30 shows that the tuning of the PID controller for Q_3 causes $u_a(t)$ to oscillate significantly. Figure 7.31 shows the EWMA residuals of the temperature of the 3 vessels when PID retuning is not implemented.

With the monitoring system implemented, Fig. 7.32 shows that r_{E,T_3} is found to start exceeding its threshold Ω_{E,T_3} at $t = 0.206$ hr. After waiting for $t_d = 0.01$ hr, r_{E,T_3} is found to have been continuously above its threshold until $t = 0.216$ hr. Because Q_3 is the only actuator that has relative degree 1 with the process state T_3, at $t = 0.216$ hr the monitoring system isolates Q_3 and declares that Q_3 is poorly tuned. As a result, at $t = 0.216$ hr, a set of PID parameters is calculated via internal model control tuning method based on the estimated transfer function of the control actuator G'_p. For a tuning with fast PID step response, internal model control tuning rule suggests the following PID parameters for processes that can be approximated with first-order-plus-dead-time transfer function [149]:

$$K_c = \frac{\tau_p}{2K_p\tau_d}, \tag{7.43a}$$

$$\tau_I = \min(\tau_p, 8\tau_d), \tag{7.43b}$$

$$\tau_D = 0, \tag{7.43c}$$

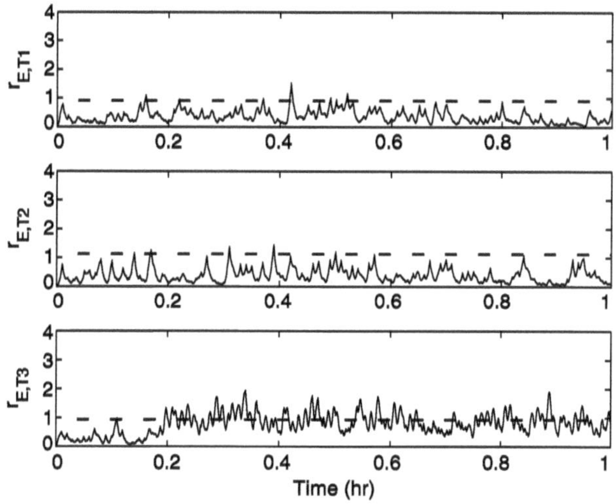

Fig. 7.31 Example 2: Temperature residuals for the 3 vessels computed via EWMA when PID retuning is not implemented. The *dashed lines* represent the EWMA residual thresholds $\Omega_{E,i}$

where K_p is the actuator's gain, τ_d is the actuator dead time, τ_p is the actuator's time constant. This results in the following PID parameters:

$$\begin{aligned} K_c &= 0.392, \\ \tau_I &= 2.82 \text{ s}, \\ \tau_D &= 0 \text{ s}. \end{aligned} \tag{7.44}$$

Figure 7.33 shows the resulting actual actuation level ($u_a(t)$) of Q_3. Though poor PID tuning is applied at $t = 0.1$ hr, its effect in terms of PID response of the control actuator is observed at $t = 0.185$ hr when the step change happens. In terms of detecting this oscillation pattern from the process sate measurements, this is detected and isolated at $t = 0.216$ hr and the PID parameters of Q_3 are retuned.

Notice in Fig. 7.31 that the magnitude of the residuals of the directly-affected process state (r_{E,T_3} in this case) is much lower than r_{E,T_1} in Example 1 (shown in Fig. 7.27). This is because the poor PID tuning problem in this example results in an actuator oscillation ($u_a(t)$) that oscillates with very high frequency around the set-point ($u_m(t)$). In terms of the process states, this leads to a smaller overall deviation of the actual process state ($x(t)$) from the expected process state ($\check{x}(t)$). This is why there is a slightly larger time lag between the initial time when $u_a(t)$ starts deviating from $u_m(t)$ and the time when the poor tuning is isolated, compared to Example 1.

Remark 7.16 While the mean and standard deviation of the residuals are calculated in the presence of process noise under normal operation at the desired steady-state, the applicability of the proposed dynamic filter for computing the residuals together with real-time state variable measurements is not limited to steady-state operation; the reason is the design of the proposed dynamic filter which can accurately predict normal evolution of the process state variables away from the steady-state in the closed-loop system, thereby leading to the computation of residual values that are

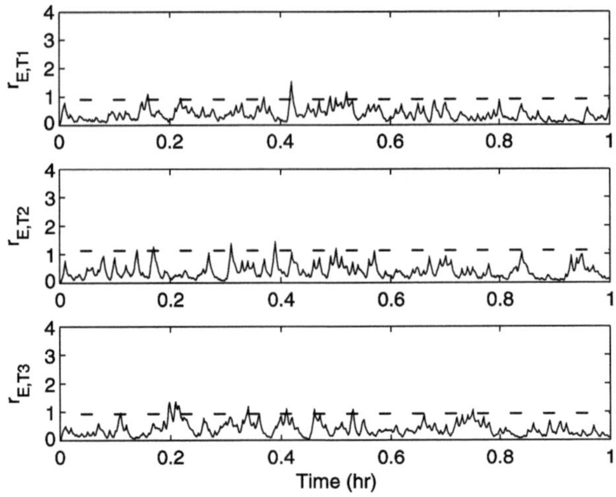

Fig. 7.32 Example 2: Temperature residuals for the 3 vessels computed via EWMA when PID retuning is implemented. The *dashed lines* represent the EWMA residual thresholds $\Omega_{E,i}$

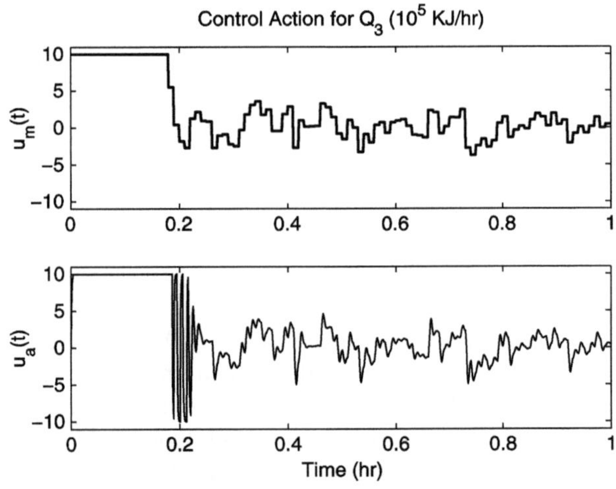

Fig. 7.33 Example 2: Requested actuation level by the MPC ($u_m(t)$) and actual actuation level ($u_a(t)$) when PID retuning is implemented

valid for process operation away from the steady-state (note that the initial condition in the example is not chosen to be the steady-state).

7.4 Conclusion

This chapter presented methods for utilizing FDI concepts for controller design as well as controller performance monitoring. The first approach strengthens existing FDI techniques by enforcing an appropriate structure on the closed-loop system that may separate regions of faulty operation in the state-space such that fault isolation may become possible. This was illustrated through two chemical process examples, a CSTR example and a polyethylene reactor example. By carefully designing the feedback controller, it was demonstrated that it is possible to enhance the isolability of particular faults. In the CSTR example, feedback linearization was used to achieve the required closed-loop system structure in order to perform fault detection and isolation, whereas in the polyethylene reactor example, a more general approach to nonlinear controller design was used in meeting the required conditions for isolability. Additionally, it was demonstrated that using a data-based method of monitoring the T^2 values of the resulting subsystems, it was possible to isolate certain faults due to the enforced closed-loop system structure. In the second part of the chapter, we focused on the problem of monitoring and retuning of low-level PID control loops used to regulate control actuators to the values computed by advanced model-based control systems like MPC. Focusing on the case where the real-time measurement of the actuation level is unavailable, we use process state measurements and process models to carry out PID controller monitoring and compute appropriate residuals. Once a poorly-tuned PID controller is detected and isolated, a PID tuning method based on the estimated transfer function of the control actuator was applied to retune PID controller. The proposed method was applied to a nonlinear reactor-separator process operating under MPC control with low-level PID controllers regulating the control actuators and its performance was successfully evaluated via extensive simulations.

Chapter 8
Isolation and Handling of Sensor Faults

8.1 Introduction

In the earlier chapters, we considered the problem of handling actuator faults. Compared to actuator faults, relatively fewer results for the problem of detecting, isolating, and handling sensor faults are available for nonlinear systems. When sensor faults are considered, observers are typically required to fully or partly recover the system state. The design of observers, however, is a challenging problem for nonlinear systems. In the context of output feedback control, high-gain observers are known to have good convergence properties and have been studied for continuous-time systems (e.g., [11, 46, 51]) and sampled-data systems with uniform [32, 120] and multiple [2] sampling rates. These results, however, rely on a special structure for the system to be in to begin with, or after an appropriate transformation. To generalize the application of this type of observers, a model predictive control formulation has been studied in [53], where the discrete nature of the control implementation is exploited to relax the relatively restrictive system structure required in the standard high-gain observer design. This generalization, however, is developed under the assumption of the locally Lipschitz continuity of the control input in the system state. Note that this assumption is hard to verify, especially under controllers such as model predictive control where the control law is not explicit but results from the solution to an optimization problem. One of the contributions of the present chapter is to generalize the design and applicability of the high-gain observers under less restrictive and easily verifiable conditions, which also helps satisfy the requirements of the filter design in the present approach.

Specifically, this chapter considers the problem of sensor FDI and FTC for nonlinear systems subject to input constraints. To this end, first results are presented that generalize the design of high-gain observers for nonlinear systems. The presented observer design is subject to less restrictive and easily verifiable assumptions compared to the results in the literature (the analysis in the standard high-gain observer design is not directly applicable to the system considered in this chapter due to the differences in the system structures or the coordinate transformations). Specifically,

P. Mhaskar et al., *Fault-Tolerant Process Control*, DOI 10.1007/978-1-4471-4808-1_8,

it expands the class of nonlinear systems to which high-gain observers can be applied, without assuming the locally Lipschitz continuity of the control input in the system state as required in [53]. Exploiting this increased applicability of the high-gain observers, a fault isolation mechanism is designed by using a bank of high-gain observers, and a novel residual generation method is proposed for FDI, which is based on the comparison between the state estimates and their predictions using previous estimates. While a similar isolation logic is presented algorithmically in [99], where the nonlinear system is first approximated by a linear parameter varying system, and there are other results that use the idea of a bank of observers in the context of linear (or linear approximations of nonlinear) systems, the present results provide a rigorous filter design and analysis that explicitly handles the presence of nonlinearities and input constraints.

8.2 Preliminaries

Consider a multi-input multi-output nonlinear system described by

$$\begin{aligned} \dot{x} &= f(x) + g(x)u, \\ y &= h(x) + v(t), \end{aligned} \tag{8.1}$$

where $x \in \mathbb{R}^n$ denotes the vector of state variables, $u \in \mathbb{R}^m$ denotes the vector of constrained input variables, taking values in a nonempty compact convex set $\mathcal{U} \subseteq \mathbb{R}^m$ that contains 0, $y = [y_1, \dots, y_p]^{\mathrm{T}} \in \mathbb{R}^p$ denotes the vector of output variables, $v = [v_1, \dots, v_p]^{\mathrm{T}} \in \mathbb{R}^p$ denotes the fault vector for the sensors, and $g(x) = [g_1(x), \dots, g_m(x)]$. In the control design, we consider the system of Eq. (8.1) under fault-free conditions (i.e., $v \equiv 0$), which satisfies Assumption 8.1 below.

Assumption 8.1 The functions $f : \mathbb{R}^n \to \mathbb{R}^n$ and $g_i : \mathbb{R}^n \to \mathbb{R}^n$, $i = 1, \dots, m$, are $\mathcal{C}^1$ functions on their domains of definition, $f(0) = 0$, and the function $h : \mathbb{R}^n \to \mathbb{R}^p$ is $\mathcal{C}^1$ on its domain of definition.

Instead of using a specific control law, the results in this chapter are developed for any control law that satisfies Assumption 8.2 below.

Assumption 8.2 For the system of Eq. (8.1), there exists a control Lyapunov function $V : \mathbb{R}^n \to \mathbb{R}$, which is a $\mathcal{C}^2$ function on its domain of definition. Let $\Omega_c = \{x \in \mathbb{R}^n : V(x) \le c\}$ denote the stability region of the closed-loop system obtained under a state feedback control law $u_c : \Omega_c \to \mathcal{U}$. Furthermore, there exists a class $\mathcal{K}$ function $\alpha : [0, c) \to [0, \infty)$ such that for any $x \in \Omega_c$, the following inequality holds:

$$L_f V(x) + L_g V(x) u_c(x) \le -\alpha\bigl(V(x)\bigr), \tag{8.2}$$

where $L_g V(x) = [L_{g_1} V(x), \dots, L_{g_m} V(x)]$.

Remark 8.1 Assumption 8.2 requires that the input prescribed by the controller should be able to make the derivative of the control Lyapunov function negative for any state, except for the origin, within the stability region obtained under the state feedback control law. It encompasses feedback controllers under which the origin is an asymptotically stable equilibrium point of the closed-loop system. It also allows inclusion of model predictive control designs, for which an inequality leading to Eq. (8.2) is used to guarantee practical stability of the closed-loop system (e.g., [93, 95, 108, 110]).

We now present an assumption on the system structure that is suitable for high-gain observer designs.

Assumption 8.3 (c.f. [53]) There exist integers $\omega_i, i = 1, \ldots, p$, with $\sum_{i=1}^{p} \omega_i = n$, and a coordinate transformation $\zeta = T(x, u)$ such that if $u = \bar{u}$, where $\bar{u} \in \mathcal{U}$ is a constant vector, then the representation of the system of Eq. (8.1) in the ζ coordinate takes the following form:

$$\begin{aligned} \dot{\zeta} &= A\zeta + B\phi(x, \bar{u}), \\ y &= C\zeta, \end{aligned} \tag{8.3}$$

where $\zeta = [\zeta_1, \ldots, \zeta_p]^{\mathrm{T}} \in \mathbb{R}^n$, $A = \text{blockdiag}[A_1, \ldots, A_p]$, $B = \text{blockdiag}[B_1, \ldots, B_p]$, $C = \text{blockdiag}[C_1, \ldots, C_p]$, $\phi = [\phi_1, \ldots, \phi_p]^{\mathrm{T}}$, $\zeta_i = [\zeta_{i,1}, \ldots, \zeta_{i,\omega_i}]^{\mathrm{T}}$, $A_i = \left[\begin{smallmatrix} 0 & I_{\omega_i - 1} \\ 0 & 0 \end{smallmatrix}\right]$, with I_{ω_i-1} being the $(\omega_i - 1) \times (\omega_i - 1)$ identity matrix, $B_i = [0^{\mathrm{T}}_{\omega_i-1}, 1]^{\mathrm{T}}$, with 0_{ω_i-1} being a vector of zeros of dimension $\omega_i - 1$, $C_i = [1, 0^{\mathrm{T}}_{\omega_i-1}]$, and $\phi_i(x, \bar{u}) = \phi_{i,\omega_i}(x, \bar{u})$, with $\phi_{i,\omega_i}(x, \bar{u})$ defined through the successive differentiation of $h_i(x)$: $\phi_{i,1}(x, \bar{u}) = h_i(x)$ and $\phi_{i,j}(x, \bar{u}) = \frac{\partial \phi_{i,j-1}}{\partial x}[f(x) + g(x)\bar{u}]$, $j = 2, \ldots, \omega_i$. Furthermore, the functions $T : \mathbb{R}^n \times \mathcal{U} \to \mathbb{R}^n$ and $T^{-1} : \mathbb{R}^n \times \mathcal{U} \to \mathbb{R}^n$ are C^1 functions on their domains of definition.

Remark 8.2 Note that the system structure requirement on high-gain observer designs as described in Assumption 8.3 is less restrictive than those in literature (e.g., [11, 46, 95]). In particular, the input information can be used in the coordinate transformation, which is assumed to be known (see [53] for a more general assumption, where the derivatives of the input variables are possibly non-zero). In contrast, the standard high-gain observer design requires that it be possible to transform the original system into a similar form to Eq. (8.3) by using a coordinate transformation without the involvement of the input.

We next present the output feedback control design, where the input is prescribed at discrete times $t_k = k\Delta$, $k = 0, \ldots, \infty$, with Δ being the hold-time of the control action. For $t \in [t_k, t_{k+1})$, consider the following output feedback controller using high-gain observers:

$$\begin{aligned}
\dot{\hat{\zeta}} &= A\hat{\zeta} + B\phi_0\big(\hat{x}, u(t_k)\big) + H(y - C\hat{\zeta}),\\
\hat{\zeta}(t_k) &= T\big(\hat{x}(t_k), u(t_k)\big),\\
u &= u_c\big(\mathrm{sat}\big(\hat{x}(t_k)\big)\big) \quad \text{for all } t \in [t_k, t_{k+1}),
\end{aligned} \tag{8.4}$$

where $\hat{x}$ and $\hat{\zeta}$ denote the estimates of x and ζ, respectively, $H = \mathrm{blockdiag}[H_1, \dots, H_p]$ is the observer gain, $H_i = [\frac{a_{i,1}}{\varepsilon}, \dots, \frac{a_{i,\omega_i}}{\varepsilon^{\omega_i}}]^{\mathrm{T}}$, with $s^{\omega_i} + a_{i,1}s^{\omega_i - 1} + \dots + a_{i,\omega_i} = 0$ being a Hurwitz polynomial and ε being a positive constant to be specified, and $\hat{x}(t_k) = T^{-1}(\hat{\zeta}(t_k^-), u(t_{k-1}))$ for $k = 1, \dots, \infty$. The initial state of the observer is denoted by $\hat{x}_0 := \hat{x}(0)$, which takes values from any compact set $\mathcal{Q} \subseteq \mathbb{R}^n$. In the transformed coordinate, the state estimate in the ζ coordinate is re-initialized at discrete times to account for the possible changes in the input. A saturation function is used to scale back the estimate (passed to the controller) to lie within the state feedback stability region (to prevent the peaking phenomenon and enable using the state feedback control law designed for the same region), which is defined as $sat(\hat{x})$:

$$\hat{x} \quad \text{for } \hat{x} \in \Omega_c, \tag{8.5}$$

$$\beta\hat{x} \quad \text{for } \hat{x} \notin \Omega_c, \tag{8.6}$$

where $\beta \in (0, 1)$ is a scaling factor such that $V(\beta\hat{x}) = c$ and the computation of β is specific to the choice of the control Lyapunov function. For a quadratic control Lyapunov function, it may be computed as follows:

$$\beta = \sqrt{\frac{c}{V(\hat{x})}}. \tag{8.7}$$

The function ϕ_0 is a nominal model of ϕ used in the observer design. The following analysis (see Proposition 8.1 below) requires the global boundedness of ϕ_0. If ϕ is known, but not globally bounded, the global boundedness of ϕ_0 can always be achieved by bounding ϕ outside a compact set of interest. To this end, the function ϕ_0 is required to satisfy the following assumption.

Assumption 8.4 $\phi_0(x, u)$ is a C^0 function on its domain of definition and globally bounded in x.

Let $D = \mathrm{blockdiag}[D_1, \dots, D_p]$, where $D_i = \mathrm{diag}[\varepsilon^{\omega_i - 1}, \dots, 1]$, and define the scaled estimation error $e = D^{-1}(\zeta - \hat{\zeta}) \in \mathbb{R}^n$. For $t \in [t_k, t_{k+1})$, the scaled estimation error evolves as follows:

$$\begin{aligned}
\varepsilon\dot{e} &= A_0 e + \varepsilon B\big[\phi\big(x, u(t_k)\big) - \phi_0\big(\hat{x}, u(t_k)\big)\big],\\
e(t_k) &= D^{-1}\big[T\big(x(t_k), u(t_k)\big) - T\big(\hat{x}(t_k), u(t_k)\big)\big],
\end{aligned} \tag{8.8}$$

where $A_0 = \text{blockdiag}[A_{0,1}, \dots, A_{0,p}]$, $A_{0,i} = \left[a_i \; \begin{smallmatrix} I_{\omega_i-1} \\ 0^T_{\omega_i-1} \end{smallmatrix} \right]$, and $a_i = [-a_{i,1}, \dots, -a_{i,\omega_i}]^T$.

Remark 8.3 The output feedback control design of Eq. (8.4) extends the class of nonlinear systems to which high-gain observers can be applied. This is achieved by utilizing the discrete nature of the control implementation under Assumption 8.3. Specifically, the control input is determined at discrete times and this information is available to the observer over each time interval. The design of high-gain observers subject to the less restrictive system structure has been studied in [53], which assumes the local Lipschitz continuity of the control input in the system state. This assumption, however, is hard to verify, in particular for model predictive control implementations, where the control input is obtained by solving a nonlinear optimization problem for a given state and an explicit control law is not available. In contrast, the assumptions used in this chapter can be verified algebraically. Therefore, one of the contributions of the present chapter is that it generalizes the design of high-gain observers subject to less restrictive and easily verifiable assumptions for constrained nonlinear systems.

Applying the change of time variable $\tau = \frac{t}{\varepsilon}$ and setting $\varepsilon = 0$, the boundary-layer system is given by

$$\frac{de}{d\tau} = A_0 e. \tag{8.9}$$

For the boundary-layer system, we define a Lyapunov function $W(e) = e^T P_0 e$, where P_0 is the symmetric positive definite solution of the Lyapunov function $A_0^T P_0 + P_0 A_0 = -I$. Let λ_{min} and λ_{max} denote the minimum and maximum eigenvalues of P_0, respectively. In preparation to the presentation of the main results, we first give the following proposition, which is similar to a result obtained in [11], and hence stated without proof.

Proposition 8.1 *Consider the system of Eq.* (8.1), *for which Assumptions* 8.1, 8.3, *and* 8.4 *hold. If* $x_0 := x(0) \in \Omega_b$, *where* $0 < b < c$, *then given* $b' \in (b, c)$, *there exists a finite time* t_e, *independent of* ε, *such that* $x(t) \in \Omega_{b'}$ *for all* $t \in [0, t_e]$. *Furthermore, there exists* $\sigma > 0$, *independent of* ε, *such that for any* $e(t) \in \mathcal{W}_o := \{e \in \mathbb{R}^n : W(e) \geq \sigma\varepsilon^2\}$ *and* $x(t) \in \Omega_c$, *the following equation holds*:

$$\dot{W} \leq -\frac{1}{2\varepsilon}\|e\|^2. \tag{8.10}$$

Remark 8.4 Proposition 8.1 establishes a finite time (t_e) such that given an initial condition within a subset of $\Omega_{b'}$, the system state continues to reside in $\Omega_{b'}$ over this time period (see Fig. 8.1). As a matter of fact, the same result holds for Ω_c. The set $\Omega_{b'}$, a subset of Ω_c, is used because in the next section, we will establish a result that if the system state is within $\Omega_{b'}$ and the scaled estimation error is sufficiently small, then the state estimate is also within Ω_c. In addition, Proposition 8.1 establishes

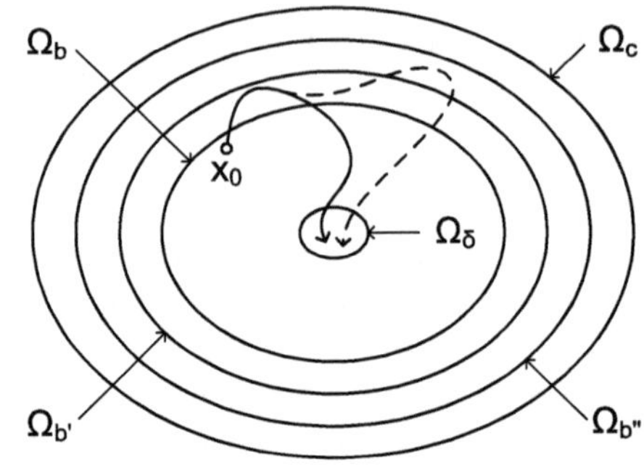

Fig. 8.1 Schematic of the stability region and the evolution of the closed-loop state trajectories under fault-free (*solid line*) and faulty (*dashed line*) conditions. The notation Ω_c denotes the stability region obtained under state feedback control. For any initial condition x_0 within Ω_b, the state estimate is guaranteed to converge before the system state goes outside $\Omega_{b'}$. Subsequently, if a fault is detected and isolated before the system state goes outside $\Omega_{b''}$ (i.e., within the FDI time window), the use of the state estimate generated using measurements from the remaining healthy sensors guarantees practical stability of the closed-loop system (i.e., the system state converges to a closed ball of radius d around the origin, which contains the set Ω_δ)

a fact that over each time interval before the system state goes outside $\Omega_{b'}$, the derivative of the Lyapunov function ($\dot{W}$) remains negative if the scaled estimation error is not within the neighborhood of the origin (or within $\mathcal{W}_o$).

8.3 Practical Stability of the Closed-Loop System Under Output Feedback Control

This section establishes the closed-loop property of the system under output feedback control and forms the basis for the sensor FDI design in the subsequent section. To this end, consider the system of Eq. (8.1), for which Assumptions 8.1–8.4 hold, under the output feedback controller of Eq. (8.4). The stability property of the closed-loop system is formalized in Theorem 8.1 below.

Theorem 8.1 *Given any* $0 < b < c$ *and* $d > 0$, *there exist* $\Delta^* > 0$ *and* $\varepsilon^* > 0$ *such that if* $\Delta \in (0, \Delta^*]$, $\varepsilon \in (0, \varepsilon^*]$, *and* $x_0 \in \Omega_b$, *then* $x(t) \in \Omega_c$ $\forall t \geq 0$ *and* $\limsup_{t\to\infty} \|x\| \leq d$.

Proof The proof is divided into two parts. In the first part, we show that given $e_b > 0$, which is to be determined in the second part, there exists $\varepsilon^* > 0$ such that if $\varepsilon \in (0, \varepsilon^*]$ and $\Delta \in (0, t_e]$, then the scaled estimation error $e(t_k^-)$ enters $\mathcal{E} := \{e \in \mathbb{R}^n : \|e\| \leq e_b\}$ no later than the time t_e, which is defined in Proposition 8.1, and stays in $\mathcal{E}$ thereafter as long as $x(t)$ remains in Ω_c. In the second part, we show that for any $d > 0$, there exist $e_b^* > 0$ and $\Delta^* > 0$ such that if $e(t_k^-) \in \mathcal{E}$ for some $t_{k'} \leq t_e$,

$e_b \in (0, e_b^*]$, and $\Delta \in (0, \Delta^*]$, then practical stability of the closed-loop system can be established.

Consider $\Delta \in (0, \Delta_1]$ and $\varepsilon \in (0, \varepsilon_1]$, where $\Delta_1 = t_e$ and $\varepsilon_1 = \sqrt{\frac{\gamma}{\sigma}}$, with $0 < \gamma < \min_{\|e\|=e_b} W(e)$. In order to show that $e(t_k^-)$ converges to $\mathcal{E}$, we only need to show that it converges to $\mathcal{W}_i := \{e \in \mathbb{R}^n : W(e) \le \sigma\varepsilon^2\}$.

Part 1: We first show that $e(t_k^-)$ reaches $\mathcal{W}_i$ no later than the time t_e. Let N be the largest integer such that $N\Delta \le t_e$. It follows from Proposition 8.1 that if $t_{k+1} \le t_e$, $k = 0, \dots, N-1$, then for any $e \in \mathcal{W}_o$ and $t \in [t_k, t_{k+1})$, we have

$$\dot{W} \le -\frac{1}{2\lambda_{\max}\varepsilon} W. \tag{8.11}$$

It follows that

$$W\big(e\big(t_{k+1}^-\big)\big) \le e^{-\frac{\Delta}{2\lambda_{\max}\varepsilon}} W\big(e(t_k)\big). \tag{8.12}$$

Since $T(x, u)$ and $T^{-1}(\zeta, u)$ are locally Lipschitz in x and ζ, respectively, and

$$\begin{aligned} e(t_k) &= D^{-1}\big[\zeta(t_k) - \hat{\zeta}(t_k)\big] \\ &= D^{-1}\big[T\big(x(t_k), u(t_k)\big) - T\big(\hat{x}(t_k), u(t_k)\big)\big], \end{aligned} \tag{8.13}$$

there exist $L_1, L_2 > 0$ such that the following equation holds:

$$\begin{aligned} \big\|e(t_k)\big\| &\le L_1 \max\big\{1, \varepsilon^{1-\omega_{\max}}\big\} \big\|x(t_k) - \hat{x}(t_k)\big\| \\ &= L_1 \max\big\{1, \varepsilon^{1-\omega_{\max}}\big\} \big\|T^{-1}\big(\zeta(t_{k-1}), u(t_{k-1})\big) - T^{-1}\big(\hat{\zeta}(t_{k-1}), u(t_{k-1})\big)\big\| \\ &\le L_1 L_2 \max\big\{1, \varepsilon^{1-\omega_{\max}}\big\} \max\big\{1, \varepsilon^{\omega_{\max}-1}\big\} \big\|e\big(t_k^-\big)\big\| \\ &= L_1 L_2 \eta_1(\varepsilon) \big\|e\big(t_k^-\big)\big\|, \end{aligned} \tag{8.14}$$

where $\omega_{\max} = \max_{i=1,\dots,p}\{\omega_i\}$ and $\eta_1(\varepsilon) = \varepsilon^{(\omega_{\max}-1)\operatorname{sgn}(\varepsilon-1)}$. Let $\tilde{L}_1 = L_1 L_2$. It follows from Eq. (8.12) and Eq. (8.14) that if $e(t) \in \mathcal{W}_o$ for all $t \in [t_k, t_{k+1})$, then the following equation holds:

$$\begin{aligned} W\big(e(t_{k+1})\big) &\le \lambda_{\max} \big\|e(t_{k+1})\big\|^2 \\ &\le \lambda_{\max} \tilde{L}_1^2 \big[\eta_1(\varepsilon)\big]^2 \big\|e\big(t_{k+1}^-\big)\big\|^2 \\ &\le \frac{\lambda_{\max}}{\lambda_{\min}} \tilde{L}_1^2 \big[\eta_1(\varepsilon)\big]^2 e^{-\frac{\Delta}{2\lambda_{\max}\varepsilon}} W\big(e(t_k)\big). \end{aligned} \tag{8.15}$$

Note that once $e(t)$ reaches $\mathcal{W}_1$, it stays there at least until the end of the same time interval. Since $T(x, u)$ is continuous, for any $x_0 \in \Omega_b$ and $\hat{x}_0 \in \mathcal{Q}$, there exists $K_1 > 0$ such that

$$\big\|e(0)\big\| \le K_1 \eta_2(\varepsilon), \tag{8.16}$$

where $\eta_2(\varepsilon) = \max\{1, \varepsilon^{1-\omega_{\max}}\}$. To guarantee that $e(t_k^-)$ reaches $\mathcal{W}_i$ by the time t_N, it is required that the following equation hold:

$$\frac{\lambda_{\max}}{\lambda_{\min}} \tilde{L}_1^2 \big[\eta_1(\varepsilon)\big]^2 e^{-\frac{\Delta}{2\lambda_{\max}\varepsilon}} \le \left\{ \frac{\sigma \varepsilon^2}{\lambda_{\max} K_1^2 [\eta_2(\varepsilon)]^2} \right\}^{\frac{1}{N}}. \tag{8.17}$$

Rearranging the above equation gives

$$\frac{[\eta_1(\varepsilon)]^{2N} [\eta_2(\varepsilon)]^2}{\varepsilon^2} e^{-\frac{N\Delta}{2\lambda_{\max}\varepsilon}} \le \frac{\sigma}{\lambda_{\max} K_1^2} \left(\frac{\lambda_{\min}}{\lambda_{\max} \tilde{L}_1^2} \right)^N. \tag{8.18}$$

Since the left-hand side of the above inequality is continuous in ε and tends to zero as ε tends to 0, there exists $\varepsilon_2 > 0$ such that if $\varepsilon \in (0, \varepsilon_2]$, then Eq. (8.17) holds.

We then show that after the scaled estimate error $e(t_k^-)$ reaches $\mathcal{W}_i$, it stays there as long as $x(t)$ stays in Ω_c. Note that given $e(t_k^-) \in \mathcal{W}_i$, it is possible that $e(t_k)$ goes outside $\mathcal{W}_i$ due to the re-initialization to the system state and its estimate in the ζ coordinate. It follows from Eq. (8.14) that if $e(t_k^-) \in \mathcal{W}_i$, then $\|e(t_k)\| \le \tilde{L}_1 \eta_1(\varepsilon) e_b$.

To guarantee that $e(t_{k+1}^-)$ stays in $\mathcal{W}_i$, it is required that the following equation hold:

$$e^{-\frac{\Delta}{2\lambda_{\max}\varepsilon}} \le \frac{\sigma \varepsilon^2}{\lambda_{\max} \tilde{L}_1^2 [\eta_1(\varepsilon)]^2 e_b^2}. \tag{8.19}$$

It can be shown that there exists $\varepsilon_3 > 0$ such that if $\varepsilon \in (0, \varepsilon_3]$, then Eq. (8.19) holds.

In the first part of the proof, it is established that for $\varepsilon \in (0, \varepsilon^*]$, where $\varepsilon^* = \min\{\varepsilon_1, \varepsilon_2, \varepsilon_3\}$, $e(t_k^-)$ enters $\mathcal{E}$ in some finite time $t_{k'} \le t_N \le t_e$, where $t_{k'}$ denotes the earliest time t_k such that $e(t_k^-) \in \mathcal{E}$, and stays in $\mathcal{E}$ thereafter as long as $x(t)$ remains in Ω_c. In addition, $x(t) \in \Omega_c$ $\forall t \in [0, t_{k'}]$.

Part 2: We first show that if the system state resides within a subset of Ω_c and the scaled estimation error is sufficiently small, then the state estimate also resides within Ω_c. It follows from the first part of the proof that we have

$$\|x - \hat{x}\| = \big\| T^{-1}(\zeta, u) - T^{-1}(\hat{\zeta}, u) \big\| \le L_2 \eta_3(\varepsilon) \|e\| \le L_2 \eta_3(\varepsilon_1) \|e\|, \tag{8.20}$$

where $\eta_3(\varepsilon) = \max\{1, \varepsilon^{\omega_{\max}-1}\}$. It can be shown that given $0 < \delta_1 < \delta_2$, there exists $\tilde{e} > 0$ such that if $e_b \in (0, \tilde{e}]$, then $V(\hat{x}) \le \delta_1$ implies $V(x) \le \delta_2$. It follows from Proposition 8.1 that given $b' \in (b, c)$, we have that $x(t_{k'}) \in \Omega_{b'}$. Therefore, there exists $e_{b,1} > 0$ such that if $e_b \in (0, e_{b,1}]$, then $\hat{x}(t_{k'}) \in \Omega_c$.

We then show the existence of $e_b^* > 0$ and $\Delta^* > 0$ such that if $e_b \in (0, e_b^*]$ and $\Delta \in (0, \Delta^*]$, then any state trajectory originating in $\Omega_{b'}$ at time $t_{k'}$ converges to a closed ball of radius d around the origin. Since $V(x)$ is a continuous function of the state, one can find a positive real number $\delta < b'$ such that $V(x) \le \delta$ implies $\|x\| \le d$. Let $\hat{\delta}$ be a positive real number such that $0 < \hat{\delta} < \delta$. If $e_b \in (0, e_{b,1}]$, the state estimate at time $t_{k'}$ can either be such that $\hat{\delta} < V(\hat{x}(t_{k'})) \le c$ or $V(\hat{x}(t_{k'})) \le \hat{\delta}$.

Case 1: Consider $\hat{x}(t_k) \in \Omega_c \backslash \Omega_{\hat{\delta}}$. For this case, we have

$$L_f V\big(\hat{x}(t_k)\big) + L_g V\big(\hat{x}(t_k)\big) u(t_k) \le -\alpha\big(V\big(\hat{x}(t_k)\big)\big) < -\alpha(\hat{\delta}).$$

It follows from the continuity properties of $f(\cdot)$, $g(\cdot)$, and $V(\cdot)$ that $L_fV(\cdot)$ and $L_gV(\cdot)$ are locally Lipschitz on the domain of interest. Therefore, there exists $L_3 > 0$ such that

$$\begin{aligned} &\left|L_fV\big(x(t_k)\big)+L_gV\big(x(t_k)\big)u(t_k)-L_fV\big(\hat{x}(t_k)\big)-L_gV\big(\hat{x}(t_k)\big)u(t_k)\right| \\ &\quad\le L_3\big\|x(t_k)-\hat{x}(t_k)\big\| \\ &\quad\le L_2L_3\eta_3(\varepsilon_1)\big\|e\big(t_k^-\big)\big\|. \end{aligned} \tag{8.21}$$

Since the functions $f(\cdot)$ and $g(\cdot)$ are continuous, u is bounded, and $\Omega_{b'}$ is bounded, one can find $K_2 > 0$ such that $\|x(t) - x(t_k)\| \le K_2\Delta$ for any $\Delta \in (0, \Delta_1]$, $x(t_k) \in \Omega_{b'}$ and $t \in [t_k, t_k + \Delta)$. It follows that $\forall t \in [t_k, t_k + \Delta)$, the following equation holds:

$$\begin{aligned} \dot{V}\big(x(t)\big) &= L_fV\big(\hat{x}(t_k)\big)+L_gV\big(\hat{x}(t_k)\big)u(t_k)+\big[L_fV\big(x(t)\big)+L_gV\big(x(t)\big)u(t_k) \\ &\quad - L_fV\big(x(t_k)\big)-L_gV\big(x(t_k)\big)u(t_k)\big]+\big[L_fV\big(x(t_k)\big)+L_gV\big(x(t_k)\big)u(t_k) \\ &\quad - L_fV\big(\hat{x}(t_k)\big)-L_gV\big(\hat{x}(t_k)\big)u(t_k)\big] \\ &< -\alpha(\hat{\delta})+L_3K_2\Delta+L_2L_3\eta_3(\varepsilon_1)\big\|e\big(t_k^-\big)\big\|. \end{aligned} \tag{8.22}$$

Consider $\Delta \in (0, \Delta_2]$, where $\Delta_2 = \frac{\alpha(\hat{\delta})}{3L_3K_2}$, and $e_b \in (0, e_{b,2}]$, where $e_{b,2} = \frac{\alpha(\hat{\delta})}{3L_2L_3\eta_3(\varepsilon_1)}$. Then, we have

$$\dot{V}\big(x(t), u(t)\big) < -\frac{1}{3}\alpha(\hat{\delta}) < 0. \tag{8.23}$$

Since $\dot{V}(x(t))$ remains negative over $[t_k, t_k + \Delta)$, $x(t)$ remains in Ω_c over the same time interval, and $V(x(t_k + \Delta)) < V(x(t_k))$.

If $\hat{x}(t_k') \in \Omega_c \backslash \Omega_{\hat{\delta}}$, we have $\dot{V}(x(t)) < 0$ over $[t_{k'}, t_{k'} + \Delta)$. It follows that $\hat{x}(t_{k'+1}) \in \Omega_c$ for $e_b \in (0, e_{b,1}]$. Similarly, it can be shown that for $t_k > t_{k'}$, $\dot{V}(x(t))$ remains negative until $\hat{x}(t_k)$ reaches $\Omega_{\hat{\delta}}$.

Case 2: Consider $\hat{x}(t_k) \in \Omega_{\hat{\delta}}$. Let δ' be a positive real number such that $\hat{\delta} < \delta' < \delta$. There exists $e_{b,3} > 0$ such that if $e_b \in (0, e_{b,3}]$, then $V(\hat{x}) \in \Omega_{\hat{\delta}}$ implies $V(x) \in \Omega_{\delta'}$ and $\{x \in \mathbb{R}^n : \|x - \hat{x}\| \le L_2\eta_3(\varepsilon_1)e_{b,3}\ \forall \hat{x} \in \Omega_{\hat{\delta}}\} \subset \Omega_\delta$. Since $V(x)$ is continuous, and x evolves continuously in time, there exists $\Delta_3 > 0$ such that for $x(t_k) \in \Omega_{\delta'}$, if $\Delta \in (0, \Delta_3]$, then $V(x(t)) \le \delta$ for any $t \in [t_k, t_k + \Delta)$. If $\Delta \in (0, \Delta_3]$, we have $x(t_{k+1}) \in \Omega_\delta$. It follows that $\hat{x}(t_{k+1}) \in \Omega_c$ for $e_b \in (0, e_{b,1}]$.

For $e_b \in (0, e_b^*]$ and $\Delta \in (0, \Delta^*]$, where $e_b^* = \min\{e_{b,1}, e_{b,2}, e_{b,3}\}$ and $\Delta^* = \min\{\Delta_1, \Delta_2, \Delta_3\}$, it can be shown by iteration that any state trajectory originating in $\Omega_{b'}$ at time $t_{k'}$ converges to the set Ω_δ, and hence converges to the closed ball of radius d around the origin.

In the second part of the proof, it is established that for any $d > 0$ there exists $e_b^* > 0$ and $\Delta^* > 0$ such that if $e(t_{k'}^-) \in \mathcal{E}$, $e_b \in (0, e_b^*]$, and $\Delta \in (0, \Delta^*]$, then $x(t) \in \Omega_c\ \forall t \ge t_{k'}$ and $\limsup_{t\to\infty} \|x\| \le d$.

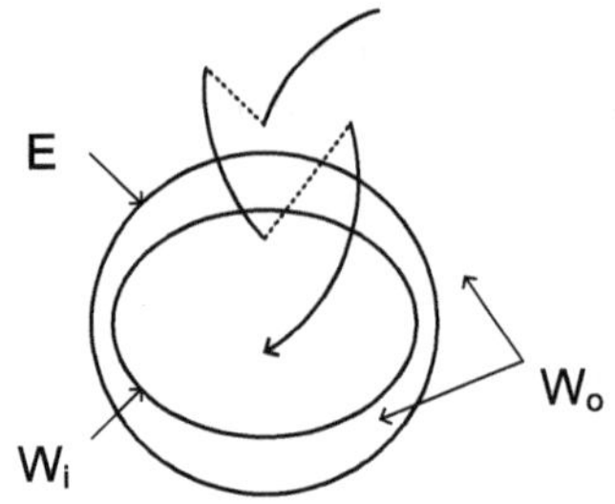

Fig. 8.2 Schematic of the evolution of the scaled estimation error. $\mathcal{E}$ is the terminal set and $\mathcal{W}_i$ is the level set of the Lyapunov function contained in $\mathcal{E}$. Note that after convergence, while jumps resulting from input changes may drive the estimation error outside $\mathcal{E}$ (see the *dotted lines*), by the end of each interval, the estimation error is guaranteed to be within $\mathcal{E}$ (see the *solid lines*)

In summary, it is shown that given any $0 < b < c$ and $d > 0$, there exist $\Delta^* > 0$ and $\varepsilon^* > 0$ such that if $\Delta \in (0, \Delta^*]$, $\varepsilon \in (0, \varepsilon^*]$, and $x_0 \in \Omega_b$, then $x(t) \in \Omega_c$ $\forall t \geq 0$ and $\limsup_{t\to\infty} \|x\| \leq d$. This concludes the proof of Theorem 8.1. □

Remark 8.5 Note that locally Lipschitz continuity of the coordinate transformation function and its inverse function is used to construct the relationship between the values of the scaled estimation error associated with different values of the input. By using this technique (not required in the standard high-gain observer design), it is shown that although the scaled estimation error may deviate from the origin due to changes in the input, a sufficiently small ε can make it be at an inner level surface at the next update time until $e(t_k^-)$ reaches the neighborhood of the origin, denoted by $\mathcal{W}_i$. Therefore, it is unnecessary to require that it converge to the neighborhood of the origin at the end of the first time interval as in [53]. More importantly, it is shown that the scaled estimation error $e(t_k^-)$ stays in the terminal set, denoted by $\mathcal{E}$, which contains the neighborhood of the origin, ultimately (see Fig. 8.2 for an illustration). Note also that the hold-time for the control implementation should be less than the time t_e (at least for the first time interval). This is because if the control input is not updated at a sufficiently fast rate, the system state may leave the stability region obtained under state feedback control, and consequently there will be no guarantee of closed-loop stability under output feedback control.

8.4 Fault Isolation and Handling Design

In this section, we present the FDI and fault-handling framechapter by utilizing the flexibility in the state observer design shown in the previous section. In particular, the fault isolation design presented in this chapter uses a bank of state observers, each of which is driven by a subset of the outputs. For each set of the outputs, we derive rigorous conditions on the faults that are detectable by the proposed method. The isolation logic is based on the assumption that only one fault takes place (see

Remark 8.12 for an extension to multiple faults). In other words, if a fault takes place in y_i, then $v_i \not\equiv 0$ and $v_j \equiv 0$ for all $j \in \{1, \dots, p\} \backslash \{i\}$ in the system of Eq. (8.1).

We first design p high-gain observers for the system of Eq. (8.1) under different sensor configurations in the same way as in Sect. 8.3. Let $y^i = h^i(x) + v^i \in \mathbb{R}^{p-1}$ denote the system output used in the design of the ith observer, where $y^i = [y_1, \dots, y_{i-1}, y_{i+1}, \dots, y_p]^{\mathrm{T}}$, $h^i(x) = [h_1(x), \dots, h_{i-1}(x), h_{i+1}(x), \dots, h_p(x)]^{\mathrm{T}}$, and $v^i = [v_1, \dots, v_{i-1}, v_{i+1}, \dots, v_p]^{\mathrm{T}}$. The FDI design relies on the satisfaction of Assumption 8.5 stated in the following.

Assumption 8.5 For the system of Eq. (8.1), Assumptions 8.3 and 8.4 hold for the ith high-gain observer design, which uses y^i as the system output, $i = 1, \dots, p$.

Remark 8.6 Assumption 8.5 dictates the system structure requirement on the fault isolation and handling design in this chapter. In particular, it implies that the system should be observable with any $p-1$ outputs. This results in a possibility of designing p observers, each of which uses $p-1$ measured outputs (in addition to the one that uses all p outputs). Note that this requirement is more general than that of physical redundancy of sensors (where multiple sensors are used to measure the same output), and can be satisfied by sensors that measure different variables, but have analytical redundancy (in the sense of enabling full-state estimation). Note that the relaxation on the system structure for the high-gain observer design presented in Sect. 8.3 aids in the ability to satisfy the above requirement, which is necessary (in some form) to be able to isolate faults in any of the p sensors. In the absence of the satisfaction of the above requirement, the key idea in the proposed method can be used to "isolate" a fault to a subset of the sensors (see Remark 8.9).

The key idea of the proposed fault detection mechanism is to monitor the error between the state estimate provided by a high-gain observer and some accurate enough predicted value. Let $\hat{x}^i$ denote the state estimate provided by the ith observer. Similarly, let $\hat{x}^0$ denote the state estimate under the nominal sensor configuration, where all the outputs are used. For the same set of the outputs used by the ith observer, let the state prediction, denoted by $\tilde{x}^i \in \mathbb{R}^n$, initially be the state estimate: $\tilde{x}^i(0) = \hat{x}^i(0)$ because no previous measurements are available. For $t_k > 0$, it is computed in the following moving horizon fashion:

$$\begin{aligned} \tilde{x}^i(t_k) &= \hat{\hat{x}}^i(t_k), \\ \dot{\hat{\hat{x}}}^i &= f\big(\hat{\hat{x}}^i, u\big), \\ \hat{\hat{x}}^i(t_{k-T}) &= \hat{x}^i(t_{k-T}), \end{aligned} \tag{8.24}$$

where $\hat{\hat{x}}^i \in \mathbb{R}^n$ denotes the state of the model used in the predictor design, and $T =$

$$1, \quad 0 < t_k \le t_{k'}, \tag{8.25}$$

$$k - k', \quad t_{k'} < t_k \le t_{k'+T'}, \tag{8.26}$$

$$T', \quad t_k > t_{k'+T'} \tag{8.27}$$

denotes the prediction horizon, with a positive integer T' being the prediction horizon after the initialization period $[0, t_{k'}]$. In Eq. (8.24), the system state at time t_k is predicted by solving the state equation with the initial condition being the state estimate at time t_{k-T}. Before time $t_{k'}$ (i.e., before the estimator convergence), a one-step prediction horizon is used. After time $t_{k'}$, the prediction horizon increases from one to T' steps as the time increases. The corresponding residual (at the discrete time t_k) is defined as follows:

$$r_i(k) = \left\|\tilde{x}^i(t_k) - \hat{x}^i(t_k)\right\|. \tag{8.28}$$

The fault isolation design is activated only after the estimation error of the initial value used for prediction is sufficiently small (i.e., after time $t_{k'}$). In the absence of faults, r_i should be below some small threshold. A fault is declared when some notable discrepancy is observed. The proposition below presents the fault detection mechanism, and explicitly characterizes the class of faults that are detectable by the proposed method. To this end, let a superscript i denote the ith sensor configuration, and t_f denote the time of fault occurrence.

Proposition 8.2 *Consider the system of Eq.* (8.1), *for which Assumptions* 8.1–8.5 *hold, under the output feedback controller of Eq.* (8.4). *Then, given any* $0 < b < c$, $d > 0$, $\delta_{0,i} > 0$, *and integer* $T > 0$, *there exist* $\tilde{\Delta}^* > 0$, $\varepsilon^{*,i} > 0$, *and* $\delta_i > 0$ *such that if* $\Delta \in (0, \tilde{\Delta}^*]$, $\varepsilon \in (0, \varepsilon^*]$, $\varepsilon^i \in (0, \varepsilon^{*,i}]$, $x_0 \in \Omega_b$, $t_{k'} \leq t_{k-T} \leq t_f$, *and* $r_i(k) > \delta_i$, *where* ε^* *is defined in Theorem* 8.1, *then* $v^i(t) \neq 0$ *for some* $t \in [t_{k'}, t_k]$. *Furthermore, for* $t_k > t_{k'}$, *if* $r_i(k-1) \leq \delta_i$ *and*

$$\left\|M_{h,i} + M_{f,i}\right\| > L_1^i \eta_2^i(\varepsilon^i)(\delta_{0,i} + \delta_i) \tag{8.29}$$

holds for all $\|\bar{e}\| \leq L_1^i \eta_2^i(\varepsilon_i)(\delta_{0,i} + \delta_i)$ *and* $\|\phi^i - \phi_0^i\| \leq k_i$, *where*

$$M_{h,i} = \exp\left(\frac{\Delta}{\varepsilon^i} A_0^i\right)\bar{e} + \int_{t_{k-1}}^{t_k} \exp\left(\frac{t_k - \tau}{\varepsilon^i} A_0^i\right) B^i(\phi^i - \phi_0^i)\, d\tau,$$

$$M_{f,i} = -\int_{t_{k-1}}^{t_k} \exp\left(\frac{t_k - \tau}{\varepsilon^i} A_0^i\right)\left[D^i\right]^{-1} H^i v^i(\tau)\, d\tau,$$

and $k_i > 0$ *is the upper bound on* $\|\phi^i - \phi_0^i\|$ *for any* $x \in \Omega_c$, *then* $r_i(k) > \delta_i$.

Proof First, we show that the system state evolves within Ω_c until time t_k. Since $V(x)$ is continuous, and x evolves continuously in time, given $b < b' < b'' < c$, there exists $\Delta_4 > 0$ such that if $x(t_k) \in \Omega_{b'}$ and $\Delta \in (0, \Delta_4]$, then $V(x(\tau)) \leq b''$ for any $\tau \in [t_k, t_k + T\Delta]$. It follows from the proof of Theorem 8.1 that there exist $\tilde{\Delta}^* = \min\{\Delta^*, \Delta_4\}$ such that if $t_f \geq t_{k-T}$, then $x(t) \in \Omega_{b''}$ for all $t \in [0, t_k]$ (see Fig. 8.1 for an illustration).

Next, we show that if the residual breaches the threshold, then a fault takes place. Since $f(x,u)$ is continuous and locally Lipschitz, given $\delta_{0,i} > 0$, there exists $e_b^{*,i} > 0$ such that if $\|\tilde{x}(t_{k-T}) - x(t_{k-T})\| < L_2^i \eta_3^i(\varepsilon^i) e_b^{*,i}$, then $\|\tilde{x}(t) - x(t)\| < \delta_{0,i}$ for any $t \in [t_{k-T}, t_k]$ (see Theorem 3.5 in [76]). It follows from the proof of Theorem 8.1 that given $e_b^{*,i} > 0$, there exists $\varepsilon^{*,i} > 0$ such that if $\varepsilon^i \in (0, \varepsilon^{*,i}]$ and $t_{k-T} \geq t_{k'}$, then $\|e^i(t_k)\| \leq e_b^{*,i}$ for any $k \geq k'$, and consequently $\|e^i(t_{k-T})\| \leq e_b^{*,i}$. In the absence of faults, the following equation holds:

$$\begin{aligned} r_i(k) &= \left\|\tilde{x}^i(t_k) - \hat{x}^i(t_k)\right\| \\ &\leq \left\|\tilde{x}^i(t_k) - x^i(t_k)\right\| + \left\|x^i(t_k) - \hat{x}^i(t_k)\right\| \\ &\leq \delta_{0,i} + L_2^i \eta_3^i(\varepsilon^i)\left\|e_i(t_k)\right\| \\ &\leq \delta_{0,i} + L_2^i \eta_3^i(\varepsilon^i) e_b^{*,i}. \end{aligned} \tag{8.30}$$

Let $\delta_i = \delta_{0,i} + L_2^i \eta_3^i(\varepsilon^i) e_b^{*,i}$. Therefore, $r_i(k) > \delta_i$ implies that $v^i(t) \neq 0$ for some $t \in [t_{k'}, t_k]$.

Finally, we show that if the residual does not breach the threshold at the previous time and Eq. (8.29) is satisfied, then the residual breaches the threshold at the current time. To this end, consider the scaled error dynamic system subject to sensor faults for $t \in [t_{k-1}, t_k)$ as follows:

$$\dot{e}^i = \frac{1}{\varepsilon^i} A_0^i e^i + B^i(\phi^i - \phi_0^i) - \left[D^i\right]^{-1} H^i v^i. \tag{8.31}$$

The solution to the above equation gives

$$\begin{aligned} e^i(t_k) = {}& \exp\left(\frac{\Delta}{\varepsilon^i} A_0^i\right) e^i(t_{k-1}) + \int_{t_{k-1}}^{t_k} \exp\left(\frac{t_k - \tau}{\varepsilon^i} A_0^i\right) B^i(\phi_i - \phi_{0,i})\, d\tau \\ & - \int_{t_{k-1}}^{t_k} \exp\left(\frac{t_k - \tau}{\varepsilon^i} A_0^i\right) \left[D^i\right]^{-1} H^i v^i(\tau)\, d\tau. \end{aligned} \tag{8.32}$$

Then, we consider two cases: (i) $t_f \geq t_{k-1}$ and (ii) $t_f < t_{k-1}$. For the first case, it follows from Eq. (8.14) that $\|e(t_{k-1})\| \leq \tilde{L}_1^i \eta_1^i(\varepsilon^i) e_b^{*,i}$. For the second case, we have $\|e(t_{k-1})\| \leq L_1^i \eta_2^i(\varepsilon^i)(\delta_{0,i} + \delta_i)$, which can be shown by a contradiction argument. Suppose $\|e(t_{k-1})\| > L_1^i \eta_2^i(\varepsilon^i)(\delta_{0,i} + \delta_i)$. Then, we have

$$\left\|x^i(t_{k-1}) - \hat{x}^i(t_{k-1})\right\| \geq \frac{1}{L_1^i \eta_2^i(\varepsilon^i)} \left\|e(t_{k-1})\right\| > \delta_{0,i} + \delta_i. \tag{8.33}$$

Because $r_i(k-1) \geq \big|\|\tilde{x}^i(t_{k-1}) - x^i(t_{k-1})\| - \|x^i(t_{k-1}) - \hat{x}^i(t_{k-1})\|\big|$ and $\|\tilde{x}^i(t_{k-1}) - x^i(t_{k-1})\| \leq \delta_{0,i}$, it follows from Eq. (8.33) that we have

$$r_i(k-1) > \delta_i. \tag{8.34}$$

The above equation contradicts the condition that $r_i(k-1) \leq \delta_i$, which shows that $\|e(t_{k-1})\| \leq L_1^i \eta_2^i(\varepsilon^i)(\delta_{0,i}+\delta_i)$. It can be shown that $\tilde{L}_1^i \eta_1^i(\varepsilon^i) = L_1^i \eta_2^i(\varepsilon^i) L_2^i \eta_3^i(\varepsilon^i)$. Consequently, we have $\tilde{L}_1^i \eta_1^i(\varepsilon^i) e_b^{*,i} < L_1^i \eta_2^i(\varepsilon^i)(\delta_{0,i}+\delta_i)$. It follows from Eq. (8.29) that for both the cases, we have

$$\left\|x^i(t_k) - \hat{x}^i(t_k)\right\| \geq \frac{1}{L_1^i \eta_2^i(\varepsilon^i)} \left\|e(t_k)\right\| > \delta_{0,i} + \delta_i. \tag{8.35}$$

By a similar argument as above, it can be shown

$$r_i(k) > \delta_i. \tag{8.36}$$

This concludes the proof of Proposition 8.2. □

Remark 8.7 Note that Proposition 8.2 establishes the fault detection property, and considers the residual for the ith sensor configuration (the one that does not contain the ith sensor). A fault is detected upon the observation of a notable discrepancy between the state estimate and prediction. This, in turn, relies on sufficient accuracy of the state estimate used for the purpose of prediction. The result of Theorem 8.1 enables achieving a desired rate of convergence of the state estimation error to facilitate good enough prediction (denoted by $\delta_{0,i}$), by using a previous estimate (an estimate after time t_k') for a given prediction horizon. It is established that, under fault-free conditions, the residual, which describes the discrepancy between the state estimate and the predicted value, is guaranteed to be below the threshold (δ_i). Therefore, the only way that the residual breaches the threshold is that the measured outputs used in the observer design are not identical to their true values, forming the basis of the fault detection mechanism.

Remark 8.8 Proposition 8.2 also establishes rigorous conditions on the class of faults that are detectable by the proposed method. In particular, it considers an interval for which no fault is detected at the end of the previous one ($r_i(k-1) \leq \delta_i$). According to these conditions, faults may not be detected at the end of the interval where it takes place if its accumulating effect is not significant enough to trigger an alarm. However, this may result in the state estimate deviating from the system state, invalidating the convergence property of the state observer established under fault-free conditions. In this way, the effect of the fault propagates and accumulates over multiple intervals, leading to possible fault detection. Note that these conditions are only sufficient for the detection of a fault. After the fault is detected, persistent detection is possible as long as the residual breaches its threshold at each instant (see Sect. 8.5 for an illustration).

With the ability of detecting a fault in a subset of the sensors, we then present a method to isolate the fault and preserve practical stability of the closed-loop system. This is formalized in Theorem 8.2 below.

Theorem 8.2 *Consider the system of Eq.* (8.1), *for which Assumptions* 8.1–8.5 *hold, under the output feedback controller of Eq.* (8.4) *and the fault detection design of Proposition* 8.2. *If* $t_{k'} \le t_{k-T} \le t_f$ *and* $r_i(k) > \delta_i$ *for all* $i \in \{1, \dots, p\} \backslash \{j\}$, *then* $v_j(t) \ne 0$ *for some* $t \in [t_{k'}, t_k]$. *Let* t_d *denote the time of fault isolation. Then, given any* $0 < b < c$ *and* $d > 0$, *there exists* $\tilde{\varepsilon}^{*,i} > 0$ *such that if* $\Delta \in (0, \tilde{\Delta}^*]$, $\varepsilon \in (0, \varepsilon^*]$, $\varepsilon^i \in (0, \tilde{\varepsilon}^{*,i}]$, $x_0 \in \Omega_b$, *where* $\tilde{\Delta}^*$ *is defined in Proposition* 8.2 *and* ε^* *defined in Theorem* 8.1, *then the control law*

$$u(t) = u_c\big(\text{sat}\big(\hat{x}^{l(t_k)}(t_k)\big)\big) \quad \textit{for all } t \in [t_k, t_{k+1}) \tag{8.37}$$

and the switching rule

$$l(t) = \begin{cases} 0, & 0 \le t < t_d, \\ j, & t_d \le t \end{cases} \tag{8.38}$$

guarantee that $x(t) \in \Omega_c$ *for all* $t \in [0, \infty)$ *and* $\limsup_{t\to\infty} \|x\| \le d$.

Proof First, we show a fault taking place in the jth sensor by a contradiction argument, using the results of Proposition 8.2. Suppose that a fault takes place in some sensor indexed by $s \in \{1, \dots, p\} \backslash \{j\}$. Since $r_s(k) > \delta_s$, a fault must have taken place in some sensor indexed by $w \in \{1, \dots, p\} \backslash \{s\}$. Note that $w \ne s$, which is contradictory to the assumption that only one sensor fault takes place. Therefore, $r_i(k) > \delta_i$ for all $i \in \{1, \dots, p\} \backslash \{j\}$ implies that a fault takes place in the jth sensor.

Then, we show practical stability of the closed-loop system under the control law of Eq. (8.37) and the switching rule of Eq. (8.38) with the focus on the analysis for the time interval after time t_d. It follows from the proof of Theorem 8.1 that there exists $\tilde{e}^i_{b,1} > 0$ such that if $x(t_k) \in \Omega_{b''}$ and $e^i_b \in (0, \tilde{e}^i_{b,1}]$, then $\hat{x}^i(t_k) \in \Omega_c$. Furthermore, given $\tilde{e}^{*,i}_b = \min\{\tilde{e}^i_{b,1}, e^i_{b,2}, e^i_{b,3}, e^{*,i}_b\}$, there exists $\tilde{\varepsilon}^{*,i} > 0$ such that if $\varepsilon^i \in (0, \tilde{\varepsilon}^{*,i}]$, then $e^i(t_k) \le \tilde{e}^{*,i}_b$ for any $k \ge k'$, and consequently $e^i(t_d) \le \tilde{e}^{*,i}_b$. It follows from the proof of Proposition 8.2 that $x(t_d) \in \Omega_{b''}$. Therefore, if $\varepsilon^i \in (0, \tilde{\varepsilon}^{*,i}]$ for all $i \in \{1, \dots, p\}$, then $\hat{x}^j(t_d) \in \Omega_c$. The rest of the proof follows from the same line of arguments as Part 2 of the proof of Theorem 8.1, and is omitted. This concludes the proof of Theorem 8.2. □

Remark 8.9 In Theorem 8.2, a fault isolation mechanism is designed by using a bank of high-gain observers, with each driven by $p - 1$ outputs. Specifically, the fault isolation logic is designed as follows: for the sensor configuration that does not use the faulty sensor, the state estimate and prediction continue to be close together because the estimate continues to be accurate enough (and close to the prediction). In contrast, for every other sensor configuration (of which the faulty sensor is a part), the state estimate and prediction diverge (subject to the satisfaction of the detectability conditions) because of the incorrect estimation of the system state. This observation forms the basis of the fault isolation mechanism. After the fault is isolated, it can be handled by simply switching to a sensor configuration that does not include the faulty sensor. Note that the key idea of the fault isolation design is to utilize the model-based sensor redundancy. If the system is observable only for certain subsets of the outputs, the proposed method can be used to narrow down the possibility of a fault to an appropriate subset of sensors. In particular, the faulty sensor can be "isolated" to be in the intersection of the subsets of the sensors corresponding to the measured outputs for which the residuals breach their thresholds.

Remark 8.10 The proposed FDI scheme remains applicable under any admissible control (possibly without using a control Lyapunov function) as long as the system state evolves within a compact set. This requirement is often satisfied for practical systems because physical variables, such as temperatures, concentrations, and pressures, typically evolve within finite ranges. The output feedback control design in Sect. 8.3 provides one way to guarantee that the state of the constrained nonlinear system evolves within a stability region, which serves as a positively invariant set for the closed-loop system under fault-free conditions. Furthermore, the fault-handling mechanism of Theorem 8.2 requires that faults be detected and isolated in a reasonably quick fashion (i.e., within a certain time window). To this end, a "cushion" (see the region $\Omega_{b''} \backslash \Omega_{b'}$ in Fig. 8.1) is built to account for possible runaway behaviors of the closed-loop system between fault occurrence and declaration within the time window dictated by the prediction horizon T'. The "cushion" is provided for the purpose of stability guarantees; in most practical situations, a sensor fault will likely cause the system state to drift (not necessarily runaway), while keeping it within the stability region (see Sect. 8.5 for an illustration), and maintaining the applicability of the proposed FDI design.

Remark 8.11 Note that the FDI scheme is presented in this chapter using high-gain observers because of their ability to deal with the system nonlinearity, and provide a convergence property at a desired rate. This property is exploited for the generation of FDI residuals. Note also that the negative impact of measurement noise on the high-gain observer can be reduced by filtering the noisy measurements before state estimation (see Sect. 8.5 for an illustration) or adopting a switched-gain approach to achieve quick convergence initially and "stable" performance later on (see, e.g., [1]). The FDI design, however, is not restricted to this particular choice of observers; any other observer that is able to provide good convergence properties and is able to handle measurement noise better can be used instead in the proposed FDI scheme.

The design and implementation of the proposed FDI and fault-handling method of Theorem 8.2 proceed as follows (see also Fig. 8.3):

1. Given the system model of Eq. (8.1), design a state feedback control law, u_c, that satisfies Assumption 8.2 and compute the stability region estimate, Ω_c, at each point of which the derivative of the control Lyapunov function, $V(x)$, can be made negative and sufficiently small by using the available input (i.e., Eq. (8.2) is satisfied).
2. Given two subsets of the stability region obtained under state feedback control, Ω_b and $\Omega_{b'}$, with $0 < b < b' < c$, compute the time t_e, by the end of which the system state remains within $\Omega_{b'}$ for any initial condition within Ω_b.

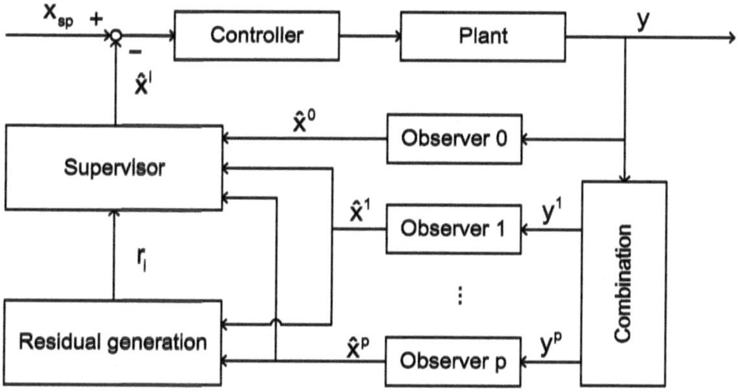

Fig. 8.3 Schematic of the FDI and fault-handling framechapter. Before FDI, the state estimate used for feedback control is generated by observer 0, which uses all the measured outputs. After a fault takes place and FDI is achieved, the supervisor switches to the observer which uses the outputs from the remaining healthy sensors

3. Given $b < b'$, and the size of the closed ball, d, to which the system state is required to converge, compute Δ^* for the system under fault-free conditions, with $\Delta^* \in (0, t_e]$, and ε^* for the high-gain observer design according to Theorem 8.1.
4. Given $b' < b'' < c$ and the prediction horizon T', compute $\tilde{\Delta}^*$ according to Proposition 8.2, and use it for the purpose of closed-loop implementation. Given the prediction error, $\delta_{0,i}$, and the size of the closed ball, d, and $b' < b''$, compute $\tilde{\varepsilon}^{*,i}$ for the ith high-gain observer design used for FDI, $i = 1, \dots, p$, according to Theorem 8.2.
5. At each time instant t_k, monitor the residuals after the scaled estimation error converges (i.e., after the time $t_{k'}$) and
   (a) If all the residuals are below their thresholds (i.e., $r_i(k) \leq \delta_i$ for all $i \in \{1, \dots, p\}$), continue to use the state estimate, $\hat{x}^0$, that is provided by the observer using all the outputs and compute the control input according to Eq. (8.37).
   (b) Otherwise, if a fault is detected and isolated (i.e., $r_i(k) > \delta_i$ for all $i \in \{1, \dots, p\} \backslash \{j\}$), switch to use the state estimate, $\hat{x}^j$, that is provided by the observer using the outputs of the remaining healthy sensors (i.e., y^j) and compute the control input according to Eq. (8.37).

Remark 8.12 The proposed methodology can be extended to detect and isolate multiple faults. To understand this point, consider the occurrence of two faults. To detect faults, we design a bank of observers, which use combinations of $p-1$ outputs. If all the residuals breach their thresholds, then at lease two faults have taken place. To isolate the faults, we design another bank of observers, which use combinations of $p-2$ outputs. If one residual does not breach its threshold and the remaining residuals do, then the two faults are isolated, which correspond to the outputs not used by that particular observer. Note that the above extension is based on the assumption that the system is observable with the chosen outputs so that the high-gain observers can be designed.

Remark 8.13 In most of the existing results on FDI of nonlinear systems, actuator and sensor faults are considered separately. With the consideration of the occurrence of one (actuator or sensor) fault, however, the proposed FDI mechanism is able to indicate whether a fault takes place in an actuator or sensor. Specifically, if only $p-1$ residuals breach their thresholds, then a sensor fault must have taken place, and that fault is also isolated. Otherwise, if all the residuals breach their thresholds, then an actuator fault must have taken place. This is because an actuator fault will not only result in possible errors in a state estimate, but also errors in the state prediction, which is used in the evaluation of all the residuals. A detailed analysis of the problem of fault isolation in this case is beyond the scope of this book.

8.5 Application to a Chemical Reactor Example

In this section, we consider a continous stirred tank reactor example, where an irreversible elementary exothermic reaction of the form $\mathrm{A} \xrightarrow{k} \mathrm{B}$ takes place. The feed to the reactor consists of reactant A at a flow rate F, concentration $C_{\mathrm{A}0}$, temperature T_0. A cooling jacket is equipped to remove heat from the reactor. The cooling stream going to the jacket is at a flow rate F_c and temperature T_{cf}. The mathematical model of this chemical reactor takes the following form:

$$\begin{aligned}
\dot{C}_{\mathrm{A}} &= \frac{F}{V}(C_{\mathrm{A}0} - C_{\mathrm{A}}) - k_0 e^{-E/RT_R} C_{\mathrm{A}}, \\
\dot{T}_R &= \frac{F}{V}(T_0 - T_R) + \frac{(-\Delta H)}{\rho c_p} k_0 e^{-E/RT_R} C_{\mathrm{A}} - \frac{UA}{\rho c_p V}(T_R - T_c), \\
\dot{T}_c &= \frac{F_c}{V_c}(T_{cf} - T_c) + \frac{UA}{\rho_c c_{pc} V_c}(T_R - T_c),
\end{aligned} \tag{8.39}$$

where C_{A} is the concentration of species A, T_R is the temperature in the reactor, T_c is the temperature in the cooling jacket, V is the volume of the reactor, k_0, E, and ΔH are the pre-exponential constant, the activation energy, and the enthalpy of the reaction, respectively, R is the ideal gas constant, ρ and c_p are the density and the heat capacity of the fluid in the reactor, respectively, U is the overall heat transfer coefficient, A is the heat transfer area of the CSTR, V_c is the volume of the cooling jacket, and ρ_c and c_{pc} are the density and the heat capacity of the cooling stream, respectively. The process parameters can be found in Table 8.1.

We first illustrate the enhanced applicability of the output feedback control design. To this end, we consider $u = [F, F_c]^{\mathrm{T}}$ and $y = [T_R, T_c]^{\mathrm{T}}$ as the input and output, respectively, where $0 \le F \le 60$ L/min and $0 \le F_c \le 10$ L/min. The control objective is to operate the process at an equilibrium point where $C_{\mathrm{A}} = 0.5$ mol/L, $T_R = 325.0$ K, and $T_c = 315.9$ K. The corresponding steady-state values of the input

Table 8.1 Process parameters for the chemical reactor example

Parameter	Value	Unit
V	100	L
k_0	7.2×10^{10}	min^{-1}
E/R	8750	K
ΔH	-5×10^4	J/mol
ρ	1000	g/L
c_p	0.239	J/g K
UA	5×10^4	J/min K
V_c	20	L
ρ_c	1000	g/L
c_{pc}	4.2	J/g K
$C_{\mathrm{A}0}$	1	mol/L
T_0	350	K
T_{cf}	293	K

variables are $F = 14.6$ L/min and $F_c = 4.7$ L/min. Note that the relative degrees for the output with respect to the input are $\omega_1 = 1$ and $\omega_2 = 1$, respectively, for the process of Eq. (8.39). Therefore, the assumption of a coordinate transformation $\zeta = T(x)$ that is required for the standard high-gain observer designs (see, e.g., [95]) is not satisfied. However, it satisfies Assumption 8.3, with the following coordinate transformation:

$$\begin{aligned}
\zeta_{1,1} &= T_R,\\
\zeta_{1,2} &= \frac{F}{V}(T_{\mathrm{A}0} - T_R) + \frac{(-\Delta H)}{\rho c_p} k_0 e^{-E/RT_R} C_{\mathrm{A}} - \frac{UA}{\rho c_p V}(T_R - T_c),\\
\zeta_{2,1} &= T_c.
\end{aligned} \tag{8.40}$$

For $t \in [t_k, t_{k+1})$, the high-gain observer is designed as follows:

$$\begin{aligned}
\dot{\hat{\zeta}}_{1,1} &= \hat{\zeta}_{1,2} + \frac{a_{1,1}}{\varepsilon}(y_1 - \hat{\zeta}_{1,1}),\\
\dot{\hat{\zeta}}_{1,2} &= \frac{a_{1,2}}{\varepsilon^2}(y_1 - \hat{\zeta}_{1,1}),\\
\dot{\hat{\zeta}}_{2,1} &= \frac{a_{2,1}}{\varepsilon}(y_2 - \hat{\zeta}_{2,1}),\\
\hat{\zeta}(t_k) &= T\big(\hat{x}(t_k), u(t_k)\big),
\end{aligned} \tag{8.41}$$

where $\varepsilon = 0.04$, $a_{1,1} = a_{2,1} = 5$, and $a_{1,2} = 10$. A Lyapunov-based model predictive controller of [93] is used to illustrate the results. The hold-time for the control action is chosen as $\Delta = 0.25$ min, the prediction horizon is chosen as 2Δ, the weighting matrices used to penalize the deviations of the state and input from their nominal

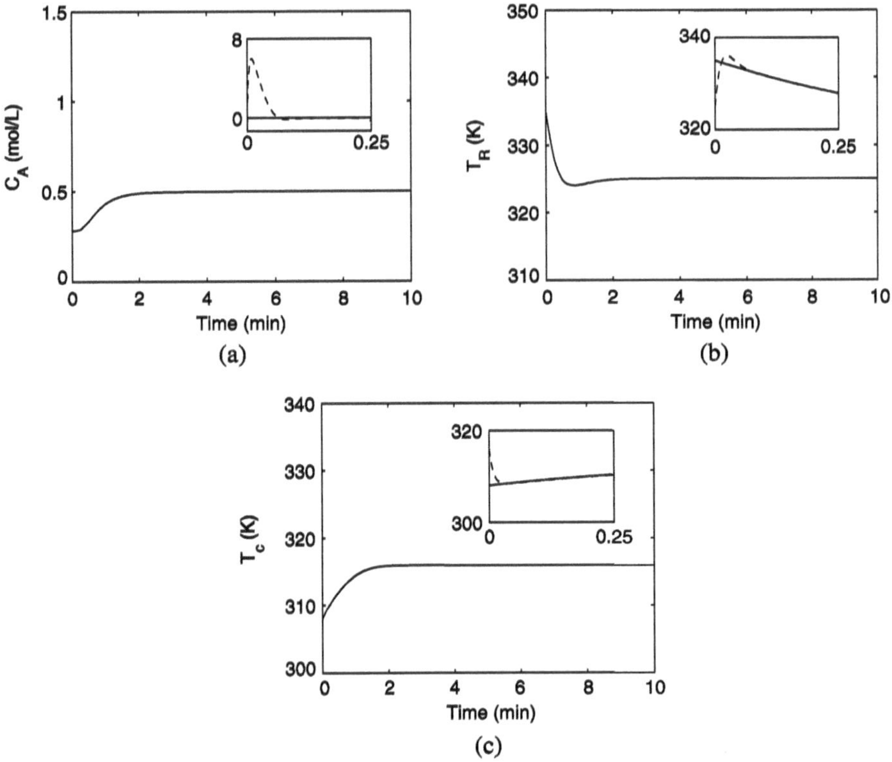

Fig. 8.4 Closed-loop state (*solid lines*) and state estimate (*dashed lines*) profiles for the chemical reactor example under fault-free conditions. The *insets* show the quick convergence of the state estimation error

values are chosen as $Q_w = \text{diag}[10^5, 10^3, 10]$ and $R_w = \text{diag}[5, 50]$, respectively, and the stability region is characterized as $\{x \in \mathbb{R}^3 : V(x) = x^{\mathrm{T}} P x \leq c\}$, where

$$P = \begin{bmatrix} 507.90 & 9.47 & 14.02 \\ 9.47 & 0.57 & 0.53 \\ 14.02 & 0.53 & 1.05 \end{bmatrix}$$

and $c = 75.5$.

To show practical stability of the closed-loop system, we consider the process starting from an initial condition $C_{\mathrm{A}} = 0.28$ mol/L, $T_R = 335$ K, and $T_c = 308$ K. The high-gain observer is initialized at the nominal equilibrium point. The closed-loop state profiles are depicted in Fig. 8.4, where the solid and dashed lines denote the process state and state estimate profiles, respectively. It can be seen that the state estimates approach the process states sufficiently fast, and the controller drives the process to the nominal equilibrium point. The corresponding input profiles are plotted in Fig. 8.5.

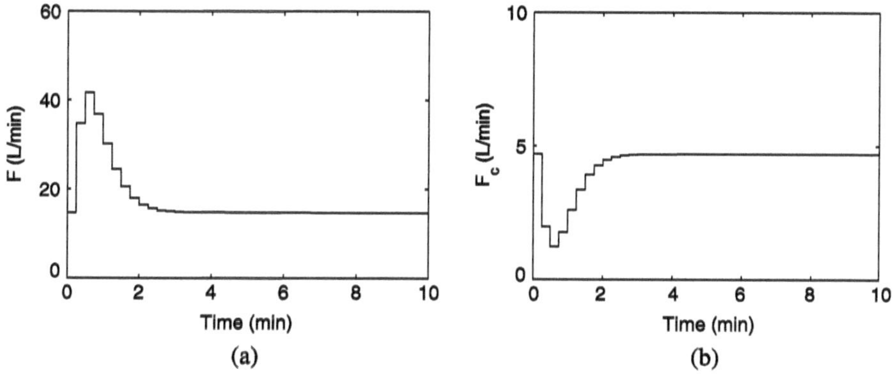

Fig. 8.5 Input profiles for the chemical reactor example under fault-free conditions

We next illustrate the FDI and fault-handling design. To this end, we first design three high-gain observers, which use outputs $y^1 = [C_A, T_R]^T$, $y^2 = [C_A, T_c]^T$, and $y^3 = [T_R, T_c]^T$, respectively. The coordinate transformations for the first and second observers are as follows:

$$\begin{aligned}
\zeta_{1,1}^1 &= C_A, \\
\zeta_{2,1}^1 &= T_R, \\
\zeta_{2,2}^1 &= \frac{F}{V}(T_{A0} - T_R) + \frac{(-\Delta H)}{\rho c_p} k_0 e^{-E/RT_R} C_A - \frac{UA}{\rho c_p V}(T_R - T_c), \\
\zeta_{1,1}^2 &= C_A, \\
\zeta_{2,1}^2 &= T_c, \\
\zeta_{2,2}^2 &= \frac{F_c}{V_c}(T_{cf} - T_c) + \frac{UA}{\rho_c c_{pc} V_c}(T_R - T_c),
\end{aligned} \tag{8.42}$$

and the corresponding observers are designed as follows ($i = 1, 2$):

$$\begin{aligned}
\dot{\hat{\zeta}}_{1,1}^i &= \frac{a_{1,1}^i}{\varepsilon}(y_1^i - \hat{\zeta}_{1,1}^i), \\
\dot{\hat{\zeta}}_{2,1}^i &= \hat{\zeta}_{2,2}^i + \frac{a_{2,1}^i}{\varepsilon}(y_2^i - \hat{\zeta}_{2,1}^i), \\
\dot{\hat{\zeta}}_{2,2}^i &= \frac{a_{2,2}^i}{\varepsilon^2}(y_2^i - \hat{\zeta}_{2,1}^i),
\end{aligned} \tag{8.43}$$

where $\varepsilon = 0.04$, $a_{1,1}^i = 5$, and $a_{2,1}^i = a_{2,2}^i = 10$. Note the third observer design is the same as the one used to show practical stability of the closed-loop system under fault free conditions (i.e., $\zeta^3 = \zeta$).

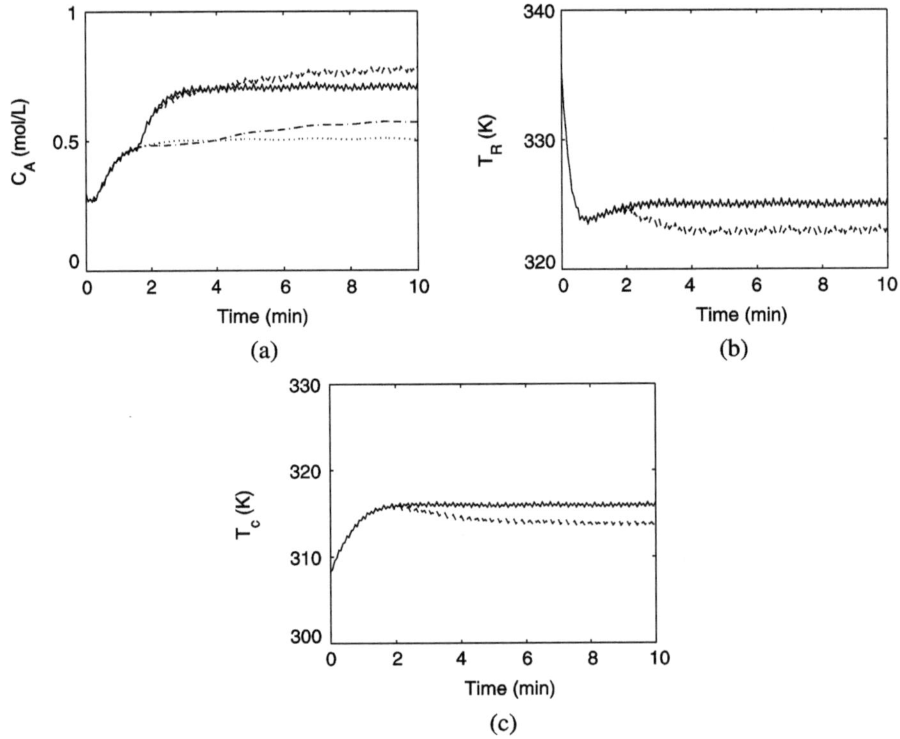

Fig. 8.6 Closed-loop measurements under faulty conditions in the presence of the proposed FDI and fault-handling method resulting in practical stability (*solid lines*) and in the absence of the proposed FDI and fault-handling method resulting in degraded control performance (*dashed lines*). The *dotted* and *dash-dotted lines* show the evolution of the state profiles for the two cases, respectively

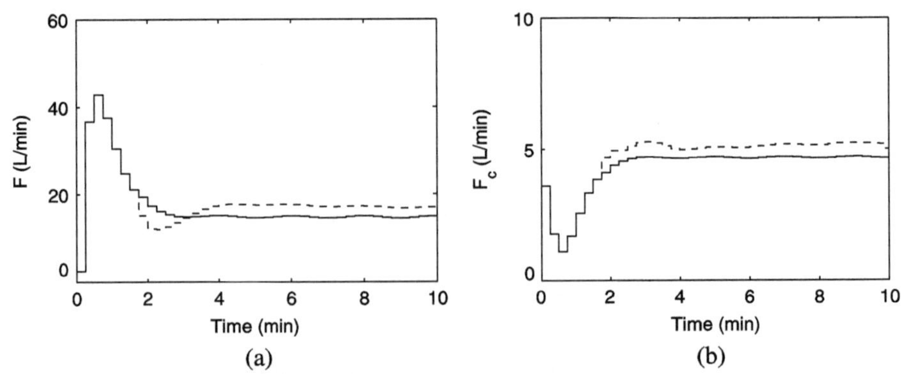

Fig. 8.7 Input profiles under faulty conditions in the presence (*solid lines*) and absence (*dashed lines*) of the proposed FDI and fault-handling method

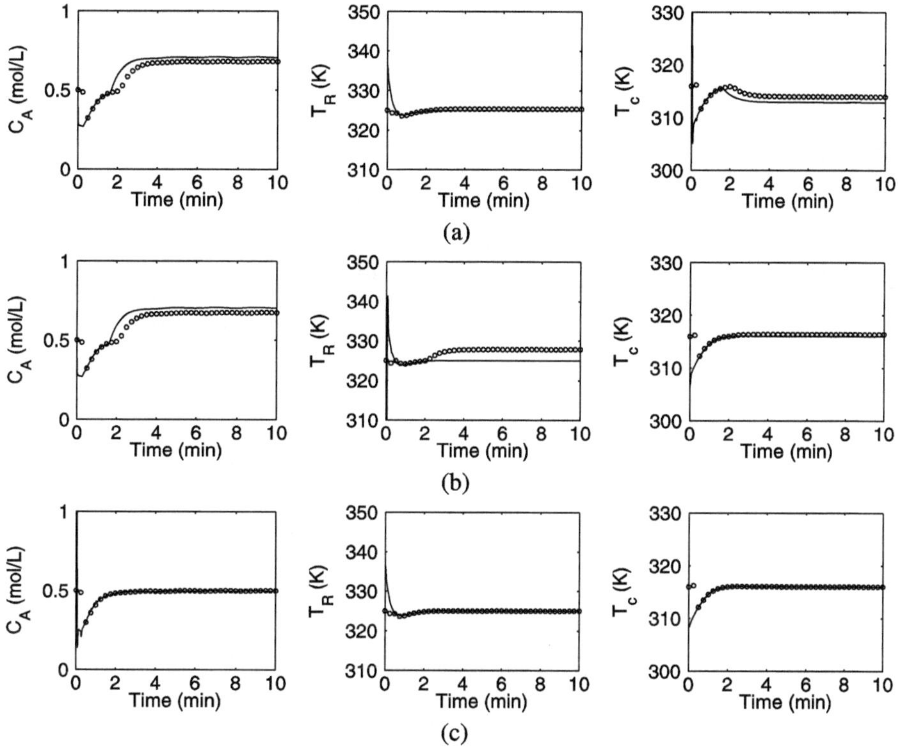

Fig. 8.8 State estimate (*solid lines*) and prediction (*circles*) profiles generated using measurements of (**a**) C_{A} and T_R, (**b**) C_{A} and T_c, and (**c**) T_R and T_c, respectively. After a fault takes place in C_{A} at time $t_f = 1.625$ min, notable discrepancies between state estimates and predictions are observed for the first two cases

To show the effectiveness of the FDI and fault-handling design subject to plant-model mismatch and measurement noise, we consider a fault that takes place in C_{A} at time $t_f = 1.625$ min by simulating a non-abrupt bias in the concentration sensor of magnitude 0.2 mol/L, which is described by $v_1 = [1 - e^{-2(t-t_f)}] \times 0.2 \times \nu(t - t_f)$ mol/L, where $\nu(t - t_f) = \begin{cases} 0 \text{ if } t<t_f \\ 1 \text{ if } t\geq t_f \end{cases}$. Furthermore, k_0 is 2 % smaller than its nominal value, and $C_{\mathrm{A}0}$ varies sinusoidally by a magnitude of 5 % about its nominal value. The concentration and temperature measurements have combinations of eleven high-frequency (about 50 Hz) sinusoidal noises with the largest of the magnitudes being 0.01 mol/L and 0.2 K, respectively. The noisy measurements are processed through a first-order low-pass filter with the filter time constant being 0.3 s. Full state feedback (i.e., the nominal sensor configuration) is used under fault-free conditions. In the FDI design, the prediction horizon after the initialization period is chosen as $T' = 2$, and the thresholds are chosen as 0.025, 0.025, and 0.05 for the three FDI filters, respectively. The thresholds are chosen by observing the normal variations of the residuals under fault-free conditions and using a conservative upper bound to account for the presence of uncertainty and measurement noise.

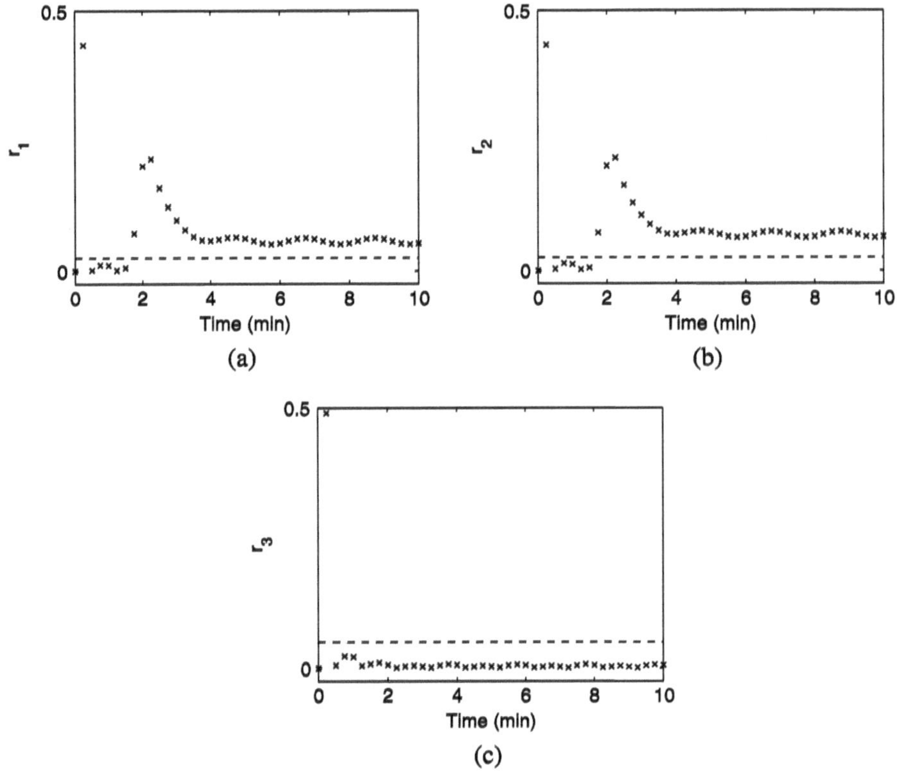

Fig. 8.9 Residuals (*crosses*) generated using measurements of (**a**) C_A and T_R, (**b**) C_A and T_c, and (**c**) T_R and T_c, respectively. The fault in C_A is detected and isolated at time $t_d = 1.75$ min via the residuals r_1 and r_2 breaching their thresholds (*dashed lines*)

As shown by the solid and dotted lines in Fig. 8.6 and Fig. 8.7, the proposed FDI and fault-handling framechapter preserves practical stability of the closed-loop system. The absence of an appropriate fault-handling mechanism, however, results in degraded control performance, as shown by the dashed and dash-dotted lines in Fig. 8.6 and Fig. 8.7. To explain the FDI mechanism, the state estimate and prediction profiles are shown in Fig. 8.8. The residuals, evaluated using the normalized state against its steady state value, and thresholds are shown by crosses and dashed lines, respectively, in Fig. 8.9. It can be seen that the residuals are above the thresholds at time 0.25 min (i.e., the second time instant) because of the initial transient in the observers for the state estimates to converge to their true values. After the state estimates converge, however, all the residuals are below the thresholds until the fault takes place. After the occurrence of the fault, residuals r_2 and r_3 breach their thresholds at the next time instant while r_1, which corresponds to the sensor configuration that does not use the faulty sensor, still stays below its threshold, resulting in detection and isolation of a fault in C_A at time $t_d = 1.75$ min. Upon FDI, the state estimate $\hat{x}^1$, which is generated by using measurements from the remaining healthy sensors, is used for feedback control, and practical stability of the closed-loop system is preserved.

8.6 Conclusions

This chapter considered the problem of sensor FDI and FTC for nonlinear systems subject to input constraints. The key idea of the proposed method is to exploit model-based sensor redundancy through state observer design. To this end, an output feedback control design using high-gain observers was first presented under less restrictive and easily verifiable assumptions. By utilizing the flexibility in the observer design, a novel FDI scheme was then proposed, which is composed of a bank of high-gain observers. Each observer uses a subset of the measured outputs to generate state estimates. Residuals are defined as the discrepancies between these state estimates and their predicted values based on previous estimates. A fault is identified when all the residuals breach their thresholds except for the one generated without using the measurements provided by the faulty sensor. Conditions characterizing the detectable faults were rigorously established. Upon FDI, the state estimate generated using measurements from the remaining healthy sensors is used to preserve practical stability of the closed-loop system. The implementation of the FDI and fault-handling framechapter subject to uncertainty and measurement noise was illustrated using a chemical reactor example.

Chapter 9
Control and Fault-Handling Subject to Asynchronous Measurements

9.1 Introduction

In this chapter, we present a control and fault-handling approach for handling sensor malfunctions. Specifically, we first modify the Lyapunov-based MPC presented in Chap. 2 to take into account sensor data losses or asynchronous measurements due to sensor malfunctions, both in the optimization problem formulation and in the controller implementation. In this LMPC scheme, when feedback is lost, instead of setting the control actuator outputs to zero or to the last available values, the actuators implement the last optimal input trajectory [119] evaluated by the controller (this requires that the actuators store in memory the last optimal input trajectory received). The LMPC is designed based on a nonlinear control law which is able to stabilize the closed-loop system and inherits the stability and robustness properties in the presence of uncertainty and sensor data losses of the nonlinear controller, while taking into account optimality considerations. Specifically, the LMPC scheme allows for an explicit characterization of the stability region, guarantees practical stability in the absence of sensor data losses or asynchronous measurements, and guarantees that the stability region is an invariant set for the closed-loop system if the maximum time in which the loop is open is shorter than a given constant that depends on the parameters of the system and the nonlinear control law that is used to formulate the optimization problem. A schematic diagram of the considered closed-loop system is shown in Fig. 9.1. The application of the LMPC to a continuous crystallization process subject to sensor malfunctions is also presented.

Subsequently, we develop an FDI scheme that will allow fault tolerant control to take place when process measurements are available at asynchronous time instants. First, an FDI scheme that employs model-based techniques is proposed that allows for the isolation of faults. This scheme employs model-based FDI filters similar to those presented in Chaps. 3 and 4 in addition to observers that estimate the fault free evolution of asynchronously measured states during time intervals in which their measurements are not available. Specifically, the presented FDI scheme provides detection and isolation of any fault that enters into the differential equation of

P. Mhaskar et al., *Fault-Tolerant Process Control*, DOI 10.1007/978-1-4471-4808-1_9,

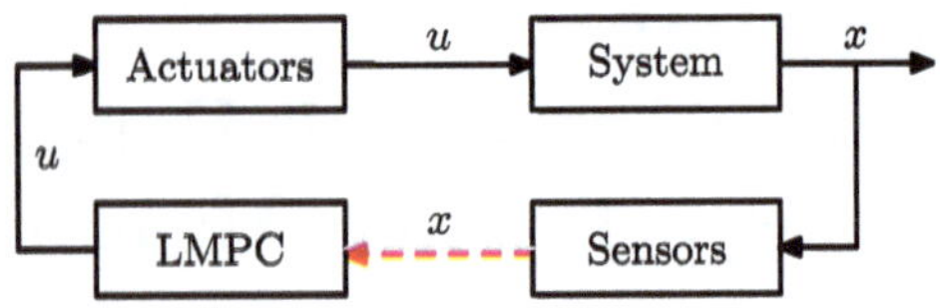

Fig. 9.1 LMPC design for systems subject to sensor data losses. *Dashed lines* denote sensor data losses/asynchronous sampling

only synchronously measured states, and grouping of faults that enter into the differential equation of any asynchronously measured state. For a fully coupled process system, fault detection occurs shortly after a fault takes place, and fault isolation, limited by the arrival of asynchronous measurements, occurs when asynchronous measurements become available. Once the FDI methodology has provided the system supervisor with a fault diagnosis, the supervisor takes appropriate action to seamlessly reconfigure the system to an alternative control configuration that will enforce the desired operation. Applications of the presented asynchronous FDI and FTC framework to a polyethylene reactor are presented.

9.2 Handling Sensor Malfunctions in the Control Design

Consider nonlinear systems described by the following state-space model:

$$\dot{x}(t) = f\bigl(x(t), u(t), w(t)\bigr), \tag{9.1}$$

where $x(t) \in \mathbb{R}^n$ denotes the vector of state variables, $u(t) \in \mathbb{R}^m$ denotes the vector of control (manipulated) input variables, $w(t) \in \mathbb{R}^w$ denotes the vector of disturbance variables, and f is a locally Lipschitz vector function on $\mathbb{R}^n \times \mathbb{R}^m \times \mathbb{R}^w$ such that $f(0, 0, 0) = 0$. This implies that the origin is an equilibrium point for the nominal system (i.e., system of Eq. (9.1) with $w(t) \equiv 0$ for all t) with $u = 0$.

The input vector is restricted to be in a nonempty convex set $U \subseteq \mathbb{R}^m$ which is defined as follows:

$$U := \bigl\{u \in \mathbb{R}^m : \|u\| \le u^{\max}\bigr\}, \tag{9.2}$$

where $u^{\max}$ is the magnitude of the input constraint.

The disturbance vector is bounded, that is, $w(t) \in W$ where

$$W := \bigl\{w \in \mathbb{R}^w : \|w\| \le \theta, \theta > 0\bigr\} \tag{9.3}$$

with θ being a known positive real number. The vector of uncertain variables, $w(t)$, is introduced into the model in order to account for the occurrence of uncertainty in the values of the process parameters and the influence of disturbances in process control applications.

9.2.1 Lyapunov-Based Control

We assume that there exists a feedback control law $u(t) = h(x(t))$ which satisfies the input constraint on u for all x inside a given stability region and renders the origin of the nominal closed-loop system asymptotically stable. This assumption is essentially equivalent to the assumption that the nominal system is stabilizable or that there exists a Lyapunov function for the nominal system or that the pair (A, B) in the case of linear systems is stabilizable. Using converse Lyapunov theorems [28, 76, 86, 97], this assumption implies that there exist functions $\alpha_i(\cdot)$, $i = 1, 2, 3, 4$ of class $\mathcal{K}$ and a continuously differentiable Lyapunov function $V(x)$ for the nominal closed-loop system that satisfy the following inequalities:

$$\alpha_1(\|x\|) \leq V(x) \leq \alpha_2(\|x\|), \tag{9.4}$$

$$\frac{\partial V(x)}{\partial x} f(x, h(x), 0) \leq -\alpha_3(\|x\|), \tag{9.5}$$

$$\left\|\frac{\partial V(x)}{\partial x}\right\| \leq \alpha_4(\|x\|), \tag{9.6}$$

$$h(x) \in U, \tag{9.7}$$

for all $x \in O \subseteq \mathbb{R}^n$ where O is an open neighborhood of the origin. We denote the region $\Omega_\rho \subseteq O$ as the stability region of the closed-loop system under the control $u = h(x)$. Note that explicit stabilizing control laws that provide explicitly defined regions of attraction for the closed-loop system have been developed using Lyapunov techniques for specific classes of nonlinear systems, particularly input-affine nonlinear systems; the reader may refer to [5, 28, 78, 150] as well as to Chap. 2 for results in this area including results on the design of bounded Lyapunov-based controllers by taking explicitly into account constraints for broad classes of nonlinear systems [45, 46, 85].

By continuity, the local Lipschitz property assumed for the vector field $f(x,u,w)$, the fact that the manipulated input u is bounded in a convex set and the continuous differentiable property of the Lyapunov function V, there exist positive constants M, L_w, L_x, and L'_x such that

$$\|f(x, u, w)\| \leq M, \tag{9.8}$$

$$\|f(x, u, w) - f(x', u, 0)\| \leq L_w\|w\| + L_x\|x - x'\|, \tag{9.9}$$

$$\left\|\frac{\partial V(x)}{\partial x} f(x, u, 0) - \frac{\partial V(x')}{\partial x} f(x', u, 0)\right\| \leq L'_x\|x - x'\|, \tag{9.10}$$

for all $x, x' \in \Omega_\rho$, $u \in U$ and $w \in W$. These constants will be used in characterizing the stability properties of the system of Eq. (9.1) under LMPC designs.

9.2.2 Modeling of Sensor Data Losses

To model sensor data losses, we assume that feedback of the state of the system of Eq. (9.1), $x(t)$, is available at asynchronous time instants t_a where $\{t_{a\geq 0}\}$ is a random increasing sequence of times, that is, the intervals between two consecutive instants are not fixed. The distribution of $\{t_{a\geq 0}\}$ characterizes the time the feedback loop is closed or the time needed to obtain a new state measurement. In general, if there exists the possibility of arbitrarily large periods of time in which feedback is not available, then it is not possible to provide guaranteed stability properties because there exists a non-zero probability that the system operates in open-loop for a period of time large enough for the state to leave the stability region. In order to study the stability properties in a deterministic framework, we assume that there exists an upper bound T_m on the interval between two successive time instants in which the feedback loop is closed or new state measurements are available, that is,

$$\max_a \{t_{a+1} - t_a\} \leq T_m. \tag{9.11}$$

This assumption is reasonable from process control and networked control systems perspectives [113, 124, 166, 167] and allows us to study deterministic notions of stability. This model of feedback/measurements is of relevance to systems subject to asynchronous measurement samplings due to sensor malfunctions.

9.2.3 LMPC Formulation with Asynchronous Feedback

When feedback is lost, most approaches set the control input to zero or to the last implemented value. Instead, in this LMPC for systems subject to sensor data losses due to sensor malfunctions, when feedback is lost, we take advantage of the MPC scheme to update the input based on a prediction obtained using the system model. This is achieved using the following implementation strategy:

1. At a sampling time, t_a, when the feedback loop is closed (i.e., the current system state $x(t_a)$ is available for the controller and the controller can send information to the actuators), the LMPC evaluates the optimal future input trajectory $u(t)$ for $t \in [t_a, t_a + N\Delta)$.
2. The LMPC sends the entire optimal input trajectory (i.e., $u(t)\ \forall t \in [t_a, t_a + N\Delta)$) to the actuators.
3. The actuators implement the input trajectory until the feedback loop is closed again at the next sampling time t_{a+1}, that is, the actuators implement $u(t)$ in $t \in [t_a, t_{a+1})$.
4. When a new measurement is received ($a \leftarrow a + 1$), go to Step 1.

In this implementation strategy, when the state is not available, the actuators keep implementing the last received optimal input trajectory. If sensor data is lost for a period larger than the prediction horizon, the actuators set the inputs to the last implemented values or to fixed values. This strategy is a receding horizon scheme, which takes into account that sensor data losses may occur. This strategy is motivated by the fact that when no feedback is available, a reasonable estimate of the future evolution of the system is given by the nominal trajectory. The LMPC design taking into account asynchronous measurements, therefore, modifies the standard implementation scheme of switching off the actuators ($u = 0$) or setting the actuators to zero or to the last computed input values. The idea of using the model to predict the evolution of the system when no feedback is possible has also been used in the context of sampled-data linear systems, see [117, 118, 121, 122]. The actuators not only receive and implement given inputs, but must also be able to store future trajectories to implement them in case sensor data losses occur. This means that to handle sensor data losses, not only the control algorithms must be modified, but also the control actuator hardware that implements the control actions.

When sensor data losses are present in the feedback loop, the existing LMPC schemes of Chap. 2 (see also [79, 108, 110, 137]) cannot guarantee the closed-loop stability no matter whether the actuators keep the inputs at the last values, set the inputs to constant values, or keep on implementing the previously evaluated input trajectories. In particular, there is no guarantee that the LMPC optimization problems will be feasible for all time, i.e., that the state will remain inside the stability region for all time. In the LMPC design of Eqs. (2.52)–(2.56), the constraint of Eq. (2.56) only takes into account the first prediction step and does not restrict the behavior of the system after the first step. If no additional constraints are included in the optimization problem, no claims on the closed-loop behavior of the system can be made. For this reason, when sensor data losses are taken into account, the constraints of the LMPC problem have to be modified. The LMPC that takes into account sensor data losses in an explicit way is based on the following finite horizon constrained optimal control problem:

$$\min_{u \in S(\Delta)} \int_{t_a}^{t_a+N\Delta} \left[\|\tilde{x}(\tau)\|_{Q_c} + \|u(\tau)\|_{R_c} \right] d\tau \tag{9.12}$$

$$\text{s.t.} \quad \dot{\tilde{x}}(t) = f\big(\tilde{x}(t), u(t), 0\big), \tag{9.13}$$

$$\dot{\hat{x}}(t) = f\big(\hat{x}(t), h\big(\hat{x}(t_a + j\Delta)\big), 0\big), \quad \forall t \in \big[t_a + j\Delta, t_a + (j+1)\Delta\big), \tag{9.14}$$

$$u(t) \in U, \tag{9.15}$$

$$\tilde{x}(t_a) = \hat{x}(t_a) = x(t_a), \tag{9.16}$$

$$V\big(\tilde{x}(t)\big) \le V\big(\hat{x}(t)\big), \quad \forall t \in [t_a, t_a + N_R\Delta), \tag{9.17}$$

where $\hat{x}(t)$ is the trajectory of the nominal system under a nonlinear control law $u = h(\hat{x}(t))$ when it is implemented in a sample-and-hold fashion, $j = 0, 1, \dots, N-1$, and N_R is the smallest integer satisfying $N_R\Delta \ge T_m$. This optimization problem does not depend on the uncertainty and assures that the LMPC inherits the properties of the nonlinear control law $h(x)$. To take full advantage of the use of the nominal model in the computation of the control action, the prediction horizon should be chosen in a way such that $N \geq N_R$.

The optimal solution to the LMPC optimization problem of Eqs. (9.12)–(9.17) is denoted as $u_a^*(t|t_a)$ which is defined for $t \in [t_a, t_a + N\Delta)$. The manipulated input of the system of Eq. (9.1) under the LMPC of Eqs. (9.12)–(9.17) is defined as follows:

$$u(t) = u_a^*(t|t_a), \quad \forall t \in [t_a, t_{a+1}), \tag{9.18}$$

where t_{a+1} is the next time instant in which the feedback loop will be closed again. This is a modified receding horizon scheme which takes advantage of the predicted input trajectory in the case of sensor data losses.

In the design of the LMPC of Eqs. (9.12)–(9.17), the constraint of Eq. (9.14) is used to generate a system state trajectory under the nonlinear control law $u = h(x)$ implemented in a sample-and-hold fashion; this trajectory is used as a reference trajectory to construct the Lyapunov-based constraint of Eq. (9.17) which is required to be satisfied for a time period which covers the maximum possible open-loop operation time T_m. This Lyapunov-based constraint allows one to prove the closed-loop stability in the presence of sensor data losses in the closed-loop system.

9.2.3.1 Stability Properties

The LMPC of Eqs. (9.12)–(9.17) computes the control input u applied to the system of Eq. (9.1) in a way such that in the closed-loop system, the value of the Lyapunov function at time instant t_a (i.e., $V(x(t_a))$) is a decreasing sequence of values with a lower bound. Following Lyapunov arguments, this property guarantees practical stability of the closed-loop system. This is achieved due to the constraint of Eq. (9.17). This property is summarized in Theorem 9.1 below. To state this theorem, we need the following propositions.

Proposition 9.1 *Consider the nominal sampled trajectory $\hat{x}(t)$ of the system of Eq. (9.1) in closed-loop for a controller $h(x)$, which satisfies the conditions of Eqs. (9.4)–(9.7), obtained by solving recursively*:

$$\dot{\hat{x}}(t) = f(\hat{x}(t), h(\hat{x}(t_k)), 0), \quad t \in [t_k, t_{k+1}), \tag{9.19}$$

where $t_k = t_0 + k\Delta$, $k = 0, 1, \ldots$. Let $\Delta, \varepsilon_s > 0$, and $\rho > \rho_s > 0$ satisfy

$$-\alpha_3(\alpha_2^{-1}(\rho_s)) + L'_x M\Delta \leq -\varepsilon_s/\Delta. \tag{9.20}$$

Then, if $\rho_{\min} < \rho$ where

$$\rho_{\min} = \max\{V(\hat{x}(t+\Delta)) : V(\hat{x}(t)) \leq \rho_s\} \tag{9.21}$$

and $\hat{x}(t_0) \in \Omega_\rho$, the following inequality holds:

$$V(\hat{x}(t)) \leq V(\hat{x}(t_k)), \quad \forall t \in [t_k, t_{k+1}), \tag{9.22}$$

$$V(\hat{x}(t_k)) \leq \max\{V(\hat{x}(t_0)) - k\varepsilon_s, \rho_{\min}\}. \tag{9.23}$$

Proof Following the definition of $\hat{x}(t)$, the time derivative of the Lyapunov function $V(x)$ along the trajectory $\hat{x}(t)$ of the system of Eq. (9.1) in $t \in [t_k, t_{k+1})$ is given by

$$\dot{V}(\hat{x}(t)) = \frac{\partial V(\hat{x}(t))}{\partial x} f(\hat{x}(t), h(\hat{x}(t_k)), 0). \tag{9.24}$$

Adding and subtracting $\frac{\partial V(\hat{x}(t_k))}{\partial x} f(\hat{x}(t_k), h(\hat{x}(t_k)), 0)$ and taking into account Eq. (9.5), we obtain

$$\begin{aligned}\dot{V}(\hat{x}(t)) \leq{}& -\alpha_3(\|\hat{x}(t_k)\|) + \frac{\partial V(\hat{x}(t))}{\partial x} f(\hat{x}(t), h(\hat{x}(t_k)), 0) \\ & - \frac{\partial V(\hat{x}(t_k))}{\partial x} f(\hat{x}(t_k), h(\hat{x}(t_k)), 0).\end{aligned} \tag{9.25}$$

From the Lipschitz property of Eq. (9.10) and the above inequality of Eq. (9.25), we have that

$$\dot{V}(\hat{x}(t)) \leq -\alpha_3(\alpha_2^{-1}(\rho_s)) + L'_x \|\hat{x}(t) - \hat{x}(t_k)\| \tag{9.26}$$

for all $\hat{x}(t_k) \in \Omega_\rho / \Omega_{\rho_s}$. Taking into account the Lipschitz property of Eq. (9.8) and the continuity of $\hat{x}(t)$, the following bound can be written for all $t \in [t_k, t_{k+1})$:

$$\|\hat{x}(t) - \hat{x}(t_k)\| \leq M\Delta. \tag{9.27}$$

Using the expression of Eq. (9.27), we obtain the following bound on the time derivative of the Lyapunov function for $t \in [t_k, t_{k+1})$, for all initial states $\hat{x}(t_k) \in \Omega_\rho / \Omega_{\rho_s}$:

$$\dot{V}(\hat{x}(t)) \leq -\alpha_3(\alpha_2^{-1}(\rho_s)) + L'_x M\Delta. \tag{9.28}$$

If the condition of Eq. (9.20) is satisfied, then $\dot{V}(\hat{x}(t)) \leq -\varepsilon_s/\Delta$. Integrating this bound on $t \in [t_k, t_{k+1})$ we obtain that the inequality of Eq. (9.22) holds. Using Eq. (9.22) recursively, it is proved that, if $x(t_0) \in \Omega_\rho / \Omega_{\rho_s}$, the state converges to Ω_{ρ_s} in a finite number of sampling times without leaving the stability region. Once the state converges to $\Omega_{\rho_s} \subseteq \Omega_{\rho_{\min}}$, it remains inside $\Omega_{\rho_{\min}}$ for all times. This statement holds because of the definition of $\rho_{\min}$ in Eq. (9.21). □

Proposition 9.1 ensures that if the nominal system under the control $u = h(x)$ implemented in a sample-and-hold fashion with state feedback every sampling time starts in the region Ω_ρ, then it is ultimately bounded in $\Omega_{\rho_{\min}}$. The following Proposition 9.2 provides an upper bound on the deviation of the system state trajectory obtained using the nominal model of Eq. (9.1), from the closed-loop state trajectory of the system of Eq. (9.1) under uncertainty (i.e., $w(t) \neq 0$) when the same control actions are applied.

Proposition 9.2 *Consider the systems*:

$$\dot{x}_a(t) = f(x_a(t), u(t), w(t)), \tag{9.29}$$

$$\dot{x}_b(t) = f\big(x_b(t), u(t), 0\big) \tag{9.30}$$

with initial states $x_a(t_0) = x_b(t_0) \in \Omega_\rho$. *There exists a class* $\mathcal{K}$ *function* $f_W(\cdot)$ *such that*

$$\big\|x_a(t) - x_b(t)\big\| \leq f_W(t - t_0), \tag{9.31}$$

for all $x_a(t), x_b(t) \in \Omega_\rho$ *and all* $w(t) \in W$ *with*

$$f_W(\tau) = \frac{L_w\theta}{L_x}\big(e^{L_x\tau} - 1\big). \tag{9.32}$$

Proof Define the error vector as $e(t) = x_a(t) - x_b(t)$. The time derivative of the error is given by

$$\dot{e}(t) = f\big(x_a(t), u(t), w(t)\big) - f\big(x_b(t), u(t), 0\big). \tag{9.33}$$

From the Lipschitz property of Eq. (9.9), the following inequality holds:

$$\big\|\dot{e}(t)\big\| \leq L_w\big\|w(t)\big\| + L_x\big\|x_a(t) - x_b(t)\big\| \leq L_w\theta + L_x\big\|e(t)\big\|, \tag{9.34}$$

for all $x_a(t), x_b(t) \in \Omega_\rho$ and $w(t) \in W$. Integrating $\|\dot{e}(t)\|$ with initial condition $e(t_0) = 0$ (recall that $x_a(t_0) = x_b(t_0)$), the following bound on the norm of the error vector is obtained:

$$\big\|e(t)\big\| \leq \frac{L_w\theta}{L_x}\big(e^{L_x(t-t_0)} - 1\big). \tag{9.35}$$

This implies that the inequality of Eq. (9.31) holds for

$$f_W(\tau) = \frac{L_w\theta}{L_x}\big(e^{L_x\tau} - 1\big), \tag{9.36}$$

which proves this proposition. □

Proposition 9.3 below bounds the difference between the magnitudes of the Lyapunov function of two states in Ω_ρ.

Proposition 9.3 *Consider the Lyapunov function* $V(\cdot)$ *of the system of Eq.* (9.1). *There exists a quadratic function* $f_V(\cdot)$ *such that*

$$V(x) \leq V\big(x'\big) + f_V\big(\big\|x - x'\big\|\big), \tag{9.37}$$

for all $x, x' \in \Omega_\rho$ *where*

$$f_V(s) = \alpha_4\big(\alpha_1^{-1}(\rho)\big)s + M_v s^2 \tag{9.38}$$

with $M_v > 0$.

Proof Since the Lyapunov function $V(x)$ is continuous and bounded on compact sets, there exists a positive constant M_v such that a Taylor series expansion of V around x' yields

$$V(x) \le V(x') + \frac{\partial V(x')}{\partial x}\|x - x'\| + M_v\|x - x'\|^2, \quad \forall x, x' \in \Omega_\rho. \tag{9.39}$$

Note that the term $M_v\|x - x'\|^2$ bounds the high order terms of the Taylor series of $V(x)$ for $x, x' \in \Omega_\rho$. Taking into account Eq. (9.6), the following bound for $V(x)$ is obtained:

$$V(x) \le V(x') + \alpha_4(\alpha_1^{-1}(\rho))\|x - x'\| + M_v\|x - x'\|^2, \quad \forall x, x' \in \Omega_\rho, \tag{9.40}$$

which proves this proposition. □

In Theorem 9.1 below, we provide sufficient conditions under which the LMPC design of Eqs. (9.12)–(9.17) guarantees that the state of the closed-loop system of Eq. (9.1) is ultimately bounded in a region that contains the origin.

Theorem 9.1 *Consider the system of Eq.* (9.1) *in closed-loop, with the loop closing at asynchronous time instants* $\{t_{a\ge 0}\}$ *that satisfy the condition of Eq.* (9.11), *under the LMPC of Eqs.* (9.12)–(9.17) *based on a controller* $h(x)$ *that satisfies the conditions of Eqs.* (9.4)–(9.7). *Let* $\Delta, \varepsilon_s > 0$, $\rho > \rho_{\min} > 0$, $\rho > \rho_s > 0$, *and* $N \ge N_R \ge 1$ *satisfy the condition of Eq.* (9.20) *and the following inequality*:

$$-N_R\varepsilon_s + f_V\big(f_W(N_R\Delta)\big) < 0 \tag{9.41}$$

with $f_V(\cdot)$ *and* $f_W(\cdot)$ *defined in Eq.* (9.38) *and Eq.* (9.32), *respectively, and* N_R *being the smallest integer satisfying* $N_R\Delta \ge T_m$. *If* $x(t_0) \in \Omega_\rho$, *then* $x(t)$ *is ultimately bounded in* $\Omega_{\rho_a} \subseteq \Omega_\rho$ *where*

$$\rho_a = \rho_{\min} + f_V\big(f_W(N_R\Delta)\big) \tag{9.42}$$

with $\rho_{\min}$ *defined as in Eq.* (9.21).

Proof In order to prove that the closed-loop system is ultimately bounded in a region that contains the origin, we prove that $V(x(t_a))$ is a decreasing sequence of values with a lower bound. The proof is divided into two parts.

Part 1: In this part, we prove that the stability results stated in Theorem 9.1 hold in the case that $t_{a+1} - t_a = T_m$ for all a and $T_m = N_R\Delta$. This case corresponds to the worst possible situation in the sense that the LMPC needs to operate in open-loop for the maximum possible amount of time. In order to simplify the notation, we assume that all the notations used in this proof refer to the final solution of the LMPC of Eqs. (9.12)–(9.17) solved at time t_a. By Proposition 9.1 and the fact that $t_{a+1} = t_a + N_R\Delta$, the following inequality can be obtained:

$$V\big(\hat{x}(t_{a+1})\big) \le \max\big\{V\big(\hat{x}(t_a)\big) - N_R\varepsilon_s, \rho_{\min}\big\}. \tag{9.43}$$

From the constraint of Eq. (9.17), the inequality of Eq. (9.43) and taking into account the fact that $\hat{x}(t_a) = \tilde{x}(t_a) = x(t_a)$, the following inequality can be written:

$$V\big(\tilde{x}(t_{a+1})\big) \leq \max\big\{V\big(x(t_a)\big) - N_R\varepsilon_s, \rho_{\min}\big\}. \tag{9.44}$$

When $x(t) \in \Omega_\rho$ for all times (this point will be proved below), we can apply Proposition 9.3 to obtain the following inequality:

$$V\big(x(t_{a+1})\big) \leq V\big(\tilde{x}(t_{a+1})\big) + f_V\big(\big\|\tilde{x}(t_{a+1}) - x(t_{a+1})\big\|\big). \tag{9.45}$$

Applying Proposition 9.2, we obtain the following upper bound on the deviation of $\tilde{x}(t)$ from $x(t)$:

$$\big\|x(t_{a+1}) - \tilde{x}(t_{a+1})\big\| \leq f_W(N_R\Delta). \tag{9.46}$$

From inequalities of Eq. (9.45) and Eq. (9.46), the following upper bound on $V(x(t_{a+1}))$ can be written:

$$V\big(x(t_{a+1})\big) \leq V\big(\tilde{x}(t_{a+1})\big) + f_V\big(f_W(N_R\Delta)\big). \tag{9.47}$$

Using the inequality of Eq. (9.44), we can rewrite the inequality of Eq. (9.47) as follows:

$$V\big(x(t_{a+1})\big) \leq \max\big\{V\big(x(t_a)\big) - N_R\varepsilon_s, \rho_{\min}\big\} + f_V\big(f_W(N_R\Delta)\big). \tag{9.48}$$

If the condition of Eq. (9.41) is satisfied, from the inequality of Eq. (9.48), we know that there exists $\varepsilon_w > 0$ such that the following inequality holds:

$$V\big(x(t_{a+1})\big) \leq \max\big\{V\big(x(t_a)\big) - \varepsilon_w, \rho_a\big\}, \tag{9.49}$$

which implies that if $x(t_a) \in \Omega_\rho/\Omega_{\rho_a}$, then $V(x(t_{a+1})) < V(x(t_a))$, and if $x(t_a) \in \Omega_{\rho_a}$, then $V(x(t_{a+1})) \leq \rho_a$.

Because $f_W(\cdot)$ and $f_V(\cdot)$ are strictly increasing functions of their arguments and $f_V(\cdot)$ is convex (see Propositions 9.2 and 9.3 for the expressions of $f_W(\cdot)$ and $f_V(\cdot)$), the inequality of Eq. (9.49) also implies that

$$V\big(x(t)\big) \leq \max\big\{V\big(x(t_a)\big), \rho_a\big\}, \quad \forall t \in [t_a, t_{a+1}). \tag{9.50}$$

Using the inequality of Eq. (9.50) recursively, it can be proved that if $x(t_0) \in \Omega_\rho$, then the closed-loop trajectories of the system of Eq. (9.1) under the LMPC of Eqs. (9.12)–(9.17) stay in Ω_ρ for all times (i.e., $x(t) \in \Omega_\rho$, $\forall t$). Moreover, it can be proved that if $x(t_0) \in \Omega_\rho$, the closed-loop trajectories of the system of Eq. (9.1) satisfy

$$\limsup_{t\to\infty} V\big(x(t)\big) \leq \rho_a.$$

This proves that $x(t) \in \Omega_\rho$ for all times and $x(t)$ is ultimately bounded in Ω_{ρ_a} for the case when $t_{a+1} - t_a = T_m$ for all a and $T_m = N_R\Delta$.

Part 2: In this part, we extend the results proved in Part 1 to the general case, that is, $t_{a+1} - t_a \leq T_m$ for all a and $T_m \leq N_R \Delta$ which implies that $t_{a+1} - t_a \leq N_R \Delta$. Because $f_W(\cdot)$ and $f_V(\cdot)$ are strictly increasing functions of their arguments and $f_V(\cdot)$ is convex, following similar steps as in Part 1, it can be shown that the inequality of Eq. (9.50) still holds. This proves that the stability results stated in Theorem 9.1 hold. □

Remark 9.1 Theorem 9.1 is important from an MPC point of view because if the maximum time without sensor data losses is smaller than the maximum time that the system can operate in open-loop without leaving the stability region, the feasibility of the optimization problem for all times is guaranteed, since each time feedback is regained, the state is guaranteed to be inside the stability region, thereby yielding a feasible optimization problem.

Remark 9.2 In the LMPC of Eqs. (9.12)–(9.17), no state constraint has been considered but the presented approach can be extended to handle state constraints by restricting the closed-loop stability region further to satisfy the state constraints.

Remark 9.3 It is also important to remark that when there are sensor data losses in the control system, standard MPC formulations do not provide guaranteed closed-loop stability results. For any MPC scheme, in order to obtain guaranteed closed-loop stability results, even in the case where initial feasibility of the optimization problem is given, the formulation of the optimization problem has to be modified accordingly to take into account sensor data losses in an explicit way.

Remark 9.4 Although the proof of Theorem 9.1 is constructive, the constants obtained are conservative. This is the case with most of the results of the type presented in this book. In practice, the different constants are better estimated through closed-loop simulations. The various inequalities provided are more useful as guidelines on the interaction between the various parameters that define the system and the controller and may be used as guidelines to design the controller and the network.

9.2.4 Application to a Continuous Crystallizer

We apply the LMPC presented in the previous section to a continuous crystallization process subject to asynchronous measurement sampling. Asynchronous measurement sampling may arise due to measurement system malfunctions or different sampling rates of the measurement sensors. In particular, a standard MPC, the LMPC presented in Chap. 2 (see also [108]), and the LMPC presented in the previous section, are applied to stabilize a continuous crystallizer at an open-loop unstable steady-state. Extensive simulations are presented to evaluate the closed-loop stability and robustness of the three control methods under three different assumptions on how the measurements from the crystallizer are obtained.

9.2.4.1 Model of a Continuous Crystallizer

In this section, the population balance model of a continuous crystallizer and the corresponding reduced-order moments model are introduced.

Population Balance Model Under the assumptions of isothermal operation, constant volume, mixed suspension, nucleation of crystals of infinitesimal size, and mixed product removal, a dynamic model for a continuous crystallizer can be derived from a population balance for the particle phase and a mass balance for the solute concentration. The resulting model has the following form [73, 84]:

$$\begin{aligned} \frac{\partial n}{\partial \bar{t}} &= -\frac{\partial (R(\bar{t})n)}{\partial r} - \frac{n}{\tau} + \delta(r-0)Q(\bar{t}), \\ \frac{dc}{d\bar{t}} &= \frac{(c_0-\rho)}{\bar{\varepsilon}\tau} + \frac{(\rho-c)}{\tau} + \frac{(\rho-c)}{\bar{\varepsilon}}\frac{d\bar{\varepsilon}}{d\bar{t}}, \end{aligned} \tag{9.51}$$

where $n(r,\bar{t})$ is the number density of crystals of radius $r \in [0,\infty)$ at time $\bar{t}$ in the suspension, τ is the residence time, c is the solute concentration in the crystallizer, c_0 is the solute concentration in the feed, $\bar{\varepsilon} = 1 - \int_0^\infty n(r,\bar{t})\frac{4}{3}\pi r^3\, dr$ is the volume of liquid per unit volume of suspension, $R(\bar{t})$ is the growth rate, $\delta(r-0)$ is the standard Dirac function, ρ is the density of crystals, and $Q(\bar{t})$ is the nucleation rate. The term $\delta(r-0)Q(\bar{t})$ accounts for the production of crystals of infinitesimal (zero) size via nucleation. $R(\bar{t})$ and $Q(\bar{t})$ are assumed to follow McCabe's growth law and Volmer's nucleation law, respectively, that is,

$$R(\bar{t}) = k_1(c-c_s), \qquad Q(\bar{t}) = \bar{\varepsilon}k_2\exp\left[-k_3/(c/c_s-1)^2\right], \tag{9.52}$$

where k_1, k_2, and k_3 are positive constants and c_s is the concentration of solute at saturation.

The values of the parameters in Eq. (9.51) and Eq. (9.52) that define the process are given in Table 9.1. The open-loop crystallizer model exhibits a highly oscillatory behavior, which is the result of the interplay between growth and nucleation caused by the relative nonlinearity of the nucleation rate as compared to the growth rate. See [25] for a detailed discussion on the nature of the oscillations exhibited by this process. The population model introduced provides a good approximation of the dynamics of a continuous crystallizer [27]. All simulations have been carried out using the model of Eq. (9.51).

Reduced-Order Moments Model The population balance model is not appropriate for synthesizing model-based low-order feedback control laws due to its distributed parameter nature. To overcome this problem, following the same approach as in [25], we derive a reduced-order moments model which accurately reproduces the dominant dynamics of the system and is suitable for directly synthesizing low-order feedback control laws.

Table 9.1 Process parameters for the continuous crystallizer

$c_s = 980.2\ \text{kg/m}^3$
$c_{0s} = 999.943\ \text{kg/m}^3$
$\rho = 1770.0\ \text{kg/m}^3$
$\tau = 1.0\ \text{hr}$
$k_1 = 5.065 \times 10^{-2}\ \text{mm}\,\text{m}^3/\text{kg}\,\text{hr}$
$k_2 = 7.958\ 1/\text{mm}^3\,\text{hr}$
$k_3 = 1.217 \times 10^{-3}$

We define the jth moment of $n(r, \bar{t})$ as

$$\mu_j = \int_0^{\infty} r^j n(r, \bar{t})\, dr, \quad j = 0, 1, \ldots, \infty. \tag{9.53}$$

Multiplying the population balance in Eq. (9.51) by r^j, integrating over all particle sizes, and introducing the following set of dimensionless variables and parameters:

$$\begin{aligned}
&\tilde{x}_0 = 8\pi\sigma^3\mu_0, \qquad \tilde{x}_1 = 8\pi\sigma^2\mu_1, \qquad \tilde{x}_2 = 4\pi\sigma\mu_2, \qquad \tilde{x}_3 = \frac{4}{3}\pi\mu_3, \ldots, \\
&t = \frac{\bar{t}}{\tau}, \qquad \sigma = k_1\tau(c_{0s} - c_s), \qquad Da = 8\pi\sigma^3 k_2 \tau, \\
&F = \frac{k_3 c_s^2}{(c_{0s} - c_s)^2}, \qquad \alpha = \frac{(\rho - c_s)}{(c_{0s} - c_s)}, \qquad \tilde{y} = \frac{(c - c_s)}{(c_{0s} - c_s)}, \qquad u = \frac{(c_0 - c_{0s})}{(c_{0s} - c_s)},
\end{aligned} \tag{9.54}$$

where c_{0s} is the steady-state solute concentration in the feed, the dominant dynamics of Eq. (9.51) can be adequately captured by the following fifth-order moments model which includes the dynamics of the first four moments and those of the solute concentration:

$$\begin{aligned}
\frac{d\tilde{x}_0}{dt} &= -\tilde{x}_0 + (1 - \tilde{x}_3) Da e^{\frac{-F}{\tilde{y}^2}}, \\
\frac{d\tilde{x}_1}{dt} &= -\tilde{x}_1 + \tilde{y}\tilde{x}_0, \\
\frac{d\tilde{x}_2}{dt} &= -\tilde{x}_2 + \tilde{y}\tilde{x}_1, \\
\frac{d\tilde{x}_3}{dt} &= -\tilde{x}_3 + \tilde{y}\tilde{x}_2, \\
\frac{d\tilde{y}}{dt} &= \frac{1 - \tilde{y} - (\alpha - \tilde{y})\tilde{y}\tilde{x}_2}{1 - \tilde{x}_3} + \frac{u}{1 - \tilde{x}_3},
\end{aligned} \tag{9.55}$$

where $\tilde{x}_\nu$, $\nu = 0, 1, 2, 3$, are dimensionless moments of the crystal size distribution, $\tilde{y}$ is the dimensionless concentration of the solute in the crystallizer, and u is a dimensionless concentration of the solute in the feed. The values of the dimensionless

Table 9.2 Dimensionless parameters for the continuous crystallizer

$\sigma = k_1\tau(c_{0s} - c_s) = 1.0$ mm
$Da = 8\pi\sigma^3 k_2\tau = 200.0$
$F = k_3 c_s^2/(c_{0s} - c_s)^2 = 3.0$
$\alpha = (\rho - c_s)/(c_{0s} - c_s) = 40.0$

model parameters in Eq. (9.54) are given in Table 9.2. Note that since the moments of order four and higher do not affect those of order three and lower, the state of the infinite dimensional system is bounded when $\tilde{x}_3$ and $\tilde{y}$ are bounded, and it converges to a globally exponentially stable equilibrium point when $\lim_{t\to\infty} \tilde{x}_3 = c_1$ and $\lim_{t\to\infty} \tilde{y} = c_2$, where c_1, c_2 are constants. In this work, the state of the crystallizer is denoted as $\tilde{x} = [\tilde{x}_0\ \tilde{x}_1\ \tilde{x}_2\ \tilde{x}_3\ \tilde{y}]^T$, and the reduced-order moments model is used to define different model predictive control strategies.

The reduced-order moments model is a very good approximation of the population balance model and is suitable for directly synthesizing model-based low-order feedback control laws. The reader may refer to [25, 49] for a detailed derivation of the moments model, and to [27] for further results and references in this area. The stability properties of the fifth-order model of Eq. (9.55) have been also studied and it has been shown [73] that the global phase space of this model has a unique unstable steady-state surrounded by a stable periodic orbit at

$$\tilde{x}_s = [\tilde{x}_{0s}\ \tilde{x}_{1s}\ \tilde{x}_{2s}\ \tilde{x}_{3s}\ \tilde{y}_s]^T = [0.0471\ 0.0283\ 0.0169\ 0.0102\ 0.5996]^T,$$

and that the linearization of Eq. (9.51) around the unstable steady-state includes two isolated complex conjugate eigenvalues with a positive real part. The control objective is to regulate the system to the unstable steady state $\tilde{x}_s$ by manipulating the solute feed concentration c_0.

We consider constraints in the input. The dimensionless solute feed concentration, u, is subject to the constraints: $-u_{\max} \le u \le u_{\max}$, where $u_{\max} = 3$. For $u_{\max} = 3$, the constraint on the inlet solute concentration corresponds to $940\ \text{kg/m}^3 \le c_0 \le 1060\ \text{kg/m}^3$.

We denote the state x as the error, that is $x = \tilde{x} - \tilde{x}_s$. Then, we can rewrite Eq. (9.55) in a more compact form:

$$\dot{x}(t) = f\big(x(t)\big) + g\big(x(t)\big)u(t), \tag{9.56}$$

where $x = [x_0\ x_1\ x_2\ x_3\ y]^T$, and f and g have the following form:

$$f(x) = \begin{bmatrix} -(x_0 + \tilde{x}_{0s}) + (1 - x_3 - \tilde{x}_{3s}) Da\, e^{\frac{-F}{(y+\tilde{y}_s)^2}} \\ -(x_1 + \tilde{x}_{1s}) + (y + \tilde{y}_s)(x_0 + \tilde{x}_{0s}) \\ -(x_2 + \tilde{x}_{2s}) + (y + \tilde{y}_s)(x_1 + \tilde{x}_{1s}) \\ -(x_3 + \tilde{x}_{3s}) + (y + \tilde{y}_s)(x_2 + \tilde{x}_{2s}) \\ \frac{1 - y - \tilde{y}_s - (\alpha - y - \tilde{y}_s)(y + \tilde{y}_s)(x_2 + \tilde{x}_{2s})}{1 - x_3 - \tilde{x}_{3s}} \end{bmatrix}, \qquad g(x) = \begin{bmatrix} 0 \\ 0 \\ 0 \\ 0 \\ \frac{1}{1 - x_3 - \tilde{x}_{3s}} \end{bmatrix}.$$

Next we are going to define a feedback control law $h_L : \mathbb{R}^n \to \mathbb{R}$ which satisfies $h_L(0) = 0$ that renders the origin $x = 0$ of the closed-loop system of Eq. (9.56) asymptotically stable under continuous measurements. Stabilizing state feedback control laws for nonlinear systems have been developed using Lyapunov techniques; the reader may refer to Chap. 2 and to [28, 78] for results on this area. In this work, we use the Lyapunov-based feedback control proposed in [85] (see also [45, 46]) which is based on a control Lyapunov function of the open-loop system.

Consider the control Lyapunov function $V(x) = x^T Px$ with $P = I$ of the system of Eq. (9.56). The following Lyapunov-based feedback control law [85] asymptotically stabilizes the open-loop unstable steady-state under continuous state feedback implementation for an appropriate set of initial conditions:

$$h_L(x) = -k(x) L_g V(x), \tag{9.57}$$

where

$$k(x) = \begin{cases} \frac{L_f V(x) + \sqrt{(L_f V(x))^2 + (u_{\max} L_g V(x))^4}}{(L_g V(x))^2 [1 + \sqrt{1 + (u_{\max} L_g V(x))^2}]}, & L_g V(x) \neq 0, \\ 0, & L_g V(x) = 0. \end{cases}$$

The feedback controller $h_L(x)$ will be used to design the contractive constraints of the two LMPCs.

9.2.4.2 Modeling Asynchronous Measurements

We assume that the sampling of the state of the continuous crystallizer of Eq. (9.51) takes at least 15 minutes, and if errors occur in the sampling system or in the communication network, it may take a much longer time. We assume that the maximum time interval (worst case occurrence) between two consecutive measurements is shorter than 2.5 hours, which is denoted as $T_{\max}$. Note that a $T_{\max}$ is needed in the present stabilization problem because the open-loop crystallizer is unstable.

To account for asynchronous sampling, the sampling times are defined by an increasing time sequence $\{t_{a \geq 0}\}$. At each sampling time t_a, a new measurement from the sensors is obtained. The interval between two consecutive samplings is not fixed. In the simulation section, we present three different ways of generating the time sequence $\{t_{a \geq 0}\}$. The only assumption made on the time sequence $\{t_{a \geq 0}\}$ is that there is an upper bound (which is $T_{\max}$) on the maximum time in which the system operates in open-loop which is needed in the present stabilization problem because the open-loop crystallizer is unstable.

We also take into account that the controller may not receive the whole state (x_0, x_1, x_2, x_3, y) at each sampling instant but just part of it, that is, the state of PSD (x_0, x_1, x_2, x_3) or the solute concentration (y) (see Figs. 9.2–9.3) may be transmitted only at a specific time instant. This is due to the fact that PSD and solute concentration are measured by different sensors with different sampling rates. At sampling time t_a, if only part of the state is available, an estimation of the current state $\hat{x}(t_a)$

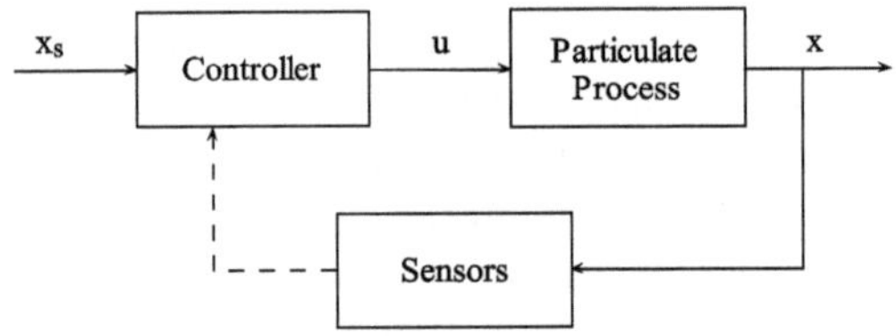

Fig. 9.2 Closed-loop system with asynchronous measurements: the whole state is sampled simultaneously

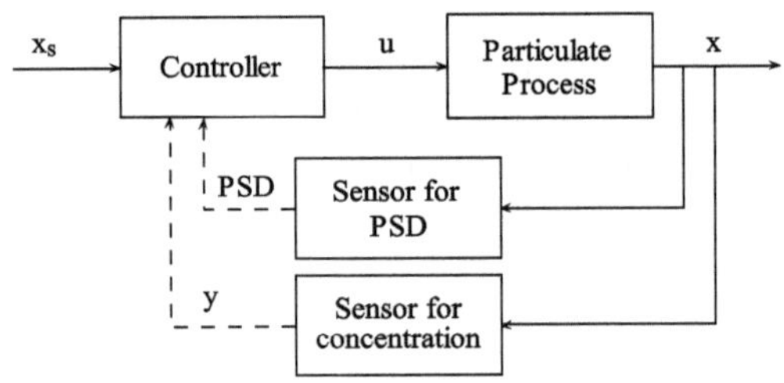

Fig. 9.3 Closed-loop system with asynchronous measurements: the states of PSD and solute concentration are sampled separately

is obtained and sent to the controller to generate a new control input. We use an auxiliary variable $s(t_a)$ to indicate what part of the process state is available at sampling time t_a as follows:

1. $s(t_a) = 1$ implies that both measurements of PSD and solute concentration are available at t_a, and $\hat{x}(t_a) = x(t_a)$.
2. $s(t_a) = 2$ implies that only the measurement of PSD is available at t_a. The corresponding value of the solute concentration at t_a is estimated by using the last available value of solute concentration, that is, $\hat{y}(t_a) = \hat{y}(t_{a-1})$.
3. $s(t_a) = 3$ implies that only the measurement of solute concentration is available at t_a. The corresponding state of PSD at t_a is estimated by the reduced-order moments model of Eq. (9.56). The last available estimated state $\hat{x}(t_{a-1})$ is taken as the initial state.

The estimated state used by the controller at each sampling time is given by the following equation:

$$\hat{x}(t_a) = \begin{cases} x(t_a) & \text{if } s(t_a) = 1, \\ \{x_0(t_a), x_1(t_a), x_2(t_a), x_3(t_a), \hat{y}(t_{a-1})\} & \text{if } s(t_a) = 2, \\ \{\hat{x}_0(t_a), \hat{x}_1(t_a), \hat{x}_2(t_a), \hat{x}_3(t_a), y(t_a)\} & \text{if } s(t_a) = 3, \end{cases} \tag{9.58}$$

where $\hat{x}_\nu$, $\nu = 0, 1, 2, 3$, are estimated by using the reduced-order moments model. Note that we have to store the implemented manipulated input trajectory.

Remark 9.5 Note that regardless of the method used to estimate the state when only partial state information is available, there exist errors between the estimated state $\hat{x}$ and the actual state of the system x, that have to be compensated by the available feedback.

In this class of processes, the solute concentration is obtained with a higher sampling rate than the crystallizer PSD. This motivates using the last available value of the solute concentration when a new PSD measurement is obtained. On the other hand, instead of using the last available values of the PSD each time we obtain a new concentration measurement, which may introduce a large error because the PSD is sampled less frequently, we use the reduced-order moments model to estimate the missing information, which increases the computational complexity but decreases the estimation error.

The controller has to take into account that the measurements arrive in an asynchronous manner and that the time in which it has to operate in open-loop may be long. In order to decide the manipulated input $u(t)$ that has to be applied at each time t, the controller uses the last estimated state $\hat{x}(t_a)$ and the corresponding sampling time t_a. We assume that each controller is defined by a function $h(\Delta, \hat{x}(t_a))$, where $\hat{x}$ is the last available estimated state and Δ is the time that has passed since that state was received. This function allows us to model different implementation strategies. For example, $h(\Delta, \hat{x}) = h_L(\hat{x})$ implements a sample-and-hold strategy based on the Lyapunov-based controller of Eq. (9.57). In this case, the input is kept constant between samples independently of the time Δ that has passed since the last measurement.

In order to consider the models in this work in a unified time scale and with the same manipulated input, we substitute Eq. (9.52), the expressions of dimensionless time t and manipulated input u into Eq. (9.51). We obtain the following asynchronous nonlinear model for the closed-loop system of the crystallizer:

$$\begin{aligned}
\frac{1}{\tau}\frac{\partial n}{\partial t} &= -k_1(c-c_s)\frac{\partial n}{\partial r} - \frac{n}{\tau} + \delta(r-0)\bar{\varepsilon}k_2\exp\bigl[-k_3/(c/c_s-1)^2\bigr],\\
\frac{1}{\tau}\frac{dc}{dt} &= \frac{(c_{0s}-\rho)}{\bar{\varepsilon}\tau} + \frac{(\rho-c)}{\tau} + \frac{(\rho-c)}{\bar{\varepsilon}\tau}\frac{d\bar{\varepsilon}}{dt} + \frac{(c_{0s}-c_s)u(t)}{\bar{\varepsilon}\tau},\\
& t\in[t_a, t_{a+1}],\\
u(t) &= h\bigl(t-t_a, \hat{x}(t_a)\bigr).
\end{aligned} \tag{9.59}$$

At time t_a, new information is available from the sensors and the content of the information is decided by the corresponding value of $s(t_a)$. The state $\hat{x}(t_a)$ is an estimation of the actual state $x(t_a)$ and it is estimated by the approach presented before in this section, see Eq. (9.58). The controller generates a future manipulated input trajectory $h(\Delta, \hat{x})$ that depends on this estimated state, where Δ is the time that has passed since t_a.

9.2.4.3 Controller Design

Three different MPC controllers, a standard MPC as presented in Sect. 2.7, the LMPC presented in Sect. 2.8 and the LMPC presented in Sect. 9.2.3, are applied to the continuous crystallizer. We denote, in the remainder of this chapter, the three model predictive controllers as MPC, LMPC I, and LMPC II, respectively. These controllers are based on the reduced-order moments model of Eq. (9.56) and the Lyapunov-based controller of Eq. (9.57). All the three model predictive controllers use the same sampling time Δ_c. In most control systems where the measurements are obtained synchronously and the communications are flawless, this sampling time is equal to the sampling time used to obtain new measurements and implement the manipulated input (sample-and-hold schemes). In this section, however, we deal with systems subject to asynchronous measurements and the time sequence that determines when new information is available is independent of Δ_c.

The cost functions of these controllers are defined by matrices $Q_c = P$ and $R_c = 4$. The weight matrices Q_c and R_c have been chosen to provide a performance similar to the Lyapunov-based controller under a sample-and-hold implementation. The sampling time of the MPC controllers is $\Delta_c = 0.25h$ which is equal to the minimum time needed to obtain a new measurement.

Through simulations, we have estimated the transition time for the crystallizer in closed-loop with the Lyapunov-based controller which is 2 hours for states x_0, y and 4 hours for states x_1, x_2, x_3. We choose the prediction horizon $N = 11$ for the model predictive controllers so that the prediction captures most of the dynamic evolution of the process.

9.2.4.4 Simulation Results

In this section, we apply the three model predictive control laws MPC, LMPC I, and LMPC II to the continuous crystallizer population balance model of Eq. (9.59) to evaluate the stability and robustness properties of the corresponding closed-loop systems in the presence of measurement unavailability and asynchronous measurements. First, we simulate with PSD and solute concentration sampled synchronously and simultaneously subject to measurement unavailability. Following that, we simulate the system with asynchronous measurements in which measurements of PSD and solute concentration come simultaneously, and then simulate with asynchronous measurements in which PSD and solute concentration are sampled separately. The control objective is to suppress the oscillatory behavior of the crystallizer and stabilize it at the open-loop unstable steady-state $\tilde{x}_s$ that corresponds to the desired PSD by manipulating the solute feed concentration. The following initial conditions are used in the simulations:

$$n(0,r) = 0.0, \qquad c(0) = 990.0\ \mathrm{kg/m^3}, \qquad \tilde{x}(0) = [0\,0\,0\,0\,0.498]^T. \tag{9.60}$$

To simulate the continuous crystallizer, we use a second-order accurate finite-difference discretization scheme. At every model evaluation step (which is different from the sampling time and should be chosen to be sufficiently small in order to get a continuous and accurate solution) of Eq. (9.59), the values of $n(t,r)$ and $c(t)$ can be obtained, so we can use them to calculate the state x at that time using Eq. (9.53) and Eq. (9.54) and the steady-state $\tilde{x}_s$.

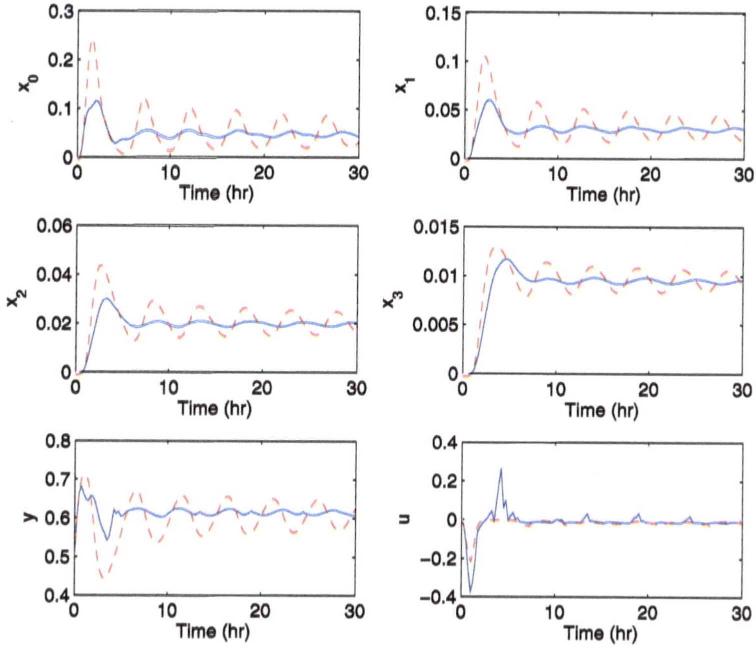

Fig. 9.4 State and manipulated input trajectories of Eq. (9.59) with PSD and solute concentration sampled synchronously and simultaneously and 95 % probability of measurement unavailability using the predicted manipulated input trajectories of LMPC II (*solid curves*) and the standard MPC (*dashed curves*)

Results of Synchronous Sampling Subject to Measurement Unavailability For this set of simulations, we assume that a new measurement of the whole state of the crystallizer is made every Δ_m, the synchronous sampling time, but the measurement might be lost due to errors in the measurement or communication systems with a probability $p \in (0, 1)$. To generate the time partition $\{t_{k\geq 0}\}$ that indicates when a new sample is available and the corresponding auxiliary variable $s(t_k)$ for a simulation of length t_{sim}, we use the following algorithm:

$t_0 = 0$, $k = 0$
while $t_k < t_{\text{sim}}$
  $t_{k+1} = t_k$, $\gamma = 0$
  while $\gamma \leq p$
    $t_{k+1} = t_{k+1} + \Delta_m$, $\gamma = rand(1)$
  end
  if $t_{k+1} > t_k + T_{\max}$ then $t_{k+1} = t_k + T_{\max}$
  $s(t_k) = 1$, $k = k + 1$
end

where t_{sim} is the simulation time, $rand(1)$ generates a uniformly distributed random value γ between 0 and 1, and $T_{\max}$ is the maximum allowable transmission interval.

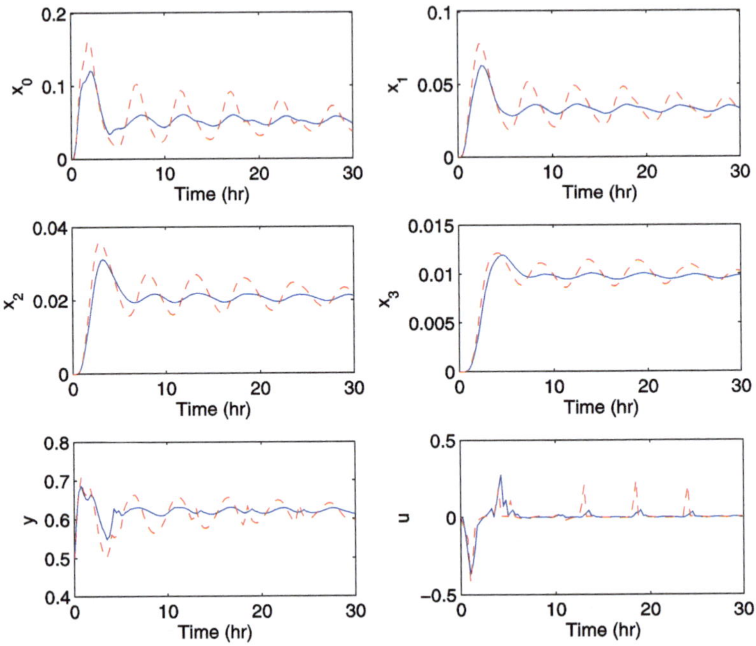

Fig. 9.5 State and manipulated input trajectories of Eq. (9.59) with PSD and solute concentration sampled synchronously and simultaneously and 95 % probability of measurement unavailability using the predicted manipulated input trajectories of LMPC II (*solid curves*) and LMPC I (*dashed curves*)

As mentioned before $T_{\max}$ is taken to be 2.5 hr, and the simulation time t_{sim} is 30 hr in this work. The sampling time Δ_m is equal to Δ_c, that is $\Delta_m = 0.25$ hr. For this sampling time, the sampled-data system in closed-loop with $u = h_L(x)$ is practically stable and its performance is similar to the closed-loop system with continuous measurements. We choose $\gamma = 95$ %, that is, there is a probability of 95 % that the measurement of the state is unavailable at every sampling time. First, we compare LMPC II with MPC. The state and manipulated input trajectories of this simulation are shown in Fig. 9.4. In this figure, it can be seen that LMPC II provides a better performance than MPC. In particular, LMPC II is able to stabilize the process at the open-loop unstable steady-state in about 5 hours while the system in closed-loop with MPC presents an oscillatory behavior indicating that the stabilization of the operating unstable steady-state has not been achieved. Second, we compare LMPC II with LMPC I. The state and manipulated input trajectories of this simulation are shown in Fig. 9.5. In this case, LMPC I is not able to regulate the system to the desired equilibrium. Finally, we compare two LMPC II controllers using the predicted manipulated input trajectory and the "last implemented manipulated input", respectively. The "last implemented manipulated input" strategy keeps constant the manipulated input, that is $h(\Delta, \tilde{x}) = u^*(0)$ for all Δ where $u^*(\cdot)$ is the optimal solution of the optimization problem that defines LMPC II with an initial

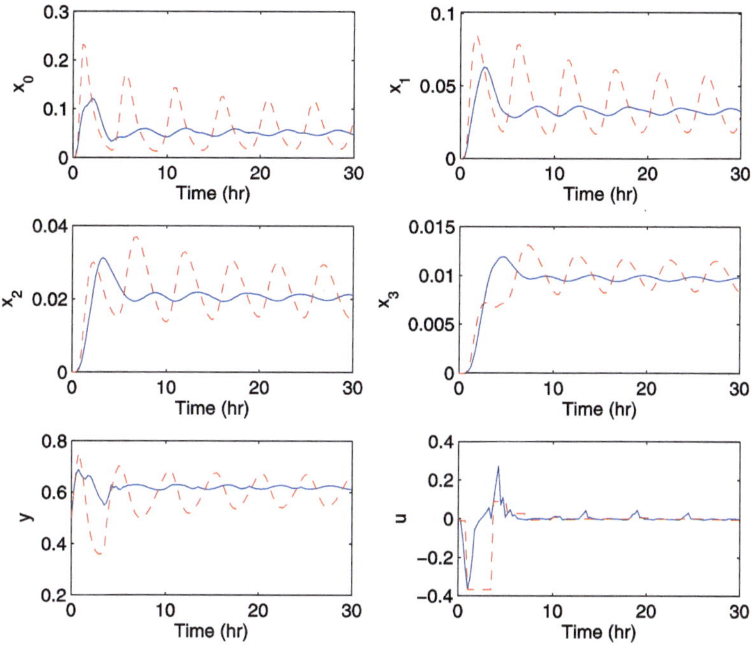

Fig. 9.6 State and manipulated input trajectories of Eq. (9.59) with PSD and solute concentration sampled synchronously and simultaneously and 95 % probability of measurement unavailability using the predicted manipulated input trajectory (*solid curves*) and the last implemented manipulated input (*dashed curves*) of LMPC II

state $\tilde{x}$. The state and manipulated input trajectories of this simulation are shown in Fig. 9.6. This simulation demonstrates that, in this case, using only the last implemented manipulated input is not possible to maintain the process at the desired steady-state.

The simulations demonstrate that LMPC II is more robust to measurements unavailability than MPC and LMPC I. This is because LMPC II is designed taking explicitly into account measurement unavailability. Moreover, we should make full use of the predicted manipulated input trajectory of LMPC II in order to get the best closed-loop system performance.

Results of Asynchronous Sampling: PSD and Solute Concentration Sampled Simultaneously For the simulations in this subsection, we assume that the time between consecutive measurements is obtained using a random process and that the PSD and solute concentration are measured simultaneously. To generate the time intervals between samples we use a random Poisson process as in [59, 110]. The Poisson process is defined by the number of events per unit time W. At a given time t, an event takes place which means that the state is sampled. The interval between two consecutive sampling times is given by $\Delta_a = \frac{-\ln \chi}{W}$, where χ is a random variable with uniform probability distribution between 0 and 1. At $t + \Delta_a$,

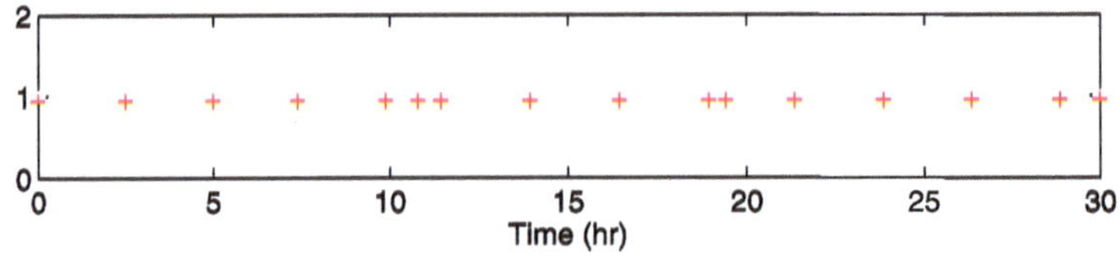

Fig. 9.7 Asynchronous sampling times for both PSD and solute concentration

another event occurs. The sequence $\{t_{k\geq 0}\}$ and the corresponding auxiliary variable $s(t_k)$ for a simulation of length t_{sim} is generated as follows:

$t_0 = 0$, $k = 0$
while $t_k < t_{\text{sim}}$
  $\chi = rand(1)$
  $t_{k+1} = t_k + \frac{-\ln\chi}{W}$
  if $t_{k+1} > t_k + T_{\max}$, then $t_{k+1} = t_k + T_{\max}$
  if $t_{k+1} < t_k + T_{\min}$, then $t_{k+1} = t_k + T_{\min}$
  $s(t_k) = 1$, $k = k + 1$
end

where $rand(1)$ generates a uniformly distributed random value χ between 0 and 1, $T_{\max}$ is the maximum allowable transmission interval and $T_{\min}$ is the minimum time interval between two consecutive samplings. Note that $T_{\min}$ should be smaller than $T_{\max}$, that is, $T_{\min} < T_{\max}$. As mentioned before $T_{\max}$ is 2.5 hr. The minimum time limit $T_{\min}$ is equal to the synchronous sampling time, that is, $T_{\min} = \Delta_m = 0.25$ hr. For the simulations carried out in this subsection we pick the value of the number of events per unit time to be $W = 0.15$. The sampling times for the simulations are shown in Fig. 9.7. Note that because the number of events is low, the time between consecutive samplings (and hence, the time in which the control system must operate in open-loop) may be large but always smaller than $T_{\max}$.

We carry out the same comparisons as we did in the previous subsection. First, we compare LMPC II with MPC. The state and manipulated input trajectories of this simulation are shown in Fig. 9.8. In this simulation, MPC can not stabilize the process, while LMPC II is able to maintain the process at the desired steady-state. Second, we compare LMPC II with LMPC I. The state and manipulated input trajectories are shown in Fig. 9.9. Though LMPC II and LMPC I can both stabilize the process, the transient of the closed-loop system under LMPC II is shorter than the transient under LMPC I and has a smaller overshoot. Finally, we compare two LMPC II controllers using the predicted manipulated input trajectory and the last implemented manipulated input, respectively; the state and manipulated input trajectories are shown in Fig. 9.10. As in the simulation of the previous subsection, the last implemented manipulated input strategy is not able to stabilize the process.

From the results of this subsection, one can also conclude that LMPC II using the predicted manipulated input trajectory is the most robust in the presence of asynchronous sampling among the three controllers.

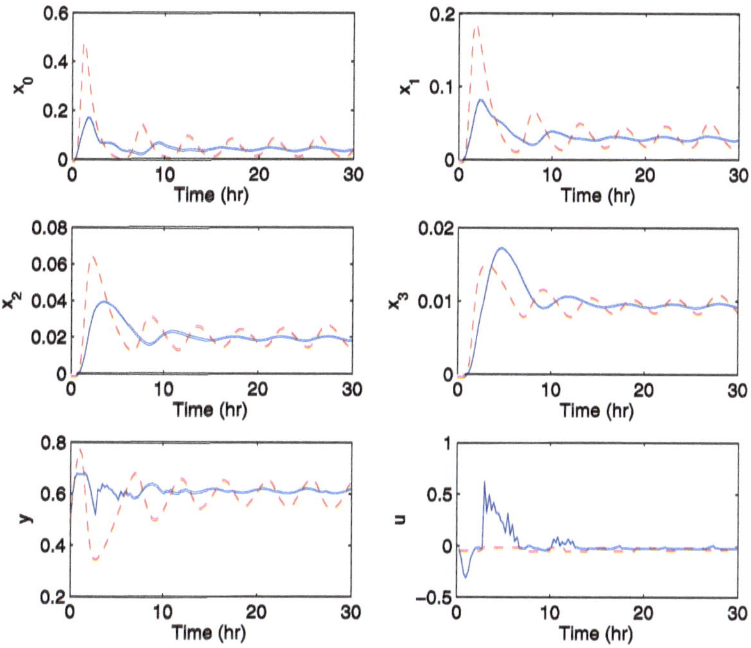

Fig. 9.8 State and manipulated input trajectories of Eq. (9.59) with PSD and solute concentration sampled asynchronously and simultaneously using the predicted manipulated input trajectories of LMPC II (*solid curves*) and the standard MPC (*dashed curves*)

Results of Asynchronous Sampling: PSD and Solute Concentration Sampled Separately For the last set of simulations, we assume that we have the measurements of PSD and solute concentration sampled separately. This implies that we may get a measurement of PSD at a sampling time but lack corresponding measurement of solute concentration; and we may have a measurement of solute concentration but lack the corresponding measurement of PSD. In addition, we have asynchronous sampling which means that the length of the time interval between two consecutive measurements is varying.

Using the same method presented in Sect. 5.2, we generate two different time sequences $\{t^p_{k\geq 0}\}$ for PSD ($s = 2$) and $\{t^c_{k\geq 0}\}$ for solute concentration ($s = 3$) using $W^p = 0.15$ and $W^c = 1$, respectively. Both time sequences are generated with the same constraints $T_{\max} = 2.5$ hr and $T_{\min} = 0.25$ hr. The choice of $W^c = 1$ for $\{t^c_{k\geq 0}\}$ is based on the fact that we can get a measurement of concentration faster. The sampling sequence $\{t^p_{k\geq 0}\}$ corresponding to the PSD measurements is shown in Fig. 9.7 and the sampling sequence $\{t^c_{k\geq 0}\}$ corresponding to the solute concentration measurements is shown in Fig. 9.11. Subsequently, the two sequences are merged into an ordered one $\{t_{k\geq 0}\}$ by increasing time and the overlapping times correspond to instants that both measurements of PSD and solute concentration can be obtained ($s = 1$). The new sequence $\{t_{k\geq 0}\}$ is shown in Fig. 9.12. Every sampling instant in

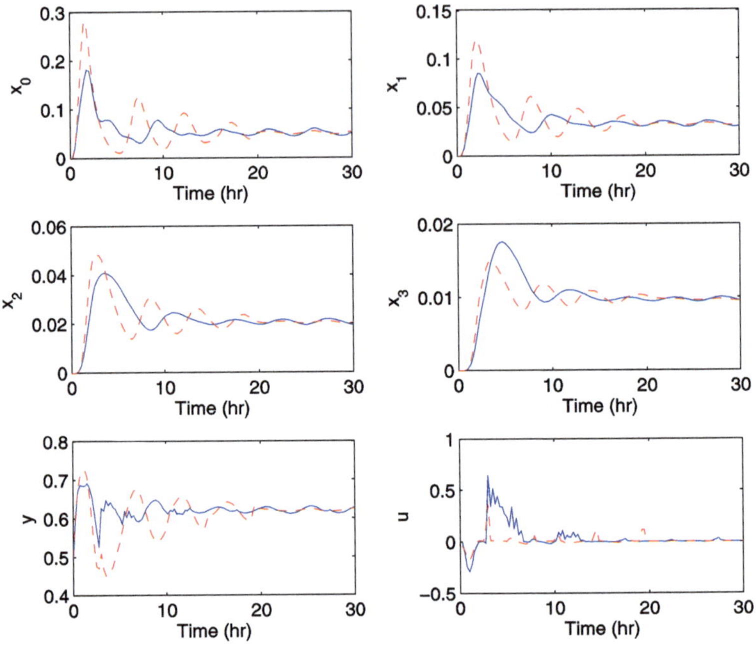

Fig. 9.9 State and manipulated input trajectories of Eq. (9.59) with PSD and solute concentration sampled asynchronously and simultaneously using the predicted manipulated input trajectories of LMPC II (*solid curves*) and LMPC I

the new sequence represents a measurement of PSD or solute concentration or both. The auxiliary variable $s(t_k)$ is defined accordingly.

We first compare LMPC II with MPC. The state and manipulated input trajectories are shown in Fig. 9.13. As expected, LMPC II is able to stabilize the process, but MPC fails. The result is consistent with the previous simulations. Following that, we compare LMPC II with LMPC I. The state and manipulated input trajectories are shown in Fig. 9.14. In this figure, it can be seen that LMPC I can also stabilize the process but it takes a longer time compared with LMPC II. Finally, we compare two LMPC II controllers using the predicted manipulated input trajectory and the last implemented input, respectively. Figure 9.15 shows the trajectories of the state and manipulated input. This simulation demonstrates that for this case, only using the last implemented manipulated input of LMPC II can not stabilize the process as in the other simulations.

In this case, the overshoots of the trajectories generated by MPC and the amplitudes of oscillations of the trajectories generated by LMPC II using the last implemented manipulated input are smaller compared with the case discussed in Sect. 5.2. This improvement is due to the decrease of the average time interval between two consecutive measurements. Despite of this decrease, the performances of LMPC I

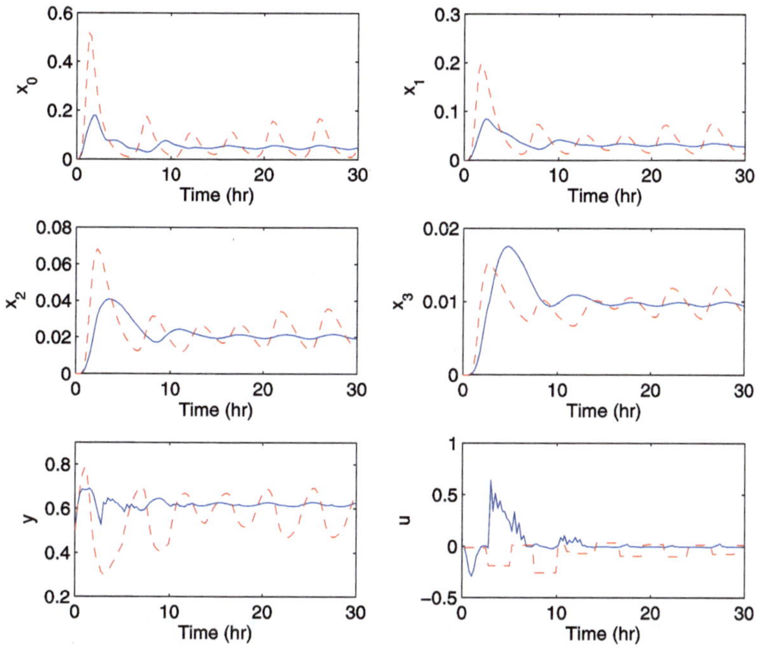

Fig. 9.10 State and manipulated input trajectories of Eq. (9.59) with PSD and solute concentration sampled asynchronously and simultaneously using the predicted manipulated input trajectory (*solid curves*) and the last implemented manipulated input (*dashed curves*) of LMPC II

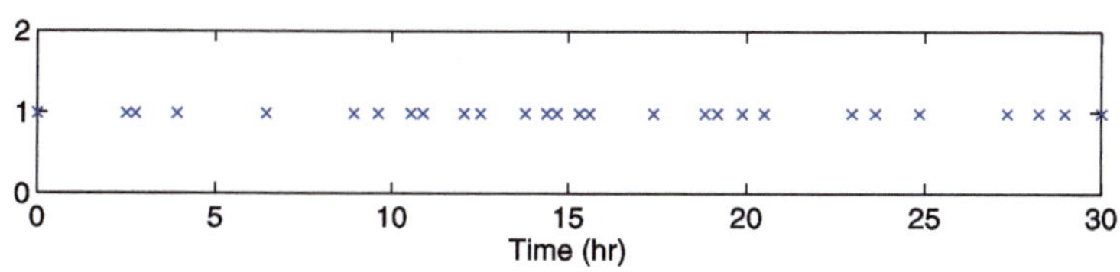

Fig. 9.11 Asynchronous sampling times for solute concentration

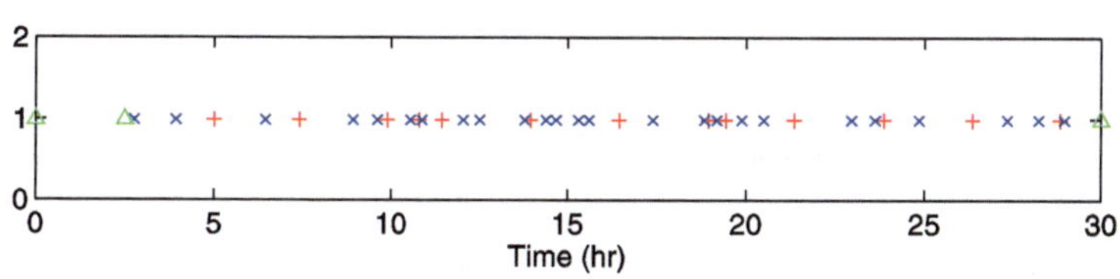

Fig. 9.12 Asynchronous sampling times, +: sampling times of PSD ($s(t_k) = 2$), ×: sampling times of solute concentration ($s(t_k) = 3$), Δ: sampling times of both PSD and solute concentration ($s(t_k) = 1$)

and LMPC II using the predicted manipulated input trajectories do not get much improvement because there still exists some large intervals between two consecutive measurements as shown in Fig. 9.12.

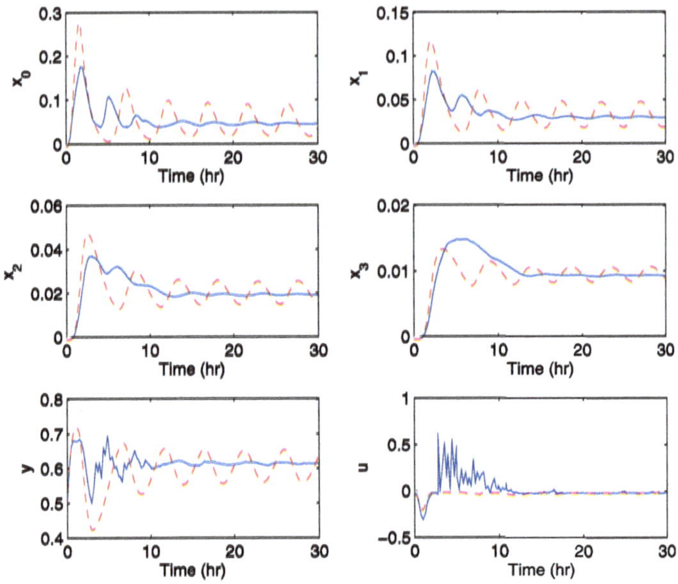

Fig. 9.13 State and manipulated input trajectories of Eq. (9.59) with PSD and solute concentration sampled asynchronously and separately using the predicted manipulated input trajectories of LMPC II (*solid curves*) and the standard MPC (*dashed curves*)

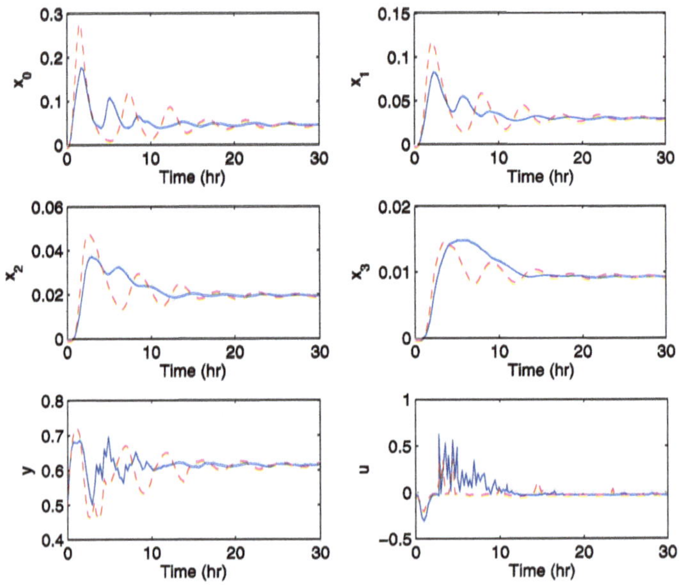

Fig. 9.14 State and manipulated input trajectories of Eq. (9.59) with PSD and solute concentration sampled asynchronously and separately using the predicted manipulated input trajectories of LMPC II (*solid curves*) and LMPC I (*dashed curves*)

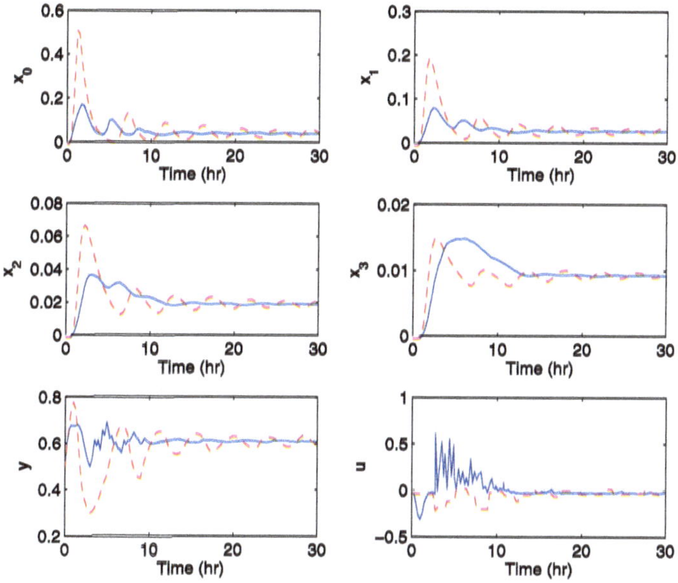

Fig. 9.15 State and manipulated input trajectories of Eq. (9.59) with PSD and solute concentration sampled asynchronously and separately using the predicted manipulated input trajectory (*solid curves*) and the last implemented manipulated input (*dashed curves*) of LMPC II

Finally, to evaluate the robustness properties of the LMPC controllers, we also carried out another set of simulations to demonstrate that LMPC II is more robust than the other two controllers when there are uncertainties in the model parameters. We assume that uncertainties are present in k_1 and k_2 of Eq. (9.59) and the actual values used to evaluate the population balance model of Eq. (9.59) are $1.1k_1$ and $1.1k_2$ (10 % uncertainty) which are different from the values (k_1 and k_2) used in the reduced-order moments model of Eq. (9.56). Figure 9.16 shows the results when MPC and LMPC II are applied and Fig. 9.17 shows the results when LMPC I and LMPC II are implemented. From the two figures, we conclude that LMPC II can stabilize the system, but both MPC and LMPC I fail.

In summary, LMPC II using the predicted manipulated input trajectory yields a more robust closed-loop performance when the process is subject to measurement unavailability, asynchronous sampling, and parametric model uncertainties.

9.3 FDI Using Asynchronous Measurements: Problem Formulation and Solution

9.3.1 Class of Nonlinear Systems

In this section, we consider nonlinear process systems described by the following state-space model:

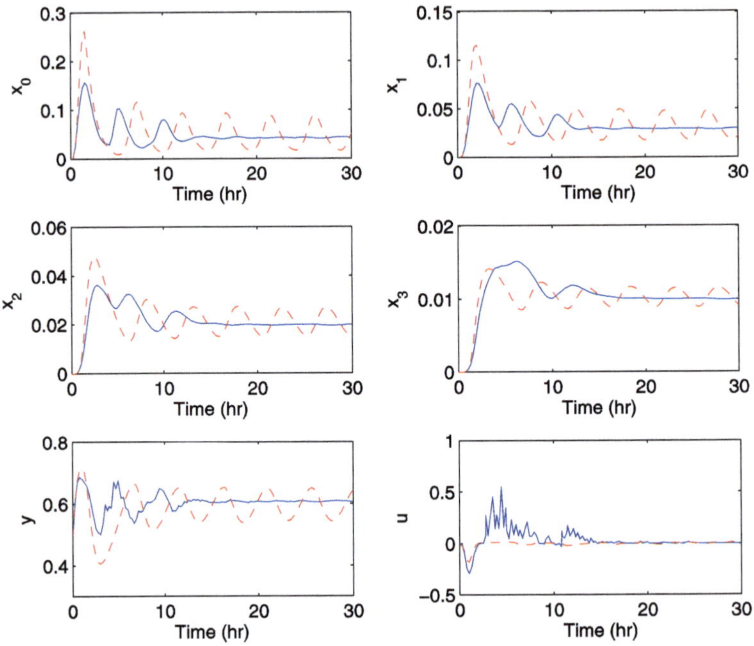

Fig. 9.16 State and manipulated input trajectories of Eq. (9.59) with 10 % uncertainty in parameters k_1 and k_2 when PSD and solute concentration are sampled asynchronously and separately using the predicted manipulated input trajectories of LMPC II (*solid curves*) and the standard MPC (*dashed curves*)

$$\begin{aligned} \dot{x}_s &= f_s(x_s, x_a, u, d), \\ \dot{x}_a &= f_a(x_s, x_a, u, d), \end{aligned} \tag{9.61}$$

where $x_s \in \mathbb{R}^{n_s}$ denotes the set of state variables that are sampled synchronously, $x_a \in \mathbb{R}^{n_a}$ denotes the set of state variables that are sampled asynchronously, $u \in \mathbb{R}^{n_u}$ denotes the input and $d \in \mathbb{R}^p$ is a model of the set of p possible faults. The faults are unknown and d_j, $j = 1, \dots, p$, can take any value. The state of the full system is given by the vector

$$x = \begin{bmatrix} x_s \\ x_a \end{bmatrix} \in \mathbb{R}^{n_s + n_a}.$$

Using this definition for x, the system of Eq. (9.61) can be written in the following equivalent compact form:

$$\dot{x} = f(x, u, d). \tag{9.62}$$

We assume that f is a locally Lipschitz vector function and that $f(0, 0, 0) = 0$. This means that the origin is an equilibrium point for the fault-free system with $u(t) \equiv 0$. Moreover, we assume that the fault-free system ($d_i(t) \equiv 0$ for all t) has an

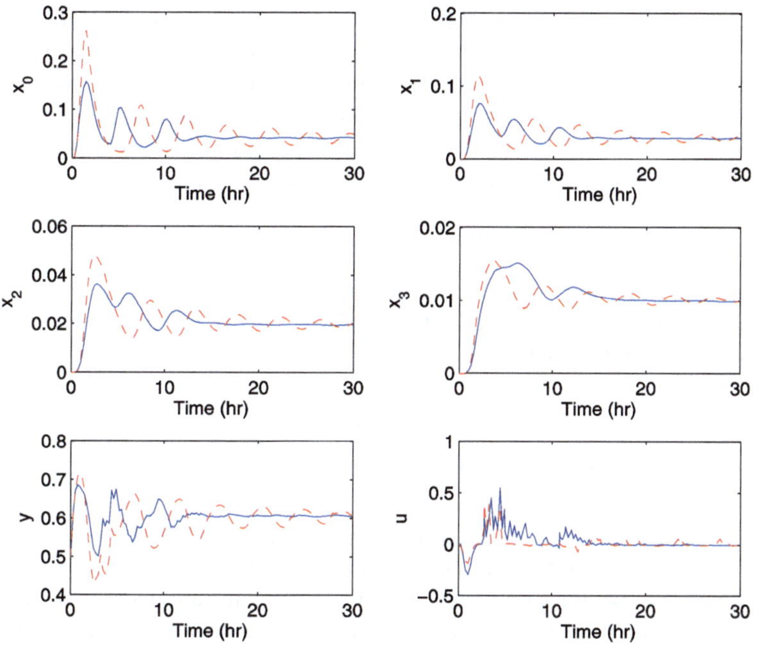

Fig. 9.17 State and manipulated input trajectories of Eq. (9.59) with 10 % uncertainty in parameters k_1 and k_2 when PSD and solute concentration are sampled asynchronously and separately using the predicted manipulated input trajectories of LMPC II (*solid curves*) and LMPC I (*dashed curves*)

asymptotically stable equilibrium at the origin $x = 0$ for a given feedback control function $h : \mathbb{R}^{n_s+n_a} \rightarrow \mathbb{R}^{n_u}$ which satisfies $h(0) = 0$.

9.3.2 Modeling of Asynchronous Measurements

The system of Eq. (9.61) is controlled using both sampled synchronous and asynchronous measurements. We assume that each state $x_{s,i}$, $i = 1, \dots, n_s$, is sampled continuously (i.e., at intervals of fixed size $\Delta > 0$ where Δ is a sufficiently small positive number). Each state $x_{a,i}$, $i = n_s + 1, \dots, n_s + n_a$, is sampled asynchronously and is only available at time instants $t_{k,i}$ where $t_{k,i}$ is a random increasing sequence of times. A controller design that takes advantage of the asynchronous measurements must take into account that it will have to operate without complete state information between asynchronous samples. This class of systems arises naturally in process control, where process variables such as temperature, flow, or concentration have to be measured. In such a case, temperature and flow measurements can be assumed to be available continuously. Concentration measurements, however, are available at an asynchronous sampling rate.

If there exists a non-zero probability that the system operates in open-loop for a period of time large enough for the state to leave the stability region or even diverge to infinity (i.e., finite escape time), it is not possible to provide guaranteed stability properties. In order to study the stability properties in a deterministic framework, we consider systems where there is a limit on the maximum number of consecutive sampling times in which measurements of $x_{a,i}$ are not available, i.e.,

$$\max(t_{k+1,i} - t_{k,i}) \leq \Delta_M.$$

This bound on the maximum period of time in which the loop is open has been also used in other works in the literature [107, 124, 167] and allows us to study deterministic notions of stability.

9.3.3 Asynchronous State Observer

An observer that takes advantage of both synchronous and asynchronous measurements can be constructed to estimate the fault-free evolution of asynchronous states between consecutive measurements. The observer states are updated by setting the observer state equal to the measurement each time a new asynchronous measurement becomes available at $t_{k,i}$. The asynchronous state observer takes the form

$$\dot{\hat{x}}_a = f_a(x_s, \hat{x}_a, u, 0) \tag{9.63}$$

with $\hat{x}_{a,i}(t_{k,i}) = x_{a,i}(t_{k,i})$ for all $t_{k,i}$, that is, each time a new asynchronous measurement is received, the estimated states $\hat{x}_{a,i}$ with $i = n_s + 1, \dots, n_s + n_a$ are reset to match the true process state. The information generated by this observer provides a fault-free estimate for each asynchronous state at any time t and allows for the design of nonlinear control laws that utilize full state information. Using the estimated states, the control input applied to the system is given by $u = h(\hat{x})$ where $\hat{x} = [x_s^T \, \hat{x}_a^T]^T$.

This control input is defined for all times because it is based on both the synchronous states and the estimated asynchronous states. We assume that Δ_M is small enough to guarantee that the system in closed-loop with this control scheme is practically stable, see [107, 124, 167] for details on similar stability results.

9.3.4 Design of Fault-Detection and Isolation Filter

In this section, we construct FDI filters that will automatically identify the source of a failure in a timely manner. Utilizing both synchronous state measurements, $\hat{x}_i(t)$, $i = 1, \dots, n_s$, and asynchronous state estimates, $\hat{x}_i(t)$, $i = n_s + 1, \dots, n_s + n_a$, the following $n_s + n_a$ filters are defined:

$$\dot{\tilde{x}}_i = f_i\big(\hat{x}_1, \dots, \tilde{x}_i, \dots, \hat{x}_{n_s+n_a}, h(\hat{x}_1, \dots, \tilde{x}_i, \dots, \hat{x}_{n_s+n_a}), 0\big), \tag{9.64}$$

where $\tilde{x}_i$ is the filter output for the ith state in $\hat{x}$ and f_i is the ith component of the vector function f. The FDI filters are only initialized at $t=0$ such that $\tilde{x}(0)=\hat{x}(0)$. For each state in $\hat{x}$, the FDI residual can be defined as

$$r_i(t)=\left|\hat{x}_i(t)-\tilde{x}_i(t)\right|, \quad i=1,\dots,n_s+n_a.$$

The synchronous residuals $r_i(t)$ with $i=1,\dots,n_s$ are computed continuously because $\hat{x}_i(t)$ with $i=1,\dots,n_s$ is known for all t. On the other hand, the asynchronous residuals $r_i(t)$, $i=n_s+1,\dots,n_s+n_a$, are computed only at times $t_{k,i}$ when a new asynchronous measurement of $\hat{x}_i(t)$, $i=n_s+1,\dots,n_s+n_a$, is received. These FDI filters operate by essentially predicting the fault-free evolution of each individual state, accounting for faults that enter the system when the predicted evolution of the state diverges from the measured evolution (see also Chap. 4).

The dynamics of the synchronous states and asynchronous observers, $\hat{x}$, and the FDI filters, $\tilde{x}_i$, are identical to those of the system of Eq. (9.61) when there are no disturbances or noise acting on the system. When the states are initialized as $\hat{x}(0)=\tilde{x}(0)=x(0)$ both the observer and filter states will track the true process states. For faults affecting the synchronous states, when a fault, d_j, occurs, only the residual corresponding to the affected state, r_i, will become nonzero. This is the case when the $f_s(x_s,x_a,h(x),d)$ vector field has a structure such that Type I faults are isolable; see Chap. 4 for a precise determination of such a structure. In the case with faults affecting asynchronously measured states, at least one r_i will become non-zero when a fault occurs. However, faults that affect asynchronous states cause the asynchronous observer $\hat{x}_a$ to diverge from the true process state x_a between consecutive measurements, and any FDI filter states that are a function of $\hat{x}_a$ will no longer accurately track the corresponding true process states. When such a fault occurs more than one residual value may become nonzero.

Continuous measurements for asynchronous states are not available, thus the FDI filters in Eq. (9.64) cannot always completely isolate all failures. We consider two classes of faults. Type I faults are faults that only affect states that are measured continuously; that is, d_j is a Type I fault if

$$\frac{\partial f_i}{\partial d_j}=0, \quad \forall i=n_s+1,\dots,n_s+n_a.$$

Type II faults affect at least one asynchronous state, that is, d_j is a Type II fault if there exists at least one $i=n_s+1,\dots,n_s+n_a$ such that

$$\frac{\partial f_i}{\partial d_j}\neq 0.$$

The FDI filter will detect and isolate a Type I fault d_j because the asynchronous state observers will track the asynchronous states accurately (i.e., the effect of the fault $d_j(t)$ on an asynchronous observer state is accounted for through the synchronous states, so $d_j(t)$ is accounted for in the observer of Eq. (9.63) and hence the FDI filter). A Type II fault enters the system in the differential equation of a state that is sampled asynchronously. The effect of Type II faults cannot be accounted for by the observer $\hat{x}_i$, and such a fault will cause $\hat{x}_i$ to no longer track x_i and will eventually affect other coupled filter states as well. Strict isolation cannot take place for a Type II fault. The FDI filter will detect and partially isolate disturbances in this case because the asynchronous state observers will diverge from the asynchronous states (i.e., the effect of the fault $d_j(t)$ on an asynchronous observer state is unmeasured and unaccounted for, thus the observer in Eq. (9.63) does not track the disturbed state). In other words, if a Type I fault occurs, then it can be detected and isolated. If a Type II fault occurs, then this fault can be grouped to the subset of Type II faults.

A fault is detected at time t_f if there exists a residual i such that $r_i(t_f) > r_{i,\max}$, where $r_{i,\max}$ is an appropriate threshold chosen to account for process and sensor noise. In order to isolate the possible source of the fault, it is necessary to wait until the residuals of all the asynchronous state filters are updated after t_f to determine if the fault is Type I or Type II. The residual of each asynchronous state filter $\tilde{x}_i$ is updated at time

$$t_i(t_f) = \min_k \{t_{k,i} \mid t_{k,i} > t_f\}.$$

If $r_i(t_i(t_f)) \le r_{i,\max}$ with $i = n_s + 1, \dots, n_s + n_a$, then the fault occurred at time t_f is a Type I fault and can be appropriately isolated. Otherwise, the fault belongs to the set of Type II faults.

Consider that a synchronous residual r_i indicates a fault at time t_f. In this case, the fault could have two possible causes, a Type I or Type II fault. In order to determine the true cause of this fault, one has to wait for the complete set of asynchronous measurements to arrive after t_f. When all the asynchronous measurements arrive and if all the residuals of the asynchronous states are smaller than the threshold, then the fault can be attributed to a Type I fault. If any asynchronous measurement arrives and the corresponding residual indicates a fault, then the fault is Type II. Note that when an asynchronous residual indicates a fault, we can also conclude that the fault is Type II. When the fault is Type II it has been detected, and it is possible to narrow the fault source down to the set of faults that enter the differential equations of asynchronous states.

When the fault can be attributed to a Type I fault and it has been detected and isolated, then automated fault tolerant (FTC) control action can be initiated. For example, when a fault event that is due to a manipulated input failure (i.e., an actuator failure) is detected and isolated, fault tolerant control methods, discussed in Chaps. 3–6, can be initiated. In general an FTC switching rule may be employed that orchestrates the reconfiguration of the control system in the event of control system failure. This rule determines which of the backup control loops can be activated, in the event that the main control loop fails, in order to preserve closed-loop stability. Owing to the limitations imposed by input constraints on the stability region for each control configuration, switching from a malfunctioning configuration to a well-functioning, but randomly selected, backup configuration will not preserve closed-loop stability if the state of the system, at the time of failure, lies outside the stability region of the chosen backup configuration. In this case, stabilization using this configuration requires more control action than is allowed by its constraints.

This observation motivates the development of switching logic, which is to switch to the control configuration for which the closed-loop state resides within the stability region at the time of control failure. Without loss of generality, let the initial actuator configuration be $k(0) = 1$ and let t_d be the time when this failure has been isolated, then the switching rule given by

$$k(t) = j, \quad \forall t \geq t_d \quad \text{if } x(t_d) \in \Omega\left(u_j^{\max}\right) \tag{9.65}$$

for some $j \in \{2, 3, \ldots, N\}$ guarantees closed-loop asymptotic stability, where $\Omega(u_j^{\max})$ is the stability region for the jth control configuration. The implementation of the above switching law requires monitoring the closed-loop state trajectory with respect to the stability regions associated with the various fall-back configurations. The reader may refer to [58] for application of FTC to a polyethylene reactor with constraints on the manipulated inputs. In this work we consider a control law without constraints on the manipulated inputs, and the primary control configuration with a faulty actuator will be deactivated in favor of a fully functional fall-back control configuration where the fall-back configuration can guarantee global stability of the closed-loop system. This integrated FDI/FTC reconfiguration allows for seamless fault-recovery in the event of an actuator failure. Section 9.3.5 demonstrates integrated FDI/FTC for the polyethylene reactor.

9.3.5 Application to a Polyethylene Reactor

9.3.5.1 Process and Measurement Modeling

The presented model-based asynchronous FDI and FTC method will be demonstrated using a model of an industrial gas phase polyethylene reactor. The feed to the reactor consists of ethylene ($[M_1]$), comonomer, hydrogen, inerts ($[In]$) and catalyst (Y). A recycle stream of unreacted gases flows from the top of the reactor and is cooled by passing through a water-cooled heat exchanger. Cooling rates in the heat exchanger are adjusted by mixing cold and warm water streams while maintaining a constant total cooling water flow rate through the heat exchanger. Mass balances on hydrogen and comonomer have not been considered in this study because hydrogen and comonomer have only mild effects on the reactor dynamics [101]. A mathematical model for this reactor has the following form [33]:

$$\begin{aligned}
\frac{d[In]}{dt} &= \frac{1}{V_g}\left(F_{In} - \frac{[\text{In}]}{[M_1]+[In]}b_t\right), \\
\frac{d[M_1]}{dt} &= \frac{1}{V_g}\left(F_{M_1} - \frac{[M_1]}{[M_1]+[In]}b_t - R_{M_1}\right) + d_4, \\
\frac{dY_1}{dt} &= F_c a_c - k_{d_1} Y_1 - \frac{R_{M_1} M_{W_1} Y_1}{B_w} + d_2,
\end{aligned}$$

$$\begin{aligned}
\frac{dY_2}{dt} &= F_c a_c - k_{d_2} Y_2 - \frac{R_{M_1} M_{W_1} Y_2}{B_w} + d_2, \\
\frac{dT}{dt} &= \frac{H_f + H_{g1} - H_{g0} - H_r - H_{\text{pol}}}{M_r C_{pr} + B_w C_{\text{ppol}}} + Q + d_1, \\
\frac{dT_{w_1}}{dt} &= \frac{F_w}{M_w}(T_{wi} - T_{w_1}) - \frac{UA}{M_w C_{pw}}(T_{w_1} - T_{g_1}), \\
\frac{dT_{g_1}}{dt} &= \frac{F_g}{M_g}(T - T_{g_1}) + \frac{UA}{M_g C_{pg}}(T_{w_1} - T_{g_1}) + d_3,
\end{aligned} \tag{9.66}$$

where

$$\begin{aligned}
b_t &= V_p C_v \sqrt{([M_1] + [In])RRT - P_v}, \\
R_{M_1} &= [M_1] k_{p0} e^{\frac{-Ea}{R}(\frac{1}{T} - \frac{1}{T_f})}(Y_1 + Y_2), \\
C_{pg} &= \frac{[M_1]}{[M_1] + [In]} C_{\text{pm1}} + \frac{[\text{In}]}{[M_1] + [In]} C_{\text{pIn}}, \\
H_f &= (F_{M_1} C_{\text{pm1}} + F_{\text{In}} C_{\text{pIn}})(T_{\text{feed}} - T_f), \\
H_{g1} &= F_g (T_{g_1} - T_f) C_{pg}, \\
H_{g0} &= (F_g + b_t)(T - T_f) C_{pg}, \\
H_r &= H_{\text{reac}} M_{W_1} R_{M_1}, \\
H_{\text{pol}} &= C_{\text{ppol}}(T - T_f) R_{M_1} M_{W_1}.
\end{aligned} \tag{9.67}$$

The definitions for all the variables used in (9.66) and (9.67) are given in Table 9.3 and their values can be found in [33] (see also [58]). Under normal operating conditions, the open-loop system behaves in an oscillatory fashion (i.e., the system possesses an open-loop unstable steady-state surrounded by a stable limit cycle). The open-loop unstable steady-state around which the system will be controlled is

$$\begin{aligned}
[In]_{ss} &= 439.7\ \text{mol/m}^3, & [M_1]_{ss} &= 326.7\ \text{mol/m}^3, \\
Y_{ss} &= 7.67\ \text{mol}, & T_{ss} &= 356.2\ \text{K}, \\
T_{g_1 ss} &= 290.4\ \text{K}, & T_{w_1 ss} &= 294.4\ \text{K},
\end{aligned}$$

where T, T_{g_1}, and T_{w_1} are the temperatures of the reactor, recycle gas after cooling, and exit-stream cooling water, respectively. In this example, we consider four possible faults, d_1, d_2, d_3, and d_4 which represent a heat jacket fault, catalyst deactivation, a change in the recycle gas flow rate, and ethylene consumption, respectively. The primary manipulated input for these studies is the heat input, Q, and the fall-back manipulated input is the feed temperature, T_{feed}. A fall-back manipulated input is required to maintain desired system performance in the presence of failure in the primary control configuration.

Table 9.3 Polyethylene reactor example process variables

a_c	active site concentration of catalyst
b_t	overhead gas bleed
B_w	mass of polymer in the fluidized bed
C_{pm1}	specific heat capacity of ethylene
C_v	vent flow coefficient
C_{pw}, C_{pIn}, C_{ppol}	specific heat capacity of water, inert gas and polymer
E_a	activation energy
F_c, F_g	flow rate of catalyst and recycle gas
F_{In}, F_{M_1}, F_w	flow rate of inert, ethylene and cooling water
H_f, H_{g0}	enthalpy of fresh feed stream, total gas outflow stream from reactor
H_{g1}	enthalpy of cooled recycle gas stream to reactor
H_{pol}	enthalpy of polymer
H_r	heat liberated by polymerization reaction
H_{reac}	heat of reaction
$[In]$	molar concentration of inerts in the gas phase
k_{d_1}, k_{d_2}	deactivation rate constant for catalyst site 1, 2
k_{p0}	pre-exponential factor for polymer propagation rate
$[M_1]$	molar concentration of ethylene in the gas phase
M_g	mass holdup of gas stream in heat exchanger
$M_r C_{pr}$	product of mass and heat capacity of reactor walls
M_w	mass holdup of cooling water in heat exchanger
M_{W_1}	molecular weight of monomer
P_v	pressure downstream of bleed vent
Q	Heat added/removed by heating jacket
R, RR	ideal gas constant, unit of J/(mol K), m^3 atm/(mol K)
T, T_f, T_{feed}	reactor, reference, feed temperature
T_{g_1}, T_{w_1}	temperature of recycle gas, cooling water stream from exchanger
T_{w_i}	inlet cooling water temperature to heat exchanger
UA	product of heat exchanger coefficient with area
V_g	volume of gas phase in the reactor
V_p	bleed stream valve position
Y_1, Y_2	moles of active site type 1, 2

Simulations have been carried out for several scenarios to demonstrate the effectiveness of the proposed FDI scheme in detecting and isolating the four faults d_1, d_2, d_3, and d_4 in the presence of asynchronous measurements. The temperature measurements (T, T_{g_1}, T_{w_1}) are all assumed to be available synchronously, while the concentration measurements ($[In]$, $[M_1]$, Y) arrive at asynchronous intervals. In all the simulations, sensor measurement and process noise are included. The sensor measurement noise trajectory was generated using a sample time of ten seconds and

Table 9.4 Polyethylene reactor noise parameters

	σ_p	σ_m	ϕ
$[In]$	1E–4	5E–2	0
$[M_1]$	1E–4	5E–2	0.7
Y	1E–4	1E–2	0.7
T	5E–3	5E–2	0.7
T_{g1}	5E–3	5E–2	0.7
T_{w1}	5E–3	5E–2	0.7

a zero-mean normal distribution with standard deviation σ_M. The autoregressive process noise was generated discretely as $w_k = \phi w_{k-1} + \xi_k$, where $k = 0, 1, \ldots$ is the discrete time step, with a sample time of ten seconds, ϕ is the autoregressive coefficient, and ξ_k is obtained at each sampling step using a zero-mean normal distribution with standard deviation σ_p. The autoregressive process noise is added to the right-hand side of the differential equations for each state and the sensor measurement noise is added to the measurements of each state. Sensor measurement noise and process noise are evaluated independently for each state variable. Table 9.4 provides the values of the noise parameters for each state of the system. The length of time between consecutive asynchronous measurements is generated randomly based on a Poisson process. The time when the system will receive the next asynchronous measurement of the ith state is given by $t_{k+1,i} = t_{k,i} + \Delta_a$ where $\Delta_a = -\ln(\xi)/W_a$ and $\xi \in (0, 1)$ is a random variable chosen from a uniform probability distribution and $W_a = 0.003\ \text{s}^{-1}$ is the mean rate of asynchronous sampling. There is an upper bound limiting the time between consecutive measurements such that $\Delta_a \leq \Delta_M = 1200$ s. This value of Δ_M is small enough to provide practical closed-loop stability around the desired equilibrium point for the polyethylene reactor. An increasing sequence of measurement arrival times is generated independently for each asynchronously measured state.

9.3.5.2 Design of the Asynchronous State Observers

To perform FDI for the polyethylene reactor system we need to construct the asynchronous state observers of the form in Eq. (9.63). The asynchronous state observers for this system have the form:

$$
\begin{aligned}
\frac{d[\hat{In}]}{dt} &= \frac{1}{V_g}\left(F_{\text{In}} - \frac{[\hat{In}]}{[\hat{M}_1] + [\hat{In}]}\hat{b}_t\right),\\
\frac{d[\hat{M}_1]}{dt} &= \frac{1}{V_g}\left(F_{M_1} - \frac{[\hat{M}_1]}{[\hat{M}_1] + [\hat{In}]}\hat{b}_t - \hat{R}_{M_1}\right),\\
\frac{d\hat{Y}}{dt} &= F_c a_c - k_{d_1}\hat{Y} - \frac{\hat{R}_{M_1} M_{W_1} Y}{B_w},
\end{aligned}
$$

$$\hat{b}_t = V_p C_v \sqrt{([\hat{M}_1] + [\hat{In}])RRT(t) - P_v}, \tag{9.68}$$

$$\hat{R}_{M_1} = [\hat{M}_1] k_{p0} e^{\frac{-Ea}{R}(\frac{1}{T(t)} - \frac{1}{T_f})} (\hat{Y}),$$

$$[\hat{In}](t_{k,[\mathrm{In}]}) = [In](t_{k,[In]}),$$

$$[\hat{M}_1](t_{k,[M_1]}) = [M_1](t_{k,[M_1]}),$$

$$\hat{Y}(t_{k,Y}) = Y(t_{k,Y}),$$

where $[\hat{In}]$, $[\hat{M}_1]$, and $\hat{Y}$ are the asynchronous observer states. Each asynchronous observer state is initialized each time new measurement information becomes available at the times $t_{k,i}$. The observer states provide estimates for the asynchronous states between consecutive measurements allowing the computation of control actions and FDI residuals at each time.

9.3.5.3 Design of the State Feedback Controller

The control objective is to stabilize the system at the open-loop unstable steady state. A nonlinear Lyapunov-based feedback controller that enforces asymptotic stability of the closed-loop system is synthesized using the method as discussed in Chap. 2. This is a single input controller that utilizes synchronous measurements as well as observer states. The polyethylene reactor dynamics belong to the following class of nonlinear systems:

$$\dot{x}(t) = f(x(t)) + g_1(x(t))u_1(t) + g_2(x(t))u_2(t) + w(x(t))d(t), \tag{9.69}$$

where

$$x(t) = \begin{bmatrix} [In] - [In]_{ss} \\ [M_1] - [M_1]_{ss} \\ Y - Y_{ss} \\ T - T_{ss} \\ T_{g_1} - T_{g_1 ss} \\ T_{w_1} - T_{w_1 ss} \end{bmatrix},$$

and

$$u_1(t) = Q, \qquad u_2(t) = T_{\mathrm{feed}}.$$

Consider the quadratic control Lyapunov function $V(x) = x^T P x$ where

$$P = 1 \times 10^{-2} diag[0.5\ 0.5\ 0.5\ 1\ 0.005\ 0.005].$$

The values of the weighting matrix P are chosen to account for the different range of numerical values for each state. The following feedback laws [150] (see also

Chap. 2) asymptotically stabilize the open-loop and possibly unstable steady-state of the nominal system (i.e., $d(t) \equiv 0$)

$$h_i(x) = \begin{cases} \frac{L_f V + \sqrt{L_f V^2 + L_{g_i} V^4}}{-L_{g_i} V} & \text{if } L_{g_i} V \neq 0, \\ 0 & \text{if } L_{g_i} V = 0, \end{cases} \quad i = 1, 2. \tag{9.70}$$

In the simulations, the primary control configuration is given by

$$u_1(t) = h_1\big(\hat{x}(t)\big),$$

and the fall-back control configuration is given by

$$u_2(t) = h_2\big(\hat{x}(t)\big),$$

where

$$\hat{x}(t) = \begin{bmatrix} [\hat{In}] - [In]_{ss} \\ [\hat{M}_1] - [M_1]_{ss} \\ \hat{Y} - Y_{ss} \\ T - T_{ss} \\ T_{g1} - T_{g1ss} \\ T_{w1} - T_{w1ss} \end{bmatrix}.$$

9.3.5.4 Design of FDI/FTC Scheme

Fault detection and isolation for the system in closed-loop with the primary configuration is accomplished by generating FDI filters as in Eq. (9.64), and for the polyethylene system the FDI filters take the following form:

$$\begin{aligned}
\frac{d[\tilde{In}]}{dt} &= \frac{1}{V_g}\left(F_{\text{In}} - \frac{[\tilde{In}]}{[\hat{M}_1] + [\tilde{In}]}\tilde{b}_t^{[\text{In}]}\right), \\
\frac{d[\tilde{M}_1]}{dt} &= \frac{1}{V_g}\left(F_{M_1} - \frac{[\tilde{M}_1]}{[\tilde{M}_1] + [\hat{In}]}\tilde{b}_t^{[M_1]} - \tilde{R}_{M_1}^{[M_1]}\right), \\
\frac{d\tilde{Y}}{dt} &= F_c a_c - k_{d_1}\tilde{Y} - \frac{\tilde{R}_{M_1}^{Y} M_{W_1}\tilde{Y}}{B_w}, \\
\frac{d\tilde{T}}{dt} &= \frac{H_f + \tilde{H}_{g1}^T - \tilde{H}_{g0}^T - \tilde{H}_r^T - \tilde{H}_{\text{pol}}^T}{M_r C_{pr} + B_w C_{\text{ppol}}} + h_1\big(\hat{x}(t)\big), \\
\frac{d\tilde{T}_{w1}}{dt} &= \frac{F_w}{M_w}(T_{w_i} - \tilde{T}_{w1}) - \frac{UA}{M_w C_{pw}}(\tilde{T}_{w1} - T_{g1}), \\
\frac{d\tilde{T}_{g1}}{dt} &= \frac{F_g}{M_g}(T - \tilde{T}_{g1}) + \frac{UA}{M_g \tilde{C}_{pg}}(T_{w1} - \tilde{T}_{g1}),
\end{aligned} \tag{9.71}$$

where

$$
\begin{aligned}
\tilde{b}_t^{[\text{In}]} &= V_p C_v \sqrt{([\hat{M}_1] + [\tilde{In}]) RR\, T - P_v}, \\
\tilde{b}_t^{[M_1]} &= V_p C_v \sqrt{([\tilde{M}_1] + [\hat{In}]) RR\, T - P_v}, \\
\tilde{b}_t^{[T]} &= V_p C_v \sqrt{([\hat{M}_1] + [\hat{In}]) RR\, \tilde{T} - P_v}, \\
\tilde{R}_{M_1}^{[M_1]} &= [\tilde{M}_1] k_{p0} e^{\frac{-E_a}{R}(\frac{1}{T} - \frac{1}{T_f})} (\hat{Y}), \\
\tilde{R}_{M_1}^{Y} &= [\hat{M}_1] k_{p0} e^{\frac{-E_a}{R}(\frac{1}{T} - \frac{1}{T_f})} (\tilde{Y}), \\
\tilde{R}_{M_1}^{T} &= [\hat{M}_1] k_{p0} e^{\frac{-E_a}{R}(\frac{1}{\tilde{T}} - \frac{1}{T_f})} (\hat{Y}), \\
\tilde{C}_{pg} &= \frac{[\hat{M}_1]}{[\hat{M}_1] + [\hat{In}]} C_{\text{pm1}} + \frac{[\hat{In}]}{[\hat{M}_1] + [\hat{In}]} C_{\text{pIn}}, \\
\tilde{H}_{g_1}^{T} &= F_g (T_{g_1} - T_f) \tilde{C}_{pg}, \\
\tilde{H}_{g_0}^{T} &= (F_g + \tilde{b}_t^{T})(\tilde{T} - T_f) \tilde{C}_{pg}, \\
\tilde{H}_r^{T} &= H_{\text{reac}} M_{W_1} \tilde{R}_{M_1}^{T}, \\
\tilde{H}_{\text{pol}}^{T} &= C_{\text{ppol}} (\tilde{T} - T_f) \tilde{R}_{M_1}^{T} M_{W_1}.
\end{aligned}
\tag{9.72}
$$

In addition, the FDI residuals take the following form:

$$
\begin{aligned}
r_{[\text{In}]} &= \left| [\hat{In}](t_k) - [\tilde{In}](t_k) \right|, \\
r_{[M_1]} &= \left| [\hat{M}_1](t_k) - [\tilde{In}](t_k) \right|, \\
r_Y &= \left| \hat{Y}(t_k) - \tilde{Y}(t_k) \right|, \\
r_T &= |T - \tilde{T}|, \\
r_{T_{g_1}} &= |T_{g_1} - \tilde{T}_{g_1}|, \\
r_{T_{w_1}} &= |T_{w_1} - \tilde{T}_{w_1}|.
\end{aligned}
\tag{9.73}
$$

In the case with measurement and process noise, the residuals will be nonzero even without a failure event. This motivates the use of detection thresholds such that a fault is declared when a residual exceeds a specific threshold value, $r_{i,\max}$ (note that a different threshold value can be used for each residual). This threshold value must be selected to avoid false alarms due to process and measurement noise, but it should also be sensitive enough (small enough) to detect faults in a timely manner so that efficient FTC action can be initiated. The threshold values used for each residual in the numerical simulations can be seen as the dashed lines in Figs. 9.20, 9.23, 9.26, and 9.29.

If the fault can be isolated to d_1 (i.e., r_T exceeds $r_{T,\max}$ at $t = t_f$, while $r_i(t_i(t_f)) \leq r_{i,\max}$ with $i = [In], [M_1], Y$), then one can invoke fault tolerant control methods to handle actuator failures by activation of a fall-back control configuration. In the simulation studies, it is assumed that a fall-back configuration, where the fall-back manipulated input is $u_2 = T_{\text{feed}}$, is available. The control law of Eq. (9.70) enforces stability when the control actuator is functioning properly, thus switching to the operational fall-back configuration will guarantee stability in the case of failure of the primary control configuration, $u_1 = Q$.

9.3.5.5 Closed-Loop Process Simulation Results

This section consists of four simulation studies, each examining one of the faults d_1, d_2, d_3, or d_4. The first simulation considers a fault, d_1, on the heating jacket which is the primary manipulated input. In this case, the simulation includes fault tolerant control that automatically reconfigures the plant so that the fall-back manipulated input, $u_2 = T_{\text{feed}}$, is activated to maintain stability. Specifically, the supervisory control element will deactivate the primary control configuration, u_1 and activate the fall-back configuration u_2 when $r_T > r_{T,\max}$ and $r_i(t_i(t_f)) \leq r_{i,\max}$ with $i = [In], [M_1], Y$. This specific fault signature corresponds to a Type I fault that can be isolated to d_1. The reader may refer to [58] to obtain more information on FTC and reconfiguration rules for a polyethylene reactor with constraints on the manipulated inputs that give rise to stability regions. This work does not consider constraints on the manipulated inputs, hence, the fall-back configuration can guarantee stability from anywhere in the state space because the closed-loop system under the fall-back control configuration is globally asymptotically stable. The remaining simulation studies explore faults that disturb the system, but do not arise from actuator failures. Since they are not caused by actuation component malfunctions these failures cannot be resolved simply by actuator reconfiguration. However, these simulations demonstrate quick detection and isolation in the presence of asynchronous measurements that enables the operator to take appropriate and focused action in a timely manner.

For the fault d_1 a simulation study has been carried out to demonstrate the proposed asynchronous fault detection and isolation and fault tolerant control method. The sequence of asynchronous measurements for this scenario is shown in Fig. 9.18. This first simulation uses the primary control configuration in which Q is the manipulated input and has a fall-back configuration, in which T_{feed} is the manipulated input, available in case of a fault in d_1. A fault takes place where $d_1 = 1$ K/s at $t = 0.5$ hr, representing a failure in the heating jacket, Q. At this time, the synchronous states in Fig. 9.19 all move away from the equilibrium point. Additionally, as asynchronous measurements become available, it is clear the asynchronous states also move away from the equilibrium point after the failure. It is unclear from the state information alone what caused this faulty behavior. However, if the FDI residuals in Fig. 9.20 are examined, it is clear that the residual r_T that is associated with the manipulated input Q, violates its threshold at $t_f = 0.5003$ hr.

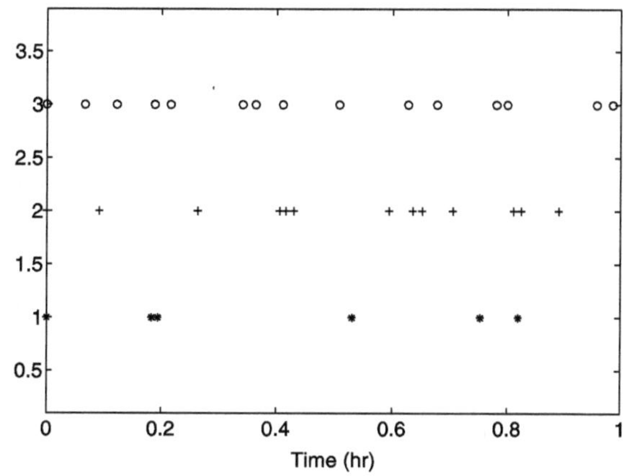

Fig. 9.18 Asynchronous sampling times $t_{k,[\mathrm{In}]}$ (*star*), $t_{k,[M_1]}$ (*cross*), and $t_{k,Y}$ (*circle*) with a fault d_1 at $t = 0.5$ hr

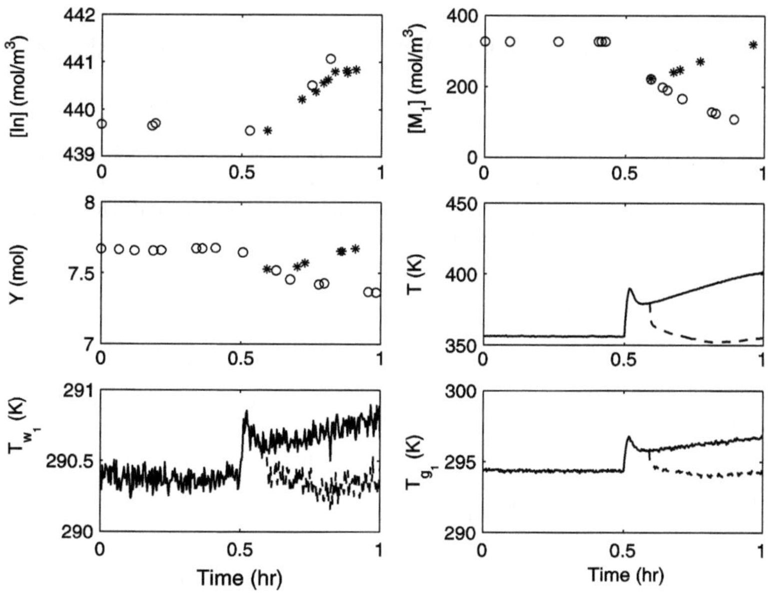

Fig. 9.19 State trajectories of the closed-loop system without fault-tolerant control (*circle/solid*) and with appropriate fault detection and isolation and fault-tolerant control where the fall-back control configuration is activated (*star/dotted*) with a fault d_1 at $t = 0.5$ hr

The fault is detected upon this threshold violation. However, isolation cannot take place until one new measurement for each asynchronous state becomes available. At $t = 0.5944$ hr all three required asynchronous measurements have arrived, and the asynchronous residuals remain below their thresholds, hence $r_i(t_i(t_f)) \leq r_{i,\max}$ with $i = [In], [M_1], Y$. This signals that this is a Type I fault that can be isolated to d_1. At this time, the system is reconfigured to the fall-back configuration where T_{feed} is the manipulated input, and the resulting state trajectory, shown as the dotted

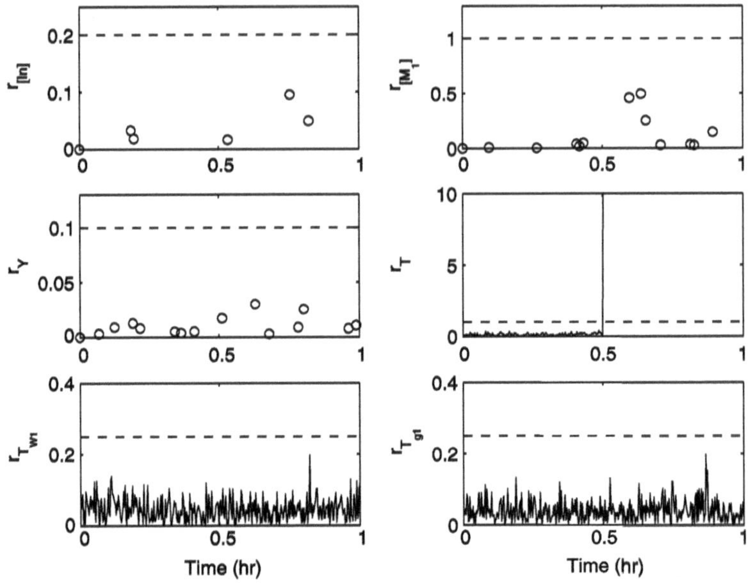

Fig. 9.20 Fault-detection and isolation residuals for the closed-loop system with a fault d_1 at $t = 0.5$ hr. The fault is detected immediately, but isolation occurs at $t = 0.59$ hr when all three asynchronous states have reported a residual below their detection threshold. This signals a Type I fault, and we can isolate the source of this fault as d_1

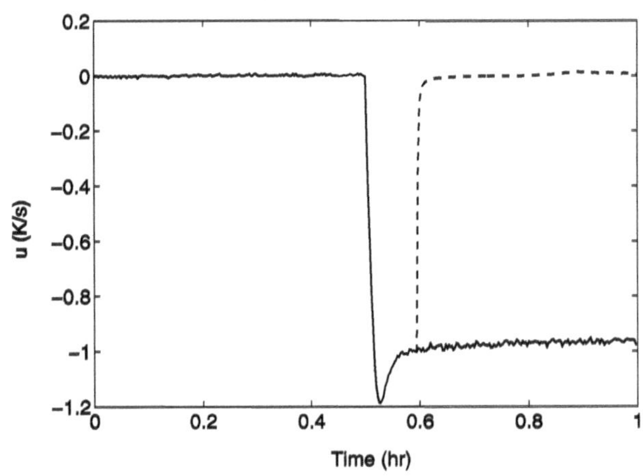

Fig. 9.21 Manipulated input for the closed-loop system without fault-tolerant control (*solid*) and with appropriate fault-tolerant control where the fall-back control configuration is activated (*dotted*) with a fault d_1 at $t = 0.5$ hr

line in Fig. 9.19, moves back to the desired operating point. The manipulated input for this scenario can be seen in Fig. 9.21 where the solid line is the manipulated input without detection and reconfiguration, and the dotted line represents the input after FDI and reconfiguration.

The second simulation demonstrates the proposed asynchronous model-based fault-detection and isolation method when a Type II fault occurs. The sequence of

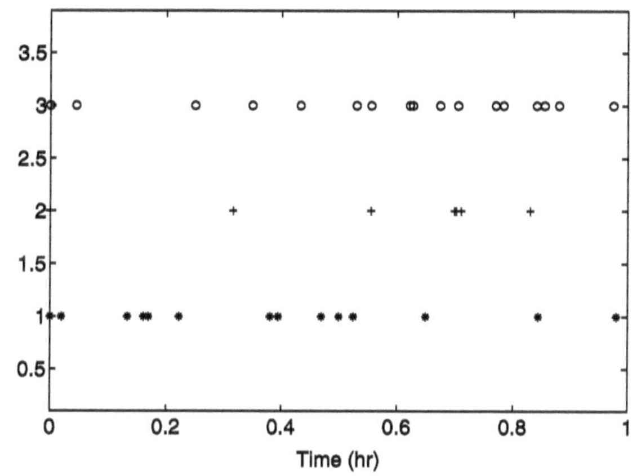

Fig. 9.22 Asynchronous sampling times $t_{k,[\mathrm{In}]}$ (*star*), $t_{k,[M_1]}$ (*cross*), and $t_{k,Y}$ (*circle*) with a fault d_2 at $t = 0.5$ hr

asynchronous measurements for this scenario are found in Fig. 9.22. This simulation uses the primary control configuration in which Q is the manipulated input. A fault takes place where $d_2 = -0.001$ mol/s at $t = 0.5$ hr, representing a catalyst deactivation event. After the failure, two synchronous states in move away from the equilibrium point (see [103] for additional figures). Additionally, as asynchronous measurements become available it can be seen that asynchronous states also move away from the equilibrium point after the failure. It is unclear from the state information alone what caused this faulty behavior. However, if the FDI residuals in Fig. 9.23 generated by (9.73) are examined, it is clear that the residuals $r_{[M_1]}$, r_Y, and r_T violate their thresholds. The fault is detected upon the first threshold violation (r_Y at $t = 0.5333$ hr). When the residual associated with Y exceeds the threshold this signals that the fault is Type II and entered the system in the differential equation of an asynchronous state. When the fault is Type II it cannot be isolated. However, such a fault can be grouped in the subset of faults that enter into the differential equation of an asynchronous state (i.e., the group of Type II faults, specifically, d_2 or d_4). At this time, the system operator can utilize the above partial isolation to examine the plant and determine the exact source of the failure. The manipulated input for this scenario can be seen in Fig. 9.24.

The third simulation study examines FDI in the presence of a Type I fault, d_3, representing a change in the recycle gas flow rate. The sequence of asynchronous measurements for this scenario are found in Fig. 9.25. This simulation study uses the primary control configuration in which Q is the manipulate input, and a fault takes place where $d_3 = 300$ K/s at $t = 0.5$ hr. At this time the synchronous states all move away from the equilibrium point (see [103] for additional figures). Additionally, as asynchronous measurements become available it is observed that the asynchronous states also move away from the equilibrium point after the failure. It is unclear from the state information alone what caused this faulty behavior. However, if the FDI residuals in Fig. 9.26 are examined, the residual associated with T_{g_1} violates its threshold at $t = 0.5003$ hr. The fault is detected upon this threshold violation. However, isolation cannot take place until one new measurement for each asynchronous

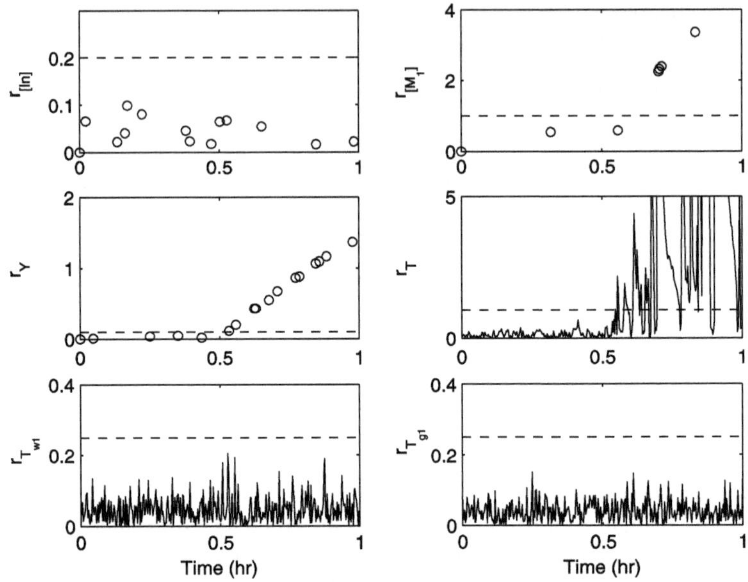

Fig. 9.23 Fault-detection and isolation residuals for the closed-loop system with a fault d_2 at $t = 0.5$ hr. The fault is detected when residual for Y exceeds the threshold. Subsequently, T and $[M_1]$ exceed their thresholds. When any asynchronous residual violates the threshold, this indicates that the fault is in the set of Type II faults, d_2 or d_4

Fig. 9.24 Manipulated input for the closed-loop system with a fault d_2 at $t = 0.5$ hr

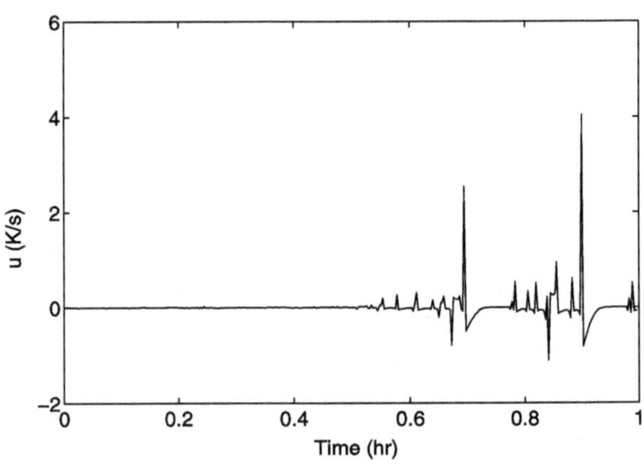

state becomes available. At $t = 0.6086$ hr, all three required asynchronous measurements have become available, and the residuals signal a Type I fault, allowing the isolation of the fault to d_3. The manipulated input for this scenario can be seen in Fig. 9.27.

The final simulation study demonstrates the proposed asynchronous model-based fault-detection and isolation method when a Type II fault occurs. The sequence of asynchronous measurements for this scenario are found in Fig. 9.28. This simula-

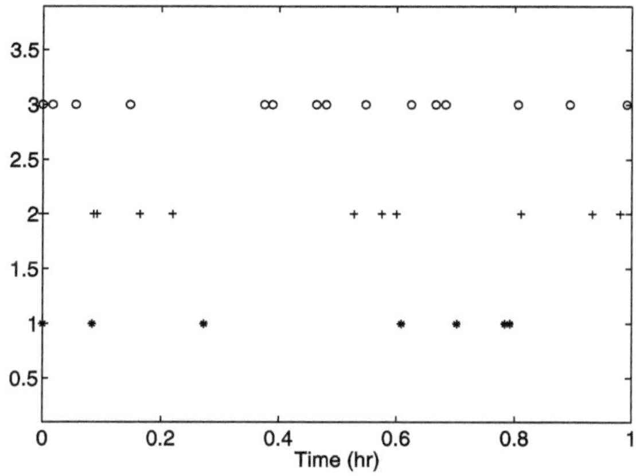

Fig. 9.25 Asynchronous sampling times $t_{k,[\mathrm{In}]}$ (*star*), $t_{k,[M_1]}$ (*cross*), and $t_{k,Y}$ (*circle*) with a fault d_3 at $t = 0.5$ hr

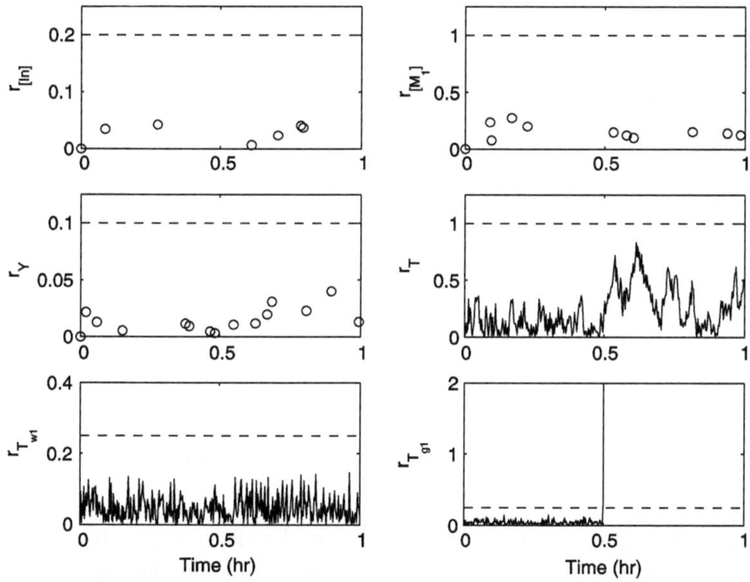

Fig. 9.26 Fault-detection and isolation residuals for the closed-loop system with a fault d_3 at $t = 0.5$ hr. A fault is detected immediately when residual for T_{g_1} exceeds the threshold. Subsequently, none of the asynchronous residuals exceed their thresholds, indicating that the fault source can be isolated as d_3

tion uses the primary control configuration in which Q is the manipulated input. A fault takes place where $d_4 = -0.2$ mol/s at $t = 0.5$ hr, representing unexpected monomer consumption. After the failure, the synchronous states diverge from their desired values (see [103] for additional figures). Additionally, as asynchronous measurements become available, it can be seen that asynchronous states also diverge after the failure. It is unclear from the state information alone what caused this faulty behavior. However, if the FDI residuals in Fig. 9.29 are examined, the residuals

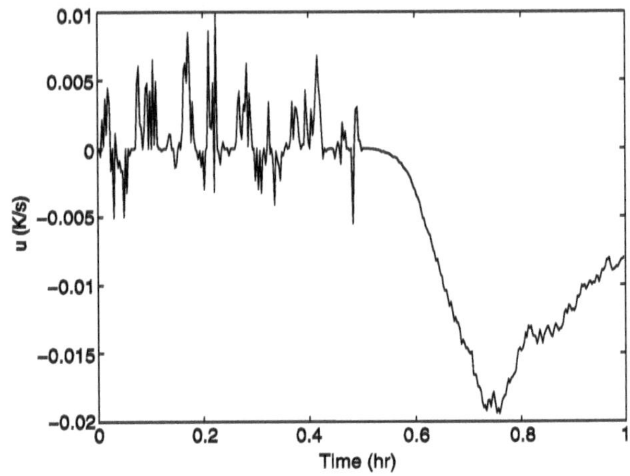

Fig. 9.27 Manipulated input for the closed-loop system with a fault d_3 at $t = 0.5$ hr

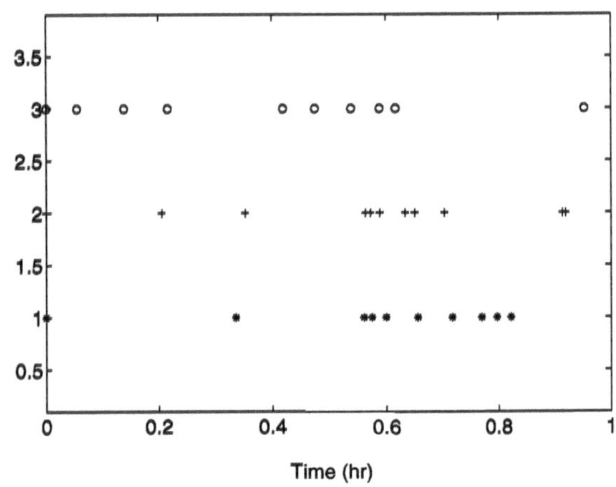

Fig. 9.28 Asynchronous sampling times $t_{k,[\mathrm{In}]}$ (*star*), $t_{k,[M_1]}$ (*cross*), and $t_{k,Y}$ (*circle*) with a fault d_4 at $t = 0.5$ hr

$r_{[\mathrm{In}]}$, $r_{[M_1]}$, r_T, and $r_{T_{g1}}$ violate their thresholds. The fault is detected upon the first threshold violation ($r_{[M_1]}$ at $t = 0.05667$ hr). When the residual $r_{[M_1]}$ exceeds the threshold this signals that a Type II fault has occurred. When a Type II fault occurs, it cannot be isolated. As in the second simulation, such a fault can be grouped in the subset of Type II faults, d_2 or d_4. At this time, the system operator can utilize the partial isolation to examine the plant and determine the exact source of the failure. The manipulated input for this scenario can be seen in Fig. 9.30.

9.4 Conclusions

This chapter presented a control and fault handling approach to handle asynchronous measurements. First, an LMPC scheme was presented where when feedback is lost, the actuators implement the last optimal input trajectory evaluated by

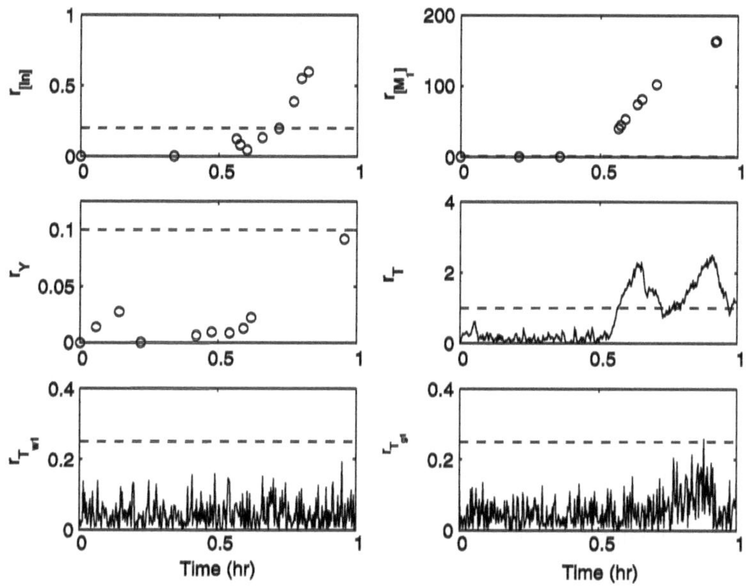

Fig. 9.29 Fault-detection and isolation residuals for the closed-loop system with a fault d_4 at $t = 0.5$ hr. The fault is detected when residual for $[M_1]$ exceeds the threshold. Subsequently, T and $[In]$ exceed their thresholds. When any asynchronous residual violates the threshold, this indicates the fault is in the set of Type II faults, d_2 or d_4

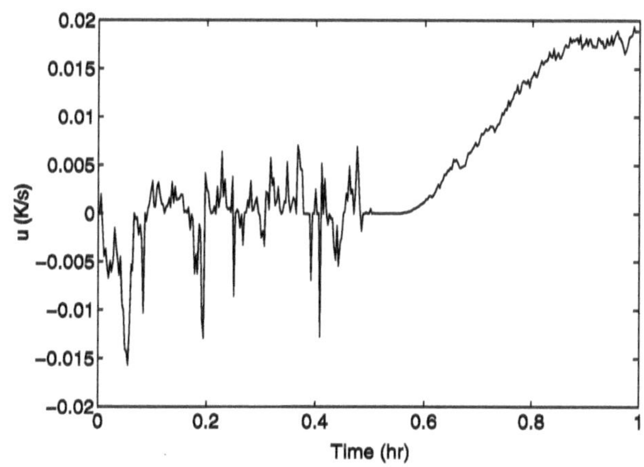

Fig. 9.30 Manipulated input for the closed-loop system with a fault d_4 at $t = 0.5$ hr

the LMPC. The LMPC scheme allows for an explicit characterization of the stability region, guarantees practical stability in the absence of sensor data losses or asynchronous measurements. Extensive simulations of the application of the LMPC to a continuous crystallization process subject to sensor malfunctions were carried out. From the simulations, we found out that the presented LMPC accounting for sensor data losses yields a more robust closed-loop performance when the process is subject to measurement unavailability, asynchronous sampling, and parametric model uncertainties. Next, an application of fault detection and isolation and fault-tolerant control to a polyethylene reactor system was presented where several process measurements were not available synchronously. First, an FDI scheme that employs model-based techniques was introduced that allowed for the isolation of faults. This scheme employed model-based FDI filters in addition to observers that estimate the fault-free evolution of asynchronously measured states during times when they are unmeasured. Specifically, the proposed FDI scheme provided detection and isolation for a Type I fault where the fault entered into the differential equation of only synchronously measured states, and grouping of Type II faults where the fault entered into the differential equation of any asynchronously measured state. The detection occurred shortly after a fault took place, and the isolation, limited by the arrival of asynchronous measurements, occurred once all of the asynchronous measurements became available. Once the FDI methodology provided the system supervisor with a fault diagnosis, the supervisor took appropriate action to seamlessly reconfigure the polyethylene reactor system to an alternative control configuration that enforced the desired operation.

References

1. Ahrens, J.H., Khalil, H.K.: High-gain observers in the presence of measurement noise: A switched-gain approach. Automatica **45**, 936–943 (2009)
2. Ahrens, J.H., Tan, X., Khalil, H.K.: Multirate sampled-data output feedback control with application to smart material actuated systems. IEEE Trans. Autom. Control **54**, 2518–2529 (2009)
3. Allgöwer, F., Chen, H.: Nonlinear model predictive control schemes with guaranteed stability. In: Berber, R., Kravaris, C. (eds.) NATO ASI on Nonlinear Model Based Process Control, pp. 465–494. Kluwer Academic, Dordrecht (1998)
4. Anderson, K.L., Blankenship, G.L., Lebow, L.G.: A rule-based adaptive PID controller. In: Proceedings of IEEE Conference on Decision and Control, Austin, Texas, pp. 564–569 (1988)
5. Antoniades, C., Christofides, P.D.: Feedback control of nonlinear differential difference equation systems. Chem. Eng. Sci. **54**, 5677–5709 (1999)
6. Aradhye, H.B., Bakshi, B.R., Strauss, R.A., Davis, J.F.: Multiscale SPC using wavelets: Theoretical analysis and properties. AIChE J. **49**, 939–958 (2003)
7. Aradhye, H.B., Bakshi, B.R., Davis, J.F., Ahalt, S.C.: Clustering in wavelet domain: A multiresolution art network for anomaly detection. AIChE J. **50**, 2455–2466 (2004)
8. Armaou, A., Demetriou, M.A.: Robust detection and accommodation of incipient component and actuator faults in nonlinear distributed processes. AIChE J. **54**, 2651–2662 (2008)
9. Artstein, Z.: Stabilization with relaxed control. Nonlinear Anal. **7**, 1163–1173 (1983)
10. Astrom, K.J., Hagglund, T., Hang, C.C., Ho, W.K.: Automatic tuning and adaptation for PID controllers—a survey. Control Eng. Pract. **1**, 699–714 (1993)
11. Atassi, A.N., Khalil, H.K.: A separation principle for the stabilization of a class of nonlinear systems. IEEE Trans. Autom. Control **44**, 1672–1687 (1999)
12. Bakshi, B.R.: Multiscale PCA with application to multivariate statistical process monitoring. AIChE J. **44**, 1596–1610 (1998)
13. Bao, J., Zhang, W.Z., Lee, P.L.: Passivity-based decentralized failure-tolerant control. Ind. Eng. Chem. Res. **41**, 5702–5715 (2002)
14. Bemporad, A., Morari, M.: Control of systems integrating logic, dynamics and constraints. Automatica **35**, 407–427 (1999)
15. Bequette, W.B.: Nonlinear control of chemical processes: A review. Ind. Eng. Chem. Res. **30**, 1391–1413 (1991)
16. Bitmead, R.R., Gevers, M., Wertz, V.: Adaptive Optimal Control—The Thinking Man's GPC. Prentice-Hall, Englewood Cliffs (1990)
17. Blanke, M., Izadi-Zamanabadi, R., Bogh, S.A., Lunau, C.P.: Fault-tolerant control systems—a holistic view. Control Eng. Pract. **5**, 693–702 (1997)

P. Mhaskar et al., *Fault-Tolerant Process Control*, DOI 10.1007/978-1-4471-4808-1,

18. Bokor, J., Szabó, Z.: Fault detection and isolation in nonlinear systems. Annu. Rev. Control **33**, 113–123 (2009)
19. Branicky, M.S.: Multiple Lyapunov functions and other analysis tools for switched and hybrid systems. IEEE Trans. Autom. Control **43**, 475–482 (1998)
20. Chand, S.: Self-monitoring tuner for feedback controller (1992)
21. Chen, W., Saif, M.: Adaptive actuator fault detection, isolation and accommodation in uncertain systems. Int. J. Control **80**, 45–63 (2007)
22. Chen, J., Patton, R.J., Zhang, H.-Y.: Design of unknown input observers and robust fault detection filters. Int. J. Control **63**, 85–105 (1996)
23. Chilin, D., Liu, J., Muñoz de la Peña, D., Christofides, P.D., Davis, J.F.: Detection, isolation and handling of actuator faults in distributed model predictive control systems. J. Process Control **20**, 1059–1075 (2010)
24. Chilin, D., Liu, J., Davis, J.F., Christofides, P.D.: Data-based monitoring and reconfiguration of a distributed model predictive control system. Int. J. Robust Nonlinear Control **22**, 68–88 (2012)
25. Chiu, T., Christofides, P.D.: Nonlinear control of particulate processes. AIChE J. **45**, 1279–1297 (1999)
26. Christofides, P.D.: Robust output feedback control of nonlinear singularly perturbed systems. Automatica **36**, 45–52 (2000)
27. Christofides, P.D.: Model-based Control of Particulate Processes. Kluwer Academic, Dordrecht (2002)
28. Christofides, P.D., El-Farra, N.H.: Control of Nonlinear and Hybrid Process Systems: Designs for Uncertainty, Constraints and Time-Delays. Springer, Berlin (2005)
29. Christofides, P.D., Teel, A.R.: Singular perturbations and input-to-state stability. IEEE Trans. Autom. Control **41**, 1645–1650 (1996)
30. Clarke, F., Ledyaev, Y., Sontag, E.: Asymptotic controllability implies feedback stabilization. IEEE Trans. Autom. Control **42**, 1394–1407 (1997)
31. Cohen, G.H., Coon, G.A.: Theoretical consideration of retarded control. Trans. Am. Soc. Mech. Eng. **75**, 827–834 (1953)
32. Dabroom, A.M., Khalil, H.K.: Output feedback sampled-data control of nonlinear systems using high-gain observers. IEEE Trans. Autom. Control **46**, 1712–1725 (2001)
33. Dadebo, S.A., Bell, M.L., McLellan, P.J., McAuley, K.B.: Temperature control of industrial gas phase polyethylene reactors. J. Process Control **7**, 83–95 (1997)
34. Daoutidis, P., Kravaris, C.: Synthesis of feedforward state feedback controllers for nonlinear processes. AIChE J. **35**, 1602–1616 (1989)
35. Daoutidis, P., Kravaris, C.: Structural evaluation of control configurations for multivariable nonlinear processes. Chem. Eng. Sci. **47**, 1091–1107 (1991)
36. Davis, J.F., Piovoso, M.L., Kosanovich, K., Bakshi, B.: Process data analysis and interpretation. Adv. Chem. Eng. **25**, 1–103 (1999)
37. DeCarlo, R.A., Branicky, M.S., Pettersson, S., Lennartson, B.: Perspectives and results on the stability and stabilizability of hybrid systems. Proc. IEEE **88**, 1069–1082 (2000)
38. DePersis, C., Isidori, A.: A geometric approach to nonlinear fault detection and isolation. IEEE Trans. Autom. Control **46**, 853–865 (2001)
39. Ding, S.X., Zhang, P., Naik, A., Ding, E.L., Huang, B.: Subspace method aided data-driven design of fault detection and isolation systems. J. Process Control **19**, 1496–1510 (2009)
40. Du, M., Gandhi, R., Mhaskar, P.: An integrated fault detection and isolation and safe-parking framework for networked process systems. Ind. Eng. Chem. Res. **50**, 5667–5679 (2011)
41. Dubljevic, S., Kazantzis, N.: A new Lyapunov design approach for nonlinear systems based on Zubov's method. Automatica **38**, 1999–2007 (2002)
42. Dunia, R., Qin, S.J.: Subspace approach to multidimensional fault identification and reconstruction. AIChE J. **44**, 1813–1831 (1998)
43. Dunia, R., Qin, S.J., Edgar, T.F., McAvoy, T.J.: Identification of faulty sensors using principal component analysis. AIChE J. **42**, 2797–2812 (1996)

44. El-Farra, N.H.: Integrated fault detection and fault-tolerant control architectures for distributed processes. Ind. Eng. Chem. Res. **45**, 8338–8351 (2006)
45. El-Farra, N.H., Christofides, P.D.: Integrating robustness, optimality and constraints in control of nonlinear processes. Chem. Eng. Sci. **56**, 1841–1868 (2001)
46. El-Farra, N.H., Christofides, P.D.: Bounded robust control of constrained multivariable nonlinear processes. Chem. Eng. Sci. **58**, 3025–3047 (2003)
47. El-Farra, N.H., Christofides, P.D.: Coordinated feedback and switching for control of hybrid nonlinear processes. AIChE J. **49**, 2079–2098 (2003)
48. El-Farra, N.H., Ghantasala, S.: Actuator fault isolation and reconfiguration in transport-reaction processes. AIChE J. **53**, 1518–1537 (2007)
49. El-Farra, N.H., Chiu, T., Christofides, P.D.: Analysis and control of particulate processes with input constraints. AIChE J. **47**, 1849–1865 (2001)
50. El-Farra, N.H., Mhaskar, P., Christofides, P.D.: Hybrid predictive control of nonlinear systems: Method and applications to chemical processes. Int. J. Robust Nonlinear Control **14**, 199–225 (2004)
51. El-Farra, N.H., Mhaskar, P., Christofides, P.D.: Output feedback control of switched nonlinear systems using multiple Lyapunov functions. Syst. Control Lett. **54**, 1163–1182 (2005)
52. Eriksson, P.-G., Isaksson, A.J.: Some aspects of control loop performance monitoring. Control Applications, vol. 2, pp. 1029–1034 (1994)
53. Findeisen, R., Imsland, L., Allgöwer, F., Foss, B.A.: Output feedback stabilization of constrained systems with nonlinear predictive control. Int. J. Robust Nonlinear Control **13**, 211–227 (2003)
54. Frank, P.M.: Fault diagnosis in dynamic systems using analytical and knowledge-based redundancy—a survey and some new results. Automatica **26**, 459–474 (1990)
55. Frank, P.M., Ding, X.: Survey of robust residual generation and evaluation methods in observer-based fault detection systems. J. Process Control **7**, 403–424 (1997)
56. Freeman, R.A., Kokotovic, P.V.: Robust Nonlinear Control Design: State-Space and Lyapunov Techniques. Birkhauser, Boston (1996)
57. Gandhi, R., Mhaskar, P.: Safe-parking of nonlinear process systems. Comput. Chem. Eng. **32**, 2113–2122 (2008)
58. Gani, A., Mhaskar, P., Christofides, P.D.: Fault-tolerant control of a polyethylene reactor. J. Process Control **17**, 439–451 (2007)
59. Gani, A., Mhaskar, P., Christofides, P.D.: Handling sensor malfunctions in control of particulate processes. Chem. Eng. Sci. **63**, 1217–1229 (2008)
60. García, C.E., Prett, D.M., Morari, M.: Model predictive control: Theory and practice—A survey. Automatica **25**, 335–348 (1989)
61. Garcia-Onorio, V., Ydstie, B.E.: Distributed, asynchronous and hybrid simulation of process networks using recording controllers. Int. J. Robust Nonlinear Control **14**, 227–248 (2004)
62. Grossmann, I.E., van den Heever, S.A., Harjukoski, I.: Discrete optimization methods and their role in the integration of planning and scheduling. In: Proceedings of 6th International Conference on Chemical Process Control, Tucson, AZ, pp. 124–152 (2001)
63. Hamelin, F., Sauter, D.: Robust fault detection in uncertain dynamic systems. Automatica **36**, 1747–1754 (2000)
64. Hang, C.C., Astrom, K.J., Ho, W.K.: Refinements of Zeigler–Nichols tuning formula. IEE Proc. Part D, Control Theory Appl. **138**, 111–118 (1991)
65. Harary, F.: Graph Theory. Perseus Books, Cambridge (1969)
66. Harris, T.J.: Assessment of control loop performance. Can. J. Chem. Eng. **67**, 856–861 (1989)
67. Henson, M.A., Seborg, D.E.: Nonlinear Process Control. Prentice-Hall, Englewood Cliffs (1997)
68. Hespanha, J.P., Morse, A.S.: Stability of switched systems with average dwell time. In: Proceedings of 38th IEEE Conference on Decision and Control, Phoenix, AZ, pp. 2655–2660 (1999)

69. Hidalgo, P.M., Brosilow, C.B.: Nonlinear model predictive control of styrene polymerization at unstable equilibrium point. Comput. Chem. Eng. **14**, 481–494 (1990)
70. Hotelling, H.: Multivariate quality control. In: Eisenhart, O. (ed.) Techniques of Statistical Analysis, pp. 113–184. McGraw-Hill, New York (1947)
71. Hu, T., Lin, Z., Qiu, L.: An explicit description of null controllable regions of linear systems with saturating actuators. Syst. Control Lett. **47**, 65–78 (2002)
72. Isidori, A.: Nonlinear Control Systems: An Introduction, 3rd edn. Springer, Berlin (1995)
73. Jerauld, G.R., Vasatis, Y., Doherty, M.F.: Simple conditions for the appearance of sustained oscillations in continuous crystallizers. Chem. Eng. Sci. **38**, 1675–1681 (1983)
74. Kazantzis, N., Kravaris, C.: Nonlinear observer design using Lyapunov's auxiliary theorem. Syst. Control Lett. **34**, 241–247 (1999)
75. Kazantzis, N., Kravaris, C., Wright, R.A.: Nonlinear observer design for process monitoring. Ind. Eng. Chem. Res. **39**, 408–419 (2000)
76. Khalil, H.K.: Nonlinear Systems, 3rd edn. Prentice Hall, Upper Saddle River (2002)
77. Khalil, H.K., Esfandiari, F.: Semiglobal stabilization of a class of nonlinear systems using output feedback. IEEE Trans. Autom. Control **38**, 1412–1415 (1993)
78. Kokotovic, P., Arcak, M.: Constructive nonlinear control: a historical perspective. Automatica **37**, 637–662 (2001)
79. Kothare, S.L.D., Morari, M.: Contractive model predictive control for constrained nonlinear systems. IEEE Trans. Autom. Control **45**, 1053–1071 (2000)
80. Kourti, T., MacGregor, J.F.: Multivariate SPC methods for process and product monitoring. J. Qual. Technol. **28**, 409–428 (1996)
81. Kravaris, C., Kantor, J.C.: Geometric methods for nonlinear process control. 2. controller synthesis. Ind. Eng. Chem. Res. **29**, 2310–2323 (1990)
82. Kresta, J.V., Macgregor, J.F., Marlin, T.E.: Multivariate statistical monitoring of process operating performance. Can. J. Chem. Eng. **69**, 35–47 (1991)
83. Krstic, N., Kanellakopoulos, I., Kokotovic, P.: Nonlinear and Adaptive Control Design, 1st edn. Wiley, New York (1995)
84. Lei, S.J., Shinnar, R., Katz, S.: The stability and dynamic behavior of a continuous crystallizer with a fines trap. AIChE J. **17**, 1459–1470 (1971)
85. Lin, Y., Sontag, E.D.: A universal formula for stabilization with bounded controls. Syst. Control Lett. **16**, 393–397 (1991)
86. Lin, Y., Sontag, E.D., Wang, Y.: A smooth converse Lyapunov theorem for robust stability. SIAM J. Control Optim. **34**, 124–160 (1996)
87. Liu, J., Muñoz de la Peña, D., Christofides, P.D.: Distributed model predictive control of nonlinear process systems. AIChE J. **55**, 1171–1184 (2009)
88. Liu, J., Chen, X., Muñoz de la Peña, D., Christofides, P.D.: Sequential and iterative architectures for distributed model predictive control of nonlinear process systems. AIChE J. **56**, 2137–2149 (2010)
89. Lucas, J.M., Saccucci, M.S.: Exponentially weighted moving average control schemes: Properties and enhancements. Technometrics **32**, 1–12 (1990)
90. MacGregor, J.F., Kourti, T.: Statistical process control of multivariate processes. J. Qual. Technol. **28**, 409–428 (1996)
91. MacGregor, J.F., Jaeckle, C., Kiparissides, C., Koutoudi, M.: Process monitoring and diagnosis by multiblock PLS methods. AIChE J. **40**, 826–838 (1994)
92. Maeder, U., Cagienard, R., Morari, M.: Explicit model predictive control. In: Tarbouriech, S., Garcia, G., Glattfelder, A.H. (eds.) Advanced Strategies in Control Systems with Input and Output Constraints. Lecture Notes in Control and Information Sciences, vol. 346, pp. 237–271. Springer, Berlin (2007)
93. Mahmood, M., Mhaskar, P.: Enhanced stability regions for model predictive control of nonlinear process systems. AIChE J. **54**, 1487–1498 (2008)
94. Mahmood, M., Mhaskar, P.: On constructing constrained control Lyapunov functions for linear systems. IEEE Trans. Autom. Control **56**, 1136–1140 (2011)

95. Mahmood, M., Gandhi, R., Mhaskar, P.: Safe-parking of nonlinear process systems: Handling uncertainty and unavailability of measurements. Chem. Eng. Sci. **63**, 5434–5446 (2008)
96. Mahmoud, N.A., Khalil, H.K.: Asymptotic regulation of minimum phase nonlinear systems using output feedback. IEEE Trans. Autom. Control **41**, 1402–1412 (1996)
97. Massera, J.L.: Contributions to stability theory. Ann. Math. **64**, 182–206 (1956)
98. Massoumnia, M., Verghese, G.C., Wilsky, A.S.: Failure detection and identification. IEEE Trans. Autom. Control **34**, 316–321 (1989)
99. Mattei, M., Paviglianiti, G., Scordamaglia, V.: Nonlinear observers with H_∞ performance for sensor fault detection and isolation: a linear matrix inequality design procedure. Control Eng. Pract. **13**, 1271–1281 (2005)
100. Mayne, D.Q., Rawlings, J.B., Rao, C.V., Scokaert, P.O.M.: Constrained model predictive control: Stability and optimality. Automatica **36**, 789–814 (2000)
101. McAuley, K.B., Macdonald, D.A., McLellan, P.J.: Effects of operating conditions on stability of gas-phase polyethylene reactors. AIChE J. **41**, 868–879 (1995)
102. McFall, C.W., Bartman, A., Christofides, P.D., Cohen, Y.: Control of monitoring of a high recovery reverse osmosis desalination process. Ind. Eng. Chem. Res. **47**, 6698–6710 (2008)
103. McFall, C.W., Muñoz de la Peña, D., Ohran, B., Christofides, P.D., Davis, J.F.: Fault detection and isolation for nonlinear process systems using asynchronous measurements. Ind. Eng. Chem. Res. **47**, 10009–10019 (2008)
104. Megretski, A.: l_2 BIBO output feedback stabilization with saturated control. In: Proceedings of the 13th IFAC World Congress, San Francisco, CA, pp. 435–440 (1996)
105. Mehranbod, N., Soroush, M., Piovoso, M., Ogunnaike, B.A.: Probabilistic model for sensor fault detection and identification. AIChE J. **49**, 1787–1802 (2003)
106. Mehranbod, N., Soroush, M., Panjapornpon, C.: A method of sensor fault detection and identification. J. Process Control **15**, 321–339 (2005)
107. Mhaskar, P., El-Farra, N.H., Christofides, P.D.: Hybrid predictive control of process systems. AIChE J. **50**, 1242–1259 (2004)
108. Mhaskar, P., El-Farra, N.H., Christofides, P.D.: Predictive control of switched nonlinear systems with scheduled mode transitions. IEEE Trans. Autom. Control **50**, 1670–1680 (2005)
109. Mhaskar, P., El-Farra, N.H., Christofides, P.D.: Robust hybrid predictive control of nonlinear systems. Automatica **41**, 209–217 (2005)
110. Mhaskar, P., El-Farra, N.H., Christofides, P.D.: Stabilization of nonlinear systems with state and control constraints using Lyapunov-based predictive control. Syst. Control Lett. **55**, 650–659 (2006)
111. Mhaskar, P., Gani, A., Christofides, P.D.: Fault-tolerant control of nonlinear processes: Performance-based reconfiguration and robustness. Int. J. Robust Nonlinear Control **16**, 91–111 (2006)
112. Mhaskar, P., Gani, A., El-Farra, N.H., McFall, C., Christofides, P.D., Davis, J.F.: Integrated fault-detection and fault-tolerant control of process systems. AIChE J. **52**, 2129–2148 (2006)
113. Mhaskar, P., Gani, A., McFall, C., Christofides, P.D., Davis, J.F.: Fault-tolerant control of nonlinear process systems subject to sensor faults. AIChE J. **53**, 654–668 (2007)
114. Mhaskar, P., El-Farra, N.H., Christofides, P.D.: Robust predictive control of switched systems: Satisfying uncertain schedules subject to state and control constraints. Int. J. Adapt. Control Signal Process. **22**, 161–179 (2008)
115. Mhaskar, P., McFall, C., Gani, A., Christofides, P.D., Davis, J.F.: Isolation and handling of actuator faults in nonlinear systems. Automatica **44**, 53–62 (2008)
116. Michalska, H., Mayne, D.Q.: Moving horizon observers and observer-based control. IEEE Trans. Autom. Control **40**, 995–1006 (1995)
117. Montestruque, L.A., Antsaklis, P.J.: On the model-based control of networked systems. Automatica **39**, 1837–1843 (2003)
118. Montestruque, L.A., Antsaklis, P.J.: Stability of model-based networked control systems with time-varying transmission times. IEEE Trans. Autom. Control **49**, 1562–1572 (2004)

119. Muñoz de la Peña, D., Christofides, P.D.: Lyapunov-based model predictive control of nonlinear systems subject to data losses. IEEE Trans. Autom. Control **53**, 2076–2089 (2008)
120. Muñoz de la Peña, D., Christofides, P.D.: Output feedback control of nonlinear systems subject to sensor data losses. Syst. Control Lett. **57**, 631–642 (2008)
121. Naghshtabrizi, P., Hespanha, J.: Designing an observer-based controller for a network control system. In: Proceedings of the 44th IEEE Conference on Decision and Control and the European Control Conference 2005, pp. 848–853, Seville, Spain (2005)
122. Naghshtabrizi, P., Hespanha, J.: Anticipative and non-anticipative controller design for network control systems. In: Networked Embedded Sensing and Control. Lecture Notes in Control and Information Sciences, vol. 331, pp. 203–218 (2006)
123. Negiz, A., Cinar, A.: Statistical monitoring of multivariable dynamic processes with state-space models. AIChE J. **43**, 2002–2020 (1997)
124. Nešić, D., Teel, A.R.: Input-to-state stability of networked control systems. Automatica **40**, 2121–2128 (2004)
125. Nešić, D., Teel, A., Kokotovic, P.: Sufficient conditions for stabilization of sampled-data nonlinear systems via discrete time approximations. Syst. Control Lett. **38**, 259–270 (1999)
126. Nijmeijer, H., van der Schaft, A.J.: Nonlinear Dynamical Control Systems. Springer, New York (1990)
127. Nishikawa, Y., Sannomiya, N., Ohta, T., Tanaka, H.: A method for auto-tuning of PID control parameters. Automatica **20**, 321–332 (1984)
128. Nomikos, P., Macgregor, J.F.: Monitoring batch processes using multiway principal component analysis. AIChE J. **40**, 1361–1375 (1994)
129. Ohran, B., Muñoz de la Peña, D., Christofides, P.D., Davis, J.F.: Enhancing data-based fault isolation through nonlinear control. AIChE J. **54**, 223–241 (2008)
130. Ohran, B., Liu, J., Munoz de la Pena, D., Christofides, P.D., Davis, J.F.: Data-based fault detection and isolation using feedback control: Output feedback and optimality. Chem. Eng. Sci. **64**, 2370–2383 (2009)
131. Patton, R.J.: Fault-tolerant control systems: The 1997 situation. In: Proceedings of the IFAC Symposium SAFEPROCESS 1997, Hull, United Kingdom, pp. 1033–1054 (1997)
132. Patton, R.J., Chen, J.: Optimal unknown input distribution matrix selection in robust fault diagnosis. Automatica **29**, 837–841 (1993)
133. Pertew, A.M., Marquez, H.J., Zhao, Q.: LMI-based sensor fault diagnosis for nonlinear Lipschitz systems. Automatica **43**, 1464–1469 (2007)
134. Pisu, P., Serrani, A., You, S., Jalics, L.: Adaptive threshold based diagnostics for steer-by-wire systems. J. Dyn. Syst. Meas. Control **128**, 428–435 (2006)
135. Prasad, P.R., Davis, J.F., Jirapinyo, Y., Bhalodia, M., Josephson, J.R.: Structuring diagnostic knowledge for large-scale process systems. Comput. Chem. Eng. **22**, 1897–1905 (1999)
136. Prasad, V., Schley, M., Russo, L.P., Bequette, B.W.: Product property and production rate control of styrene polymerization. J. Process Control **12**, 353–372 (2002)
137. Primbs, J.A., Nevistic, V., Doyle, J.C.: A receding horizon generalization of pointwise min-norm controllers. IEEE Trans. Autom. Control **45**, 898–909 (2000)
138. Qin, S.J.: Control performance monitoring—a review and assessment. Comput. Chem. Eng. **23**, 173–186 (1998)
139. Raich, A., Çinar, A.: Statistical process monitoring and disturbance diagnosis in multivariable continuous processes. AIChE J. **42**, 995–1009 (1996)
140. Rajamani, R., Ganguli, A.: Sensor fault diagnostics for a class of non-linear systems using linear matrix inequalities. Int. J. Control **77**, 920–930 (2004)
141. Rao, C.V., Rawlings, J.B.: Constrained process monitoring: Moving-horizon approach. AIChE J. **48**, 97–109 (2002)
142. Rollins, D.R., Davis, J.F.: An unbiased estimation technique when gross errors exist in process measurements. AIChE J. **38**, 563–572 (1992)
143. Rollins, D.R., Davis, J.F.: Unbiased estimation of gross errors when the covariance matrix is unknown. AIChE J. **39**, 1335–1341 (1993)

144. Romagnoli, J.A., Palazoglu, A.: Introduction to Process Control. CRC Press, Boca Raton (2006)
145. Rugh, W.J.: Analytical framework for gain scheduling. IEEE Control Syst. **11**, 79–84 (1991)
146. Saberi, A., Stoorvogel, A.A., Sannuti, P., Niemann, H.: Fundamental problems in fault detection and identification. Int. J. Robust Nonlinear Control **10**, 1209–1236 (2000)
147. Saito, T.: PID controller system (1990)
148. Shi, D., Tsung, F.: Modeling and diagnosis of feedback-controlled process using dynamic PCA and neural networks. Int. J. Prod. Res. **41**, 365–379 (2003)
149. Skogestad, S.: Simple analytic rules for model reduction and PID controller tuning. J. Process Control **13**, 291–309 (2003)
150. Sontag, E.: A 'universal' construction of Artstein's theorem on nonlinear stabilization. Syst. Control Lett. **13**, 117–123 (1989)
151. Sontag, E.D.: Smooth stabilization implies coprime factorization. IEEE Trans. Autom. Control **34**, 435–443 (1989)
152. Soroush, M., Zambare, N.: Nonlinear output feedback control of a class of polymerization reactors. IEEE Trans. Control Syst. Technol. **8**, 310–320 (2000)
153. Staroswiecki, M., Comtet-Varga, G.: Analytical redundancy relations for fault detection and isolation in algebraic dynamic systems. Automatica **37**, 687–699 (2001)
154. Sung, S., Lee, I., Lee, B.: On-line process identification and automatic tuning method for PID controllers. Chem. Eng. Sci. **53**, 1847–1859 (1998)
155. Tatara, E., Birol, I., Teymor, F., Cinar, A.: Agent-based control of autocatalytic replicators in network of reactors. Comput. Chem. Eng. **29**, 807–815 (2005)
156. Teel, A.: Global stabilization and restricted tracking for multiple integrators with bounded controls. Syst. Control Lett. **18**, 165–171 (1992)
157. Teng, F.C., Lotfi, A., Tsoi, A.C.: Novel fuzzy logic controllers with self-tuning capability. J. Comput. **3**, 9–16 (2008)
158. Tracy, N.D., Young, J.C., Mason, R.L.: Multivariate control charts for individual observations. J. Qual. Technol. **24**, 88–95 (1992)
159. Tsung, F.: Statistical monitoring and diagnosis of automatic control processes using dynamic PCA. Int. J. Prod. Res. **38**, 625–637 (2000)
160. Tsung, F., Shi, J.: Integrated design of run-to-run PID controller and SPC monitoring for process disturbance rejection. IIE Trans. **31**, 517–527 (1999)
161. Tsung, F., Shi, J., Wu, C.F.J.: Joint monitoring of PID-controlled process. J. Qual. Technol. **31**, 275–285 (1999)
162. Vemuri, A.T.: Sensor bias fault diagnosis in a class of nonlinear systems. IEEE Trans. Autom. Control **46**, 949–954 (2001)
163. Venkatasubramanian, V., Rengaswamy, R., Kavuri, S.N., Yin, K.: A review of process fault detection and diagnosis. Part III: Process history based methods. Comput. Chem. Eng. **27**, 327–346 (2003)
164. Venkatasubramanian, V., Rengaswamy, R., Yin, K., Kavuri, S.N.: A review of process fault detection and diagnosis. Part I: Quantitative model-based methods. Comput. Chem. Eng. **27**, 293–311 (2003)
165. Veronesi, M., Visioli, A.: Performance assessment and retuning of PID controllers. Ind. Eng. Chem. Res. **48**, 2616–2623 (2009)
166. Walsh, G., Beldiman, O., Bushnell, L.: Asymptotic behavior of nonlinear networked control systems. IEEE Trans. Autom. Control **46**, 1093–1097 (2001)
167. Walsh, G., Ye, H., Bushnell, L.: Stability analysis of networked control systems. IEEE Trans. Control Syst. Technol. **10**, 438–446 (2002)
168. Wang, Q.G., Zhang, Z., Chek, L.S., Astrom, K.J.: Guaranteed dominant pole placement with PID controllers. J. Process Control **19**, 349–352 (2009)
169. Whiteley, J.R., Davis, J.F.: Knowledge-based interpretation of sensor patterns. Comput. Chem. Eng. **16**, 329–346 (1992)
170. Whiteley, J.R., Davis, J.F.: Qualitative interpretation of sensor patterns. IEEE Expert **8**, 54–63 (1992)

171. Wise, B.M., Gallagher, N.B.: The process chemometrics approach to monitoring and fault detection. J. Process Control **6**, 329–348 (1996)
172. Yan, X.-G., Edwards, C.: Sensor fault detection and isolation for nonlinear systems based on a sliding mode observer. Int. J. Adapt. Control Signal Process. **21**, 657–673 (2007)
173. Yoon, S., MacGregor, J.F.: Statistical and causal model-based approaches to fault detection and isolation. AIChE J. **46**, 1813–1824 (2000)
174. Yoon, S., MacGregor, J.F.: Fault diagnosis with multivariate statistical models part I: using steady state fault signatures. J. Process Control **11**, 387–400 (2001)
175. Zayed, A., El-Fallah, A., El-Fandi, M., Hussain, A.: A novel implicit adaptive pole-placement PID controller. In: Proceedings of the IASTED International Conference on Modeling and Simulation, Banff, Canada, vol. 12, pp. 296–300 (2009)
176. Zhang, X.: Sensor bias fault detection and isolation in a class of nonlinear uncertain systems using adaptive estimation. IEEE Trans. Autom. Control **56**, 1220–1226 (2011)
177. Zhang, X., Polycarpou, M.M., Parisini, T.: A robust detection and isolation scheme for abrupt and incipient faults in nonlinear systems. IEEE Trans. Autom. Control **47**, 576–593 (2002)
178. Zhang, X.D., Parisini, T., Polycarpou, M.M.: Adaptive fault-tolerant control of nonlinear uncertain systems: An information-based diagnostic approach. IEEE Trans. Autom. Control **49**, 1259–1274 (2004)
179. Zhang, X., Parisini, T., Polycarpou, M.M.: Sensor bias fault isolation in a class of nonlinear systems. IEEE Trans. Autom. Control **50**, 370–376 (2005)
180. Zhang, X., Polycarpou, M.M., Parisini, T.: Fault diagnosis of a class of nonlinear uncertain systems with Lipschitz nonlinearities using adaptive estimation. Automatica **46**, 290–299 (2010)
181. Zhao, Z., Tomizuka, M., Isaka, S.: Fuzzy gain scheduling of PID controllers. IEEE Trans. Syst. Man Cybern. **23**, 1392–1398 (1993)
182. Zhou, D.H., Frank, P.M.: Fault diagnostics and fault tolerant control. IEEE Trans. Aerosp. Electron. Syst. **34**, 420–427 (1998)
183. Zhuang, M., Atherton, D.P.: Automatic tuning of optimum PID controllers. IEE Proc. Part D, Control Theory Appl. **140**, 216–224 (1993)
184. Ziegler, J.G., Nichols, N.B.: Optimum settings for automatic controllers. Trans. Am. Soc. Mech. Eng. **64**, 759–768 (1942)

Index

P. Mhaskar et al., *Fault-Tolerant Process Control*, DOI 10.1007/978-1-4471-4808-1,

MIX
Papier aus verantwortungsvollen Quellen
Paper from responsible sources
FSC® C105338

If you have any concerns about our products,
you can contact us on
ProductSafety@springernature.com

In case Publisher is established outside the EU,
the EU authorized representative is:
Springer Nature Customer Service Center GmbH
Europaplatz 3, 69115 Heidelberg, Germany

Printed by Libri Plureos GmbH
in Hamburg, Germany